DIE GRUNDLEHREN DER

MATHEMATISCHEN WISSENSCHAFTEN

IN EINZELDARSTELLUNGEN MIT BESONDERER BERÜCKSICHTIGUNG DER ANWENDUNGSGEBIETE

HERAUSGEGEBEN VON

J. L. DOOB · R. GRAMMEL · E. HEINZ · F. HIRZEBRUCH
E. HOPF · H. HOPF · W. MAAK · W. MAGNUS
F. K. SCHMIDT · K. STEIN

GESCHÄFTSFÜHRENDE HERAUSGEBER

B. ECKMANN UND B. L. VAN DER WAERDEN
ZÜRICH

BAND 119

SPRINGER-VERLAG
BERLIN · GÖTTINGEN · HEIDELBERG
1963

THE THEORY OF BRANCHING PROCESSES

BY

THEODORE E. HARRIS

DEPARTMENT OF MATHEMATICS,
THE RAND CORPORATION, SANTA MONICA, CALIFORNIA

WITH 6 FIGURES

SPRINGER-VERLAG
BERLIN · GÖTTINGEN · HEIDELBERG
1963

Library of Congress Catalog Card Number 63-19966

Printed in Germany

Druck der Universitätsdruckerei H. Stürtz AG., Würzburg

TO MY MOTHER AND MY FATHER

Preface

It was about ninety years ago that GALTON and WATSON, in treating the problem of the extinction of family names, showed how probability theory could be applied to study the effects of chance on the development of families or populations. They formulated a mathematical model, which was neglected for many years after their original work, but was studied again in isolated papers in the twenties and thirties of this century.

During the past fifteen or twenty years, the model and its generalizations have been treated extensively, for their mathematical interest and as a theoretical basis for studies of populations of such objects as genes, neutrons, or cosmic rays. The generalizations of the Galton-Watson model to be studied in this book can appropriately be called *branching processes*; the term has become common since its use in a more restricted sense in a paper by KOLMOGOROV and DMITRIEV in 1947 (see Chapter II). We may think of a branching process as a mathematical representation of the development of a population whose members reproduce and die, subject to laws of chance. The objects may be of different types, depending on their age, energy, position, or other factors. However, they must not interfere with one another. This assumption, which unifies the mathematical theory, seems justified for some populations of physical particles such as neutrons or cosmic rays, but only under very restricted circumstances for biological populations.

Chapter I studies the original model of GALTON and WATSON, which was designed to answer the following question: If a man has probabilities $p_0, p_1, p_2, \dots$ for having 0, 1, 2, ... sons, if each son has the same probabilities for sons of his own, and so on, what is the probability that the family will eventually become extinct, and more generally, what is the probability of a given number of male descendants in a given generation? Chapter II deals with a natural generalization, where each object may be one of several types, and Chapter III carries on the generalization, so that one can deal with objects described by continuous variables such as age, energy, etc. The theory is then applied in Chapter IV to some of the simpler mathematical models for neutron chain reactions. Chapter V treats the model of GALTON and WATSON in cases where the development of a family is traced continuously in time, rather than by generations, and Chapter VI describes the most natural way of treating populations whose objects are subject to aging effects. Finally, Chapter VII describes a mathematical theory of the electron-photon cascade, one of the components of cosmic radiation.

In this book, the emphasis is on a systematic development of the mathematical theory, but I have described briefly the more important applications, indicating in a general way the weak points of the assumptions underlying some of them. The mathematical level of the treatment varies. I believe that most of Chapters I, II, and V can be mastered by anyone with a working knowledge of Markov chains and continuous probability distributions, at the level of FELLER'S *Probability Theory* and PARZEN'S *Modern Probability Theory*, respectively. I hope that such readers can at least follow the main results in the remainder of the book, whose detailed reading requires an amount of measure-theoretic probability about equal to that in KOLMOGOROV'S basic monograph *Foundations of Probability Theory*, plus a few results from the more advanced treatises of DOOB and LOÈVE. Occasional use is made of matrix theory, the theory of analytic functions, and the theory of Fourier and Laplace integrals.

Although I have tried to give rigorous proofs of the more basic results, I have not hesitated to include a heuristic proof (so labeled) when I did not know a rigorous one, or when the length of a rigorous one seemed out of proportion to its importance.

Thanks are due D. A. DARLING and RUPERT MILLER, who read the entire manuscript and suggested numerous improvements, and also my colleague RICHARD BELLMAN, who made many suggestions about the presentation. I wish to thank my former teacher, S. S. WILKS, who introduced me to this problem, and J. L. DOOB, who encouraged me to write the book. I appreciate the excellent work of MARGARET WRAY, who typed several versions of the manuscript, and of ELEANOR HARRIS, who prepared it for the printer.

I am indebted to The RAND Corporation for support of this work under a broad research contract with the United States Air Force.

Finally, I want to thank my wife for her patience during the many evenings when I was busy with this book.

Santa Monica, California T. E. H.

July, 1963

Contents

THE THEORY
OF BRANCHING PROCESSES

Chapter I

The Galton-Watson branching process

1. Historical remarks

The decay of the families of men who occupied conspicuous positions in past times has been a subject of frequent remark, and has given rise to various conjectures ... The instances are very numerous in which surnames that were once common have since become scarce or have wholly disappeared. The tendency is universal, and, in explanation of it, the conclusion has been hastily drawn that a rise in physical comfort and intellectual capacity is necessarily accompanied by diminution in "fertility" ... On the other hand, M. ALPHONSE DE CANDOLLE has directed attention to the fact that, by the ordinary law of chances, a large proportion of families are continually dying out, and it evidently follows that, until we know what that proportion is, we cannot estimate whether any observed diminution of surnames among the families whose history we can trace, is or is not a sign of their diminished "fertility".

These remarks of FRANCIS GALTON were prefaced to the solution by the Reverend H. W. WATSON of the "problem of the extinction of families", which appeared in 1874[1]. Not willing to accept uncritically the hypothesis that distinguished families are more likely to die out than ordinary ones, GALTON recognized that a first step in studying the hypothesis would be to determine the probability that an ordinary family will disappear, using fertility data for the whole population. Accordingly, he formulated the problem of the extinction of families as follows:

Let $p_0, p_1, p_2, \ldots$ be the respective probabilities that a man has 0, 1, 2, ... sons, let each son have the same probability for sons of his own, and so on. What is the probability that the male line is extinct after r generations, and more generally what is the probability for any given number of descendants in the male line in any given generation?

WATSON's ingenious solution of this problem used a device that has been basic in most subsequent treatments. However, because of a purely algebraic oversight, WATSON concluded erroneously that every family will die out, even when the population size, on the average, increases from one generation to the next.

Although we shall not be concerned with questions of demography, let us note at this point that GALTON (1891) studied statistics on the reproductive rates of English peers, coming to the interesting conclusion

[1] WATSON and GALTON (1874). Essentially the same discussion was given in an appendix to GALTON's book (1889). GALTON originally posed the problem in the pages of the *Educational Times.*

that one factor in lowering the rates was the tendency of peers to marry heiresses. An heiress, coming from a family with no sons, would be expected to have, by inheritance, a lower-than-ordinary fertility, and GALTON's data bore out this expectation.

The mathematical model of GALTON and WATSON (we shall call it the *Galton-Watson process*) appears to have been neglected for many years after its creation, the next treatment known to the author being that of R. A. FISHER (1922, 1930a, 1930b). FISHER used a mathematical model identical with that of GALTON and WATSON to study the survival of the progeny of a mutant gene and to study random variations in the frequencies of genes. J. B. S. HALDANE (1927) likewise applied the model to genetics.

The first complete and correct determination of the probability of extinction for the Galton-Watson process was given by J. F. STEFFENSEN (1930, 1932). The problem was also treated by A. KOLMOGOROV (1938), who determined the asymptotic form of the probability that the family is still in existence after a large finite number of generations.

A. J. LOTKA (1931a, 1931b, 1939a) carried out GALTON's idea, using American fertility data, to determine the probability of extinction of a male line of descent. N. SEMENOFF (1935, Chapter III) used the Galton-Watson model in the elementary stages of his theoretical treatise on chemical (as opposed to nuclear) chain reactions, and W. SHOCKLEY and J. R. PIERCE (1938) employed the model to study the multiplication of electrons in an electronic detection device (the electron multiplier).

After 1940 interest in the model increased, partly because of the analogy between the growth of families and nuclear chain reactions, and partly because of the increased general interest in applications of probability theory. Early work stimulated by the nuclear analogy included that of D. HAWKINS and S. ULAM (1944) and C. J. EVERETT and S. ULAM (1948a, b, c, d). During the past 15 years the model has been the subject of numerous papers in Britain, the Soviet Union, and the United States.

The original Galton-Watson process and its generalizations are connected with work dating back to NIELS ABEL[1] on functional equations and the iteration of functions, and with various lines of development in the theory of stochastic processes. For example, there is an interesting connection between the Galton-Watson process and the so-called birth-and-death processes, introduced in a special form by G. U. YULE (1924) in a study of the rate of formation of new species. The species, rather than individual animals, are the multiplying objects. There are also connections with the theory of cosmic radiation formulated independently

[1] ABEL (1881). This posthumous paper appears in ABEL's collected works.

by H. J. BHABHA and W. HEITLER (1937) and by J. F. CARLSON and J. R. OPPENHEIMER (1937).

These biological and physical problems have required treatment by mathematical models more elaborate than the Galton-Watson process, which is the subject of the present chapter. Although some of these later models are only remotely related to the Galton-Watson process, others can justifiably be considered its direct descendants. It is with these that the chapters after the first will be principally concerned.

With few exceptions we shall treat only processes in which it is assumed that different objects reproduce *independently* of one another. This is a severe limitation for any application to biological problems, although there are situations, which we shall point out, where the assumption of independence seems reasonable. For many processes of interest in physics the assumption of independence seems realistic, although, of course, the models are always imperfect in other ways.

2. Definition of the Galton-Watson process

Let us imagine objects that can generate additional objects of the same kind; they may be men or bacteria reproducing by familiar biological methods, or neutrons in a chain reaction. An initial set of objects, which we call the 0-th generation, have children that are called the first generation; their children are the second generation, and so on. The process is affected by chance events.

In this chapter we choose the simplest possible mathematical description of such a situation, corresponding to the model of GALTON and WATSON. First of all we keep track only of the sizes of the successive generations, not the times at which individual objects are born or their individual family relationships. We denote by $Z_0, Z_1, Z_2, \ldots$ the numbers in the 0-th, first, second, ... generations. (We can sometimes interpret $Z_0, Z_1, \ldots$ as the sizes of a population at a sequence of points in time; see Secs. V.5 and VI.27.) Furthermore, we make the two following assumptions.

(1) If the size of the n-th generation is known, then the probability law governing later generations does not depend on the sizes of generations preceding the n-th; in other words, $Z_0, Z_1, \ldots$ form a *Markov chain*. We shall nearly always make the additional assumption that the transition probabilities for the chain do not vary with time.

(2) The Markov chains considered in this chapter have a very special property, corresponding to the assumption that different objects do not interfere with one another: The number of children born to an object does not depend on how many other objects are present.

Assumption (1) could fail, for example, if a man with few brothers tends to have fewer sons than a man with many brothers. In this case it would help us to know whether a generation comprising six men were all brothers or came from three different fathers. We might restore the Markovian nature of the mathematical model by introducing different types corresponding to different fertilities. This would lead to the models of Chapters II and III.

Assumption (2) fails if the different objects interact with one another. Some discussion of this point for biological populations is given in Secs. 7.3, V.2, and VI.23. The assumption is supposed to be good for particles such as those of the neutron processes of Chapter IV and the electron-photon cascades of Chapter VII.

The author will occasionally remind the reader of the weak points in applications of the various mathematical models to be introduced. However, there will be no systematic attempt to evaluate the worth of the various assumptions.

2.1. Mathematical description of the Galton-Watson process. Let $Z_0, Z_1, Z_2, \ldots$ denote the successive random variables in our Markov process (more particularly, Markov chain, since the states in the process are nonnegative integers). We interpret Z_n as the number of objects in the n-th generation of a population or family. *We shall always assume that $Z_0=1$, unless the contrary is stated.* The appropriate adjustments if $Z_0\neq 1$ are easily made, because we assume that the families of the initial objects develop independently of one another.

We denote by P the probability measure for our process. The probability distribution of Z_1 is prescribed by putting $P(Z_1=k)=p_k$, $k=0, 1, 2, \ldots, \Sigma p_k=1$, where p_k is interpreted as the probability that an object existing in the n-th generation has k children in the $(n+1)$-th generation. It is assumed that p_k does not depend on the generation number n.

The conditional distribution of Z_{n+1}, given $Z_n=k$, is appropriate to the assumption that different objects reproduce independently; that is, Z_{n+1} is distributed as the sum of k independent random variables, each distributed like Z_1. If $Z_n=0$, then Z_{n+1} has probability 1 of being 0. Thus we have defined the *transition probabilities* of our Markov process, denoted by

$$P_{ij}=P(Z_{n+1}=j|Z_n=i), \qquad i, j, n=0, 1, \ldots. \tag{2.1}$$

These transition probabilities are defined for each i and j even though, strictly speaking, the right side of (2.1) is not defined as a conditional probability if $P(Z_n=i)=0$.

Having defined the process, we shall want to know some of its properties: the probability distribution and moments of Z_n; the probability that the random sequence $Z_0, Z_1, Z_2, \ldots$ eventually goes to zero; and the behavior of the sequence in case it does not go to zero.

2.2. Generating functions. We shall make repeated use of the *probability generating function*

$$f(s)=\sum_{k=0}^{\infty} p_k s^k, \qquad |s|\leqq 1, \tag{2.2}$$

where s is a complex variable.

The *iterates* of the generating function $f(s)$ will be defined by

$$f_0(s)=s, \qquad f_1(s)=f(s), \tag{2.3}$$

$$f_{n+1}(s)=f[f_n(s)], \qquad n=1, 2, \ldots. \tag{2.4}$$

The reader can verify that each of the iterates is a probability generating function, and that the following relations are a consequence of (2.3) and (2.4):

$$f_{m+n}(s)=f_m[f_n(s)], \qquad m, n=0, 1, \ldots, \tag{2.5}$$

and in particular,

$$f_{n+1}(s)=f_n[f(s)]. \tag{2.6}$$

3. Basic assumptions

Throughout this chapter we shall, without further mention, make the following assumptions, unless the contrary is stated.

(a) None of the probabilities $p_0, p_1, \ldots$ is equal to 1, and $p_0+p_1<1$. Thus f is strictly convex on the unit interval.

(b) The expected value $\mathscr{E} Z_1=\sum_{k=0}^{\infty} k p_k$ is finite. This implies that the derivative $f'(1)$ is finite. The symbols $f'(1)$, $f''(1)$, etc., will usually refer to left-hand derivatives at $s=1$, since we usually suppose $|s|\leqq 1$.

4. The generating function of Z_n

The following basic result was discovered by WATSON (1874) and has been rediscovered a number of times since. The Basic Assumptions are not required for this result.

Theorem 4.1. *The generating function of Z_n is the n-th iterate $f_n(s)$.*

Proof. Let $f_{(n)}(s)$ designate the generating function of Z_n, $n=0, 1, \ldots$. Under the condition that $Z_n=k$, the distribution of Z_{n+1} has the generating function $[f(s)]^k$, $k=0, 1, \ldots$. Accordingly the generating function of Z_{n+1} is

$$f_{(n+1)}(s)=\sum_{k=0}^{\infty} P(Z_n=k)[f(s)]^k=f_{(n)}[f(s)], \qquad n=0, 1, \ldots. \tag{4.1}$$

From the definitions of $f_{(0)}$ and f_0, we see that they are equal. Using (2.6) and (4.1), we then see by induction that $f_{(n)}(s)=f_n(s)$, $n=1, 2, \ldots$. □

Theorem 4.1 enables us to calculate the generating function, and hence the probability distribution, of Z_n in a routine manner by simply computing the iterates of f, although only rarely can the n-th iterate be found in a simple explicit form. From the point of view of probability theory, the main value of Theorem 4.1 is that it enables us to calculate the moments of Z_n and to obtain various asymptotic laws of behavior for Z_n when n is large.

Remark. We leave it to the reader to show that if k is a positive integer, then $Z_0, Z_k, Z_{2k}, Z_{3k}, \ldots$ is a Galton-Watson process with the generating function $f_k(s)$.

We now consider the moments of Z_n.

5. Moments of Z_n

Definitions 5.1. Let

$$m = \mathscr{E} Z_1, \qquad \sigma^2 = \text{Variance } Z_1 = \mathscr{E} Z_1^2 - m^2.$$

Note that $m = f'(1)$ and $\sigma^2 = f''(1) + m - m^2$.

We can obtain the moments of Z_n by differentiating either (2.4) or (2.6) at $s=1$. Thus differentiating (2.4) yields

$$f'_{n+1}(1) = f'[f_n(1)]\, f'_n(1) = f'(1)\, f'_n(1), \tag{5.1}$$

whence by induction $f'_n(1) = m^n$, $n = 0, 1, \ldots$. If $f''(1) < \infty$, we can differentiate (2.4) again, obtaining

$$f''_{n+1}(1) = f'(1)\, f''_n(1) + f''(1)\, [f'_n(1)]^2. \tag{5.2}$$

We obtain $f''_n(1)$ by repeated application of (5.2) with $n = 0, 1, 2, \ldots$; thus

$$\left.\begin{aligned} \text{Variance } Z_n = \mathscr{E} Z_n^2 - (\mathscr{E} Z_n)^2 &= \frac{\sigma^2 m^n (m^n - 1)}{m^2 - m}, \qquad m \neq 1; \\ &= n\sigma^2, \qquad m = 1. \end{aligned}\right\} \tag{5.3}$$

We thus have the following results[1].

Theorem 5.1. *The expected value $\mathscr{E} Z_n$ is m^n, $n = 0, 1, \ldots$. If $\sigma^2 =$ Variance $Z_1 < \infty$, then the variance of Z_n is given by* (5.3).

If higher moments of Z_1 exist, then higher moments of Z_n can be found in a similar fashion.

---

[1] Steffensen (1932) showed how to obtain the moments in essentially this manner.

6. The probability of extinction

We now consider the problem originally posed by GALTON: find the probability of extinction of a family.

Definition 6.1. By *extinction* we mean the event that the random sequence $\{Z_n\}$ consists of zeros for all but a finite number of values of n.

Since Z_n is integer-valued, extinction is also the event that $Z_n \to 0$. Moreover, since $P(Z_{n+1}=0|Z_n=0)=1$, we have the equalities

$$\left.\begin{aligned} P(Z_n \to 0) &= P(Z_n=0 \text{ for some } n) \\ &= P[(Z_1=0) \cup (Z_2=0) \cup \ldots] \\ &= \lim_{n\to\infty} P[(Z_1=0) \cup \ldots \cup (Z_n=0)] \\ &= \lim P(Z_n=0) = \lim f_n(0). \end{aligned}\right\} \tag{6.1}$$

It is obvious that $f_n(0)$ is a nondecreasing function of n.

Definition 6.2. Let q be the probability of extinction, i.e.,

$$q = P(Z_n \to 0) = \lim f_n(0). \tag{6.2}$$

Theorem 6.1.[1] *If* $m=\mathscr{E}Z_1 \leqq 1$, *the extinction probability* q *is* 1. *If* $m>1$, *the extinction probability is the unique nonnegative solution less than* 1 *of the equation*

$$s = f(s). \tag{6.3}$$

Proof. We see by induction that $f_n(0)<1$, $n=0, 1, \ldots$.

We have already observed that $0=f_0(0) \leqq f_1(0) \leqq f_2(0) \leqq \cdots \leqq q = \lim f_n(0)$. Since $f_{n+1}(0)=f[f_n(0)]$, and since $\lim f_n(0)=\lim f_{n+1}(0)=q$, we see that $q=f(q)$, and of course $0 \leqq q \leqq 1$. If $m \leqq 1$, then $f'(s)<1$ for $0 \leqq s < 1$. Using the law of the mean to express $f(s)$ in terms of $f(1)$, we see that $f(s)>s$ for $0 \leqq s<1$ if $m \leqq 1$. Hence in this case $q=1$.

If $m>1$, then $f(s)<s$ when s is slightly less than 1, but $f(0) \geqq 0$. Hence (6.3) has at least one solution in the half-open interval $[0, 1)$. If there were two solutions, say s_0 and t_0 with $0 \leqq s_0 < t_0 < 1$, then ROLLE's theorem would imply the existence of ξ and η, $s_0 < \xi < t_0 < \eta < 1$, such that $f'(\xi)=f'(\eta)=1$. But this is impossible because f is strictly convex. Now $\lim f_n(0)$ cannot be 1 because $\{f_n(0)\}$ is a nondecreasing sequence, while $f_{n+1}(0)=f[f_n(0)]$ would be less than $f_n(0)$ if $f_n(0)$ were slightly less than 1. Hence q must be the unique solution of (6.3) in $[0, 1)$. □

WATSON (1874, 1889) deduced that q is a root of (6.3), but failed to notice that if $m>1$, the relevant root is less than 1.

Remark. The reader can show that $s \leqq f(s) \leqq f(q)=q$ when $0 \leqq s \leqq q$. Then by induction $f(s) \leqq f_2(s) \leqq f_3(s) \leqq \cdots \leqq q$ when $0 \leqq s \leqq q$, and hence

---

[1] First proved in complete generality by STEFFENSEN (1930, 1932).

$f_n(s) \to q$ since $f_n(s) \geqq f_n(0)$. Furthermore, in case $m > 1$, if $q < s < 1$, then $1 > s \geqq f(s) \geqq f_2(s) \geqq \ldots$, and we see that $f_n(s) \to q$ in this case also. Hence, whatever the value of m,

$$\lim_{n\to\infty} f_n(s) = q, \qquad 0 \leqq s < 1. \tag{6.4}$$

(See Fig. 1.)

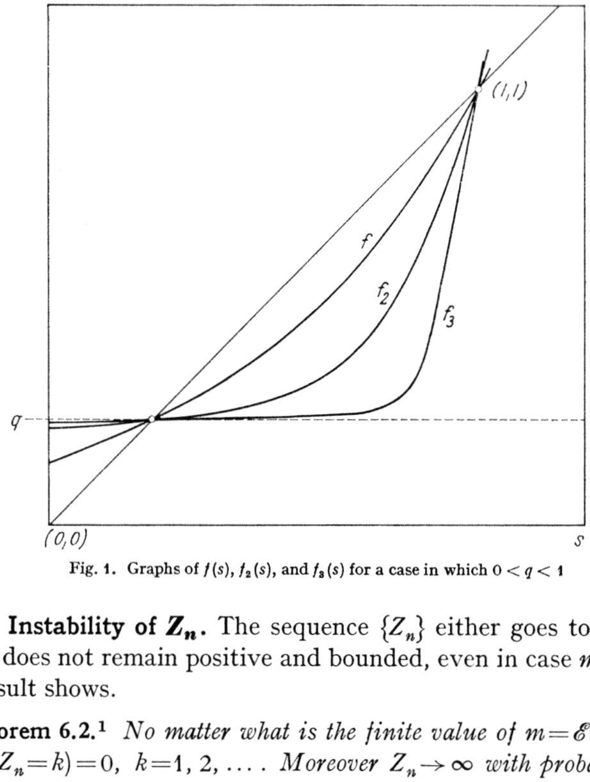

Fig. 1. Graphs of $f(s)$, $f_2(s)$, and $f_3(s)$ for a case in which $0 < q < 1$

6.1. Instability of Z_n. The sequence $\{Z_n\}$ either goes to ∞ or goes to 0; it does not remain positive and bounded, even in case $m = 1$, as the next result shows.

Theorem 6.2.[1] *No matter what is the finite value of $m = \mathscr{E} Z_1$, we have $\lim_{n\to\infty} P(Z_n = k) = 0$, $k = 1, 2, \ldots$. Moreover $Z_n \to \infty$ with probability $1 - q$ and $Z_n \to 0$ with probability q.*

Proof. We shall first show that each of the states $k = 1, 2, \ldots$ is *transient*; that is, if we put $R_k = P(Z_{n+j} = k$ for some $j \geqq 1 | Z_n = k)$, then $R_k < 1$. To see this, observe that if $p_0 = 0$, then $R_k = p_1^k < 1$, while if $p_0 > 0$, then $R_k \leqq 1 - P_{k0} = 1 - p_0^k < 1$. Hence for each $k = 1, 2, \ldots$ we have $\lim_{n\to\infty} P(Z_n = k) = 0$ and $P(Z_n = k$ for infinitely many values of $n) = 0$. (See FELLER (1957, pp. 283–286, 352–353).) Since Z_n does not take the same positive value infinitely often, it must go to 0 or ∞. From

[1] The first sentence of the theorem is implicit in the work of STEFFENSEN (1930, 1932). FELLER (1951) pointed out the applicability of the theory of Markov chains.

Theorem 6.1 we know that q is the probability that $Z_n \to 0$. This completes the proof. □

The unstable behavior depicted by Theorem 6.2 is different from the behavior of biological populations, which, typically, expand to the limit supportable by the environment and then level off more or less. However, the Galton-Watson process has been used to represent the early stages of the development of a family, before the environment is saturated. An example is given in Sec. 7.3.

In Sec. 11 we shall see that there is a generalized sense in which stability *is* possible.

Remark. If $m \leqq 1$, it is obvious that $P(Z_n = k) \to 0$ as $n \to \infty$, uniformly for $k = 1, 2, \ldots$. If $m > 1$, this uniformity still holds, at least if $f''(1) < \infty$[1].

7. Examples

Before taking up a more detailed study of the Galton-Watson process, we consider several examples of calculations and applications.

7.1. Fractional linear generating functions[2]. Suppose the probabilities $p_1, p_2, \ldots$ form a geometric series: $p_k = bc^{k-1}$, $k = 1, 2, \ldots$, $0 < b, c$; $b \leqq 1 - c$; while $p_0 = 1 - p_1 - p_2 - \cdots$. Then the generating function $f(s)$ is a fractional linear function, $f(s) = 1 - b/(1-c) + bs/(1-cs)$, and it can be verified by induction that the iterates $f_n(s)$ are the same kind of function with different constants b_n and c_n. The first moment is

$$m = f'(1) = \frac{b}{(1-c)^2}. \tag{7.1}$$

The equation $s = f(s)$ has the nonnegative root s_0 given by

$$s_0 = \frac{1-b-c}{c(1-c)}. \tag{7.2}$$

This root is 1 if $m = 1$ and is, if $m \neq 1$, the only nonnegative root different from 1. Thus $s_0 = q$ if $m \geqq 1$.

The iterates $f_n(s)$ can be written conveniently in the form

$$f_n(s) = 1 - m^n \left(\frac{1-s_0}{m^n - s_0}\right) + \frac{m^n \left(\dfrac{1-s_0}{m^n - s_0}\right)^2 s}{1 - \left(\dfrac{m^n - 1}{m^n - s_0}\right) s}, \qquad m \neq 1;\; n = 1, 2, \ldots; \tag{7.3}$$

$$f_n(s) = \frac{nc - (nc + c - 1)s}{1 - c + nc - ncs}, \qquad 0 < c < 1;\; m = 1;\; n = 1, 2, \ldots. \tag{7.4}$$

For each n, the probabilities $P(Z_n = k)$ can be found from (7.3) or (7.4), and are again a geometric series, excluding possibly the first term.

---

[1] Harris (1948, p. 481).

[2] This family of functions was considered in nineteenth-century works on functional iteration, such as that of Schroeder (1871).

LOTKA (1931a, 1931b, 1939a) applied Theorem 6.1 to determine the extinction probability for American male lines of descent. He found that the probabilities p_k, $k \geqq 1$ were well represented by the geometric series $p_k = bc^{k-1}$, with $b = 0.2126$, $c = 0.5893$, and $p_0 = 0.4825$. The extinction probability is thus $q = 0.819$. (The numerical values here, based on the 1939 paper, are slightly different from those in the earlier papers. The numbers are based on the Census of 1920.)

7.2. Another example. Suppose $f(s)$ has the fractional linear form given above. If h is a function such that $h^{-1}[f(h(s))]$ is a generating function $g(s)$, then the iterates $g_n(s)$ are generating functions of the form[1] $h^{-1}[f_n(h(s))]$. For example, suppose $f(s) = s/[m - (m-1)s]$, $m > 1$, and let $h(s) = s^k$, where k is a positive integer. Then

$$f_n(s) = \frac{s}{m^n - (m^n - 1)s}, \quad g_n(s) = \frac{s}{[m^n - (m^n - 1)s^k]^{1/k}}, \quad n = 1, 2, \ldots. \tag{7.5}$$

We shall find an interesting interpretation for the forms (7.3), (7.4), and (7.5) when we study branching processes with a continuous time parameter (Sec. V.13).

7.3. Survival of animal families or genes. Imagine a large biological population that has reached equilibrium with its environment, so that each mature animal (or plant) has an average of one descendant reaching maturity in the next generation. Let us now fix our attention on a family descended from some one animal. Suppose that the family has an average multiplication rate m, which, because of a mutation or other reasons, may be different from 1. We are interested in the probability that this family survives indefinitely, assuming that we know the probabilities for the number of mature children of one animal in the family. Let us suppose that as long as the family remains small relative to the rest of the population, the assumptions of GALTON and WATSON are applicable; in particular we assume that mature members of the family are scattered through the population and do not compete with one another appreciably. However, if the family grows very large, then its members may saturate the environment and compete with one another, and the assumptions of GALTON and WATSON would no longer hold.

Suppose we want to find the probability of extinction of the family. If $m \leqq 1$, then it seems reasonable (or at least internally consistent) to apply the Galton-Watson model and deduce that the probability of extinction is 1, since the model implies that the family is never large. On the other hand, if $m > 1$ it may appear inconsistent to apply the model, because it implies that the family can grow very large. However,

---

[1] The usefulness of this consideration in the present context was pointed out by HAWKINS and ULAM (1944). The form $h = s^k$ was used by HARRIS (1948).

the following argument suggests that we may sometimes apply the model to calculate survival probabilities even if $m>1$.

Let us suppose that there is a large number N such that if the size of the family reaches N, then the family will very probably survive and flourish. (This is a mathematical truth about the Galton-Watson process if $m>1$, and we now suppose it to be true about the real family as well.) If N is small compared to the total population size, then the probability P_1 that the family reaches the size N may reasonably be calculated according to the assumptions of GALTON and WATSON. The probability that the family then survives indefinitely is very close to 1, and since P_1 is close to the value q calculated according to the Galton-Watson model, we may suppose that the survival probability is very close to q.

R. A. FISHER (1922, 1930a, Chapter 4; 1930b) applied the Galton-Watson model to the survival of a family of *genes*; see also Sec. 12.

If each animal or gene produces a large number of progeny, only a few of which survive to reproduce, it is reasonable that the generating function f should have a Poisson form. FISHER (1930a) used this form to calculate the chances for survival of the family of a gene when m is near 1. For example, if $m=1.01$, the chance of indefinite survival is 0.0197. The chance that the 127-th generation is not empty is 0.0271 if $m=1.01$ and is 0.0153 if $m=1$.

7.4. Electron multipliers[1]**.** An electron multiplier is a device for amplifying a weak current of electrons. Each electron, as it strikes the first in a series of plates, gives rise to a random number of electrons, which strike the next plate and produce more electrons, etc. The numbers of electrons produced at successive plates have been treated as a Galton-Watson process.

8. Asymptotic results when $m>1$

In Secs. 8—10, we shall study the limiting distribution of Z_n when n is large, and the behavior of the random sequence $Z_1, Z_2, \ldots$. The limiting distribution cannot in general be obtained explicitly, except when $m=1$. However, in Chapter V we shall see that rather explicit results can be obtained for those processes that can be imbedded in a scheme with a continuous time parameter.

The usual notions of convergence of a sequence of random variables $\{W_n\}$ to a random variable W are defined in treatises on probability theory such as KOLMOGOROV (1933) or LOÈVE (1960). Briefly, convergence in probability means that for each $\varepsilon>0$ we have

[1] SHOCKLEY and PIERCE (1938) and WOODWARD (1948) gave early treatments. The former reference treats also the complications due to random noise in the instrument.

$\lim_{n\to\infty} \text{Prob}\{|W_n - W| > \varepsilon\} = 0$; convergence in mean square means that $\lim_{n\to\infty} \mathscr{E}|W_n - W|^2 = 0$; convergence with probability 1 means that with probability 1 the limit $\lim_{n\to\infty} W_n$ exists and is equal to W. The second mode of convergence implies the first; the third also implies the first. Any of the three modes of convergence implies that the distribution of W_n converges to the distribution of W at any continuity point of the latter.

8.1. Convergence of the sequence $\{Z_n/m^n\}$. We have shown that when $\mathscr{E}Z_1 = m > 1$, the family of one ancestor has a positive probability

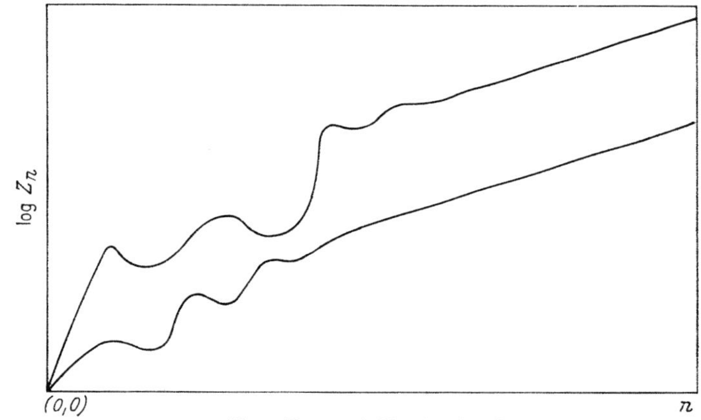

Fig. 2. Two sample histories of log Z_n

of surviving indefinitely. In case it does survive, then from Theorem 6.2 we know that the size increases indefinitely. Hence we should expect the population eventually to increase at a geometric rate, in accordance with the Malthusian law of growth. That this is true is a consequence of Theorem 8.1 and Remark 1 below, which together imply that $Z_n \sim Wm^n$, where W is a random variable that is 0 when and only when Z_n goes to 0. If we plot $\log Z_n$ as a function of n, pretending for convenience that n is a continuous variable, we obtain a curve that eventually is almost a straight line with the slope $\log m$, provided Z_n does not go to 0. The intercept of the limiting line on the vertical axis is the random variable $\log W$. The *randomness* of W should be stressed. It arises because the size of the family may undergo proportionally large random fluctuations in the early generations, these fluctuations acting as multiplicative factors whose effect persists in later generations. The situation is illustrated in Fig. 2, which shows how two different families, both with the same probabilities p_k, might develop.

Definition 8.1. Let $W_n = Z_n/m^n$, $n=0, 1, \ldots$.

Theorem 8.1.[1] *If $m>1$ and $\mathscr{E}Z_1^2 < \infty$, then the random variables W_n converge with probability 1, and in mean square (i.e., in the mean, order two), to a random variable W, and*

$$\mathscr{E}W=1, \qquad \text{Variance } W = \frac{\text{Variance } Z_1}{m^2-m} > 0. \tag{8.1}$$

Corollary. *The conditional variance of W, given $W>0$, is positive; i.e., $\mathscr{E}(W^2|W>0)-[\mathscr{E}(W|W>0)]^2>0$.*

Proof of Theorem 8.1. From the definition of Z_n, we have, with probability 1, $\mathscr{E}(Z_{n+1}|Z_n)=mZ_n$, $n=0, 1, \ldots$. Using this relation repeatedly, we find that with probability 1 (only the final equality is meaningful if $k=0$)

$$\left.\begin{aligned}\mathscr{E}(Z_{n+k}|Z_n) &= \mathscr{E}[\mathscr{E}(Z_{n+k}|Z_{n+k-1}, Z_{n+k-2}, \ldots, Z_n)|Z_n] \\ &= \mathscr{E}[\mathscr{E}(Z_{n+k}|Z_{n+k-1})|Z_n] = \mathscr{E}[mZ_{n+k-1}|Z_n] \\ &= \cdots = m^k Z_n, \qquad n, k=0, 1, 2, \ldots.\end{aligned}\right\} \tag{8.2}$$

Dividing both sides of (8.2) by m^{n+k} we obtain the important relation

$$\mathscr{E}(W_{n+k}|W_n) = \mathscr{E}(W_{n+k}|W_n, W_{n-1}, \ldots, W_0) = W_n, \quad n, k=0, 1, 2, \ldots, \tag{8.3}$$

the first equality in (8.3) holding because $W_0, W_1, \ldots$ is a Markov process. From (5.3) and (8.3), we have

$$\left.\begin{aligned}\mathscr{E}W_n^2 &= 1 + \frac{\sigma^2}{m^2-m}(1-m^{-n}), \\ \mathscr{E}(W_{n+k}-W_n)^2 &= \frac{\sigma^2 m^{-n}}{m^2-m}(1-m^{-k}), \qquad n, k=0, 1, \ldots.\end{aligned}\right\} \tag{8.4}$$

It then follows[2] that the W_n converge in mean square to a random variable W and that the mean and variance of W are as given in the theorem. Moreover, letting $k\to\infty$ in (8.4), we see that $\mathscr{E}(W-W_n)^2 = O(m^{-n})$, and hence $\sum_{n=0}^{\infty}\mathscr{E}(W-W_n)^2 = \mathscr{E}\sum_{n=0}^{\infty}(W-W_n)^2 < \infty$. Since the infinite series has a finite expectation and is thus finite with probability 1, it must be true with probability 1 that $(W-W_n)^2 \to 0$; i.e., that $W_n \to W$. □

---

[1] Convergence in distribution of Z_n/m^n, for the case in which all moments of Z_1 are finite, was proved by Hawkins and Ulam (1944) and independently by Yaglom (1947) under the present conditions. Convergence in mean square was shown by Harris (1948) from (8.3), and as pointed out by J. L. Doob, (8.3) implies that the sequence Z_n/m^n is a martingale and hence converges with probability 1 if $\mathscr{E}Z_1 < \infty$. However, see Remark 3 following the theorem.

[2] See, e.g., Loève (1960, p. 161).

To prove the corollary, consider the generating function[1]

$$G(s) = \frac{f[s(1-q)+q]-q}{1-q}.$$

Note that $G(0)=0$ and $G^{(r)}(0)=(1-q)^{r-1}f^{(r)}(q)$, $r=1,2,\ldots$. The reader can verify that $G''(1)<\infty$ and that G satisfies the Basic Assumptions of Sec. 3. Hence the variance $G''(1)+G'(1)-(G'(1))^2$ is positive, which is equivalent to the assertion that $(1-q)f''(1)-m^2+m>0$. Since $P(W>0)=1-q$ (see Remark 1 below), this is equivalent to the assertion of the corollary.

Remark 1. It is obvious that $W=0$ whenever $Z_n\to 0$. However, it is conceivable that the conditional probability that $W=0$, given $Z_n\to\infty$, could be positive. This would be the case if Z_n had a positive probability of growing at a less-than-geometric rate. However, we can show[2] that $P(W=0|Z_n\to\infty)=0$, as follows. In Sec. 13.1, we shall see that under the condition $Z_1=k$, the random vector $(Z_2, Z_3, \ldots, Z_n)$ is distributed as the sum of k independent vectors, each having the same distribution as the vector $(Z_1, Z_2, \ldots, Z_{n-1})$. Hence

$$P(W=0|Z_1=k)=P(W_n\to 0|Z_1=k)=\big(P(W_n\to 0)\big)^k,$$

and thus, if $q^*=P(W=0)$, then $q^*=f(q^*)$. But q^* cannot be 1, since Variance $W>0$; hence $q^*=q$.

Since we are thus assured that, with probability 1, W is 0 only when $Z_n\to 0$, Theorem 8.1 implies the assertion made earlier in this section that $Z_n\sim Wm^n$.

Remark 2. If $1/m$ is small, formula (8.1) implies that the coefficient of variation of W (i.e., the ratio standard deviation/mean) is approximately equal to the coefficient of variation of Z_1. In Chapter VI we shall see that an analogous relation holds for certain "age-dependent" branching processes.

Remark 3. From (8.3) (see also the footnote to Theorem 8.1), the random variables W_n are a *martingale* and hence, since $\mathscr{E}|W_n|=1$, the W_n converge to a random variable W whether or not $\mathscr{E}Z_1^2$ is finite and whether or not $m>1$. This random variable is obviously identically 0 if $m\leq 1$. An example due to LEVINSON (1959) shows that if $\mathscr{E}Z_1^2=\infty$, we may have $W\equiv 0$ even when $m>1$. However, under an additional condition that is satisfied if, e.g., $p_n=O(n^{-2-\varepsilon})$ for some $\varepsilon>0$, LEVINSON has shown that $W\not\equiv 0$.

---

[1] See HARRIS (1948, p. 477) for another use of the function G. (The notation k is used in that paper.)

[2] I am indebted to J. L. DOOB for this argument, which is simpler than my original one. (HARRIS (1948, p. 480).)

Since the W_n form a martingale, the oscillations of the sequence $W_1, W_2, \ldots$ are subject to general theorems about martingale oscillations. See DOOB (1953) and DUBINS (1962).

8.2. The distribution of W. In view of Theorem 8.1, we can use the distribution of W to study the distribution of Z_n for large n. In order to study the distribution of W, we shall derive from (2.4) a functional equation for its moment-generating function[1] $\varphi(s)=\mathscr{E}e^{-sW}$.

Definitions 8.2. Let $\varphi_n(s)$, $n=0, 1, 2, \ldots$, and $\varphi(s)$ denote respectively the moment-generating functions of W_n and W,

$$\varphi_n(s)=\mathscr{E}e^{-sW_n}=f_n(e^{-s/m^n}), \qquad \varphi(s)=\mathscr{E}e^{-sW}, \qquad Re(s)\geqq 0. \tag{8.5}$$

Let $K(u)=P(W\leqq u)$ be the distribution of W.

Equation (2.4) then becomes

$$\varphi_{n+1}(ms)=f[\varphi_n(s)], \qquad Re(s)\geqq 0;\ n=0, 1, \ldots. \tag{8.6}$$

Since the random variables W_n converge in probability to W, their distributions converge to that of W and $\varphi_n(s)\to\varphi(s)$ when $Re(s)\geqq 0$. Since $f(s)$ is continuous for $|s|\leqq 1$, we see that $f[\varphi_n(s)]\to f[\varphi(s)]$. We then have the following result, which was first proved in somewhat less generality by HAWKINS and ULAM (1944).

Theorem 8.2.[2] *Under the conditions of Theorem* 8.1, *the moment-generating function* $\varphi(s)=\mathscr{E}e^{-sW}$ *satisfies the relations*

$$\varphi(ms)=f[\varphi(s)], \qquad Re(s)\geqq 0, \tag{8.7}$$

with $\varphi'(0)=-1$. *The characteristic function* $\varphi(-it)$, $-\infty<t<\infty$, *is the only characteristic function satisfying* (8.7) *corresponding to a distribution with first moment* 1.

Proof. The equality $\varphi(ms)=f[\varphi(s)]$ is a consequence of (8.6) and the existence of $\lim\varphi_n$ and $\lim f(\varphi_n)$. The SCHWARZ inequality implies $(\mathscr{E}|W-W_n|)^2\leqq\mathscr{E}|W-W_n|^2$, and hence $\mathscr{E}W_n$ converges to $\mathscr{E}W$, which thus must equal 1, since $\mathscr{E}W_n=1$. Hence $\varphi'(0)=-1$.

To prove uniqueness, let $\Psi_1(t)$ and $\Psi_2(t)$ be characteristic functions satisfying $\Psi_r(mt)=f[\Psi_r(t)]$, $r=1, 2$, with $\Psi_1'(0)=\Psi_2'(0)=i$. Then

---

[1] It is customary to define the moment-generating function as (in our notation) $\varphi(-s)$, but this would force us to deal with the half plane $Re(s)\leqq 0$, which is inconvenient.

[2] HARRIS (1947, 1948). LEVINSON (1959) has pointed out that the uniqueness proof does not essentially use the finiteness of the second moments. The fact that φ satisfies (8.7) was first proved, for the case in which all moments of Z_1 are finite, by HAWKINS and ULAM (1944), and (8.7) was also proved by YAGLOM (1947). Such equations are closely related to those studied by ABEL (posthumous 1881), SCHROEDER (1871), KOENIGS (1884), and others.

$\Psi_1(t)-\Psi_2(t)=t\gamma(t)$, where $\gamma(t)$ is a continuous function such that $\gamma(0)=0$ (LOÈVE (1960, p. 199)). Since $|f'(s)|\leqq m$ for $|s|\leqq 1$, we have $|mt\gamma(mt)|=|f(\Psi_1(t))-f(\Psi_2(t))|\leqq m|\Psi_1(t)-\Psi_2(t)|=m|t\gamma(t)|$. Hence $|\gamma(t)|\geqq|\gamma(mt)|$, or, by induction, $|\gamma(t)|\leqq|\gamma(t/m^n)|$, $n=1, 2, \ldots$. Since γ is continuous, we have, for each t, $|\gamma(t)|\leqq \lim_{n\to\infty}|\gamma(t/m^n)|=|\gamma(0)|=0$. □

The following result is an example of the use that can be made of (8.7).

Theorem 8.3.[1] *Under the conditions of Theorem 8.1 the distribution $K(u)=P(W\leqq u)$ is absolutely continuous except for a jump of magnitude q at $u=0$.*

Sketch of proof. If $f(0)=0$, then it can be shown from (8.7) that $\varphi(it)\to 0$ as $t\to\pm\infty$ and that $d\varphi(it)/dt$ is absolutely integrable on $(-\infty,\infty)$. It follows from this that $K(u)$ has a density function given by

$$K'(u)=\frac{1}{2\pi i u}\int_{-\infty}^{\infty} e^{-itu}\,\frac{d\varphi(-it)}{dt}\,dt, \qquad u>0.$$

If $f(0)>0$, the argument needs only a slight modification. □

Other uses of (8.7) are the following (see HARRIS (1948)). We can determine the moments of W by differentiating (8.7) at $s=0$, provided f has a sufficient number of finite derivatives at $s=1$. We can determine φ numerically from (8.7) and then find K by using the Fourier inversion formula. We can determine the behavior of φ as $s\to\infty$ (and as $s\to-\infty$ if f is an entire function) and make some deductions about $K(u)$ near $u=0$ (or as $u\to\infty$).

Recent results of RAMACHANDRAN (1962) help describe the behavior of K when f is entire.

8.3. Asymptotic form of $P(Z_n=0)$. The asymptotic form of the probabilities $f_n(0)=P(Z_n=0)$ as $n\to\infty$ can be determined using a result on the iteration of functions that goes back at least to KOENIGS (1884). (The finiteness of $\mathscr{E}Z_1^2$ is not required.)

Theorem 8.4. *If $m>1$ and $q>0$, then*

$$f_n(0)=q-d\,[f'(q)]^n+O\big([f'(q)]^{2n}\big), \tag{8.8}$$

where d is a positive constant.

It does not seem possible to find d explicitly in general. Notice that $0<f'(q)<1$ because of the convexity of f.

Since (8.8) is so basic, we sketch the proof. The method is essentially that of KOENIGS.

[1] HARRIS (1947, 1948) contains the proof and further study of K. The continuity of K except at 0 was proved by YAGLOM (1947).

Proof. Let Δ be a positive number whose size will be determined later. In the proof c, c_1, and c_2 will denote constants independent of n and s.

If Δ is sufficiently small, then

$$f(s)-q=f'(q)(s-q)[1+(s-q)h(s)], \qquad |s-q|\leqq\Delta, \tag{8.9}$$

where h is continuous. Now suppose $f'(q)<c<1$ and choose Δ small enough so that $|f(s)-q|\leqq c|s-q|$, provided $|s-q|\leqq\Delta$; it is obvious from (8.9) that we can do this. Then if $|s-q|\leqq\Delta$, we have also $|f(s)-q|\leqq c\Delta$, and by induction we have $|f_n(s)-q|\leqq c^n\Delta$, $n=1, 2, \ldots$.

Replacing s by $f_{n-1}(s)$ in (8.9) and using repeatedly the resulting recurrence relation, we obtain, keeping $|s-q|\leqq\Delta$,

$$f_n(s)-q=(f'(q))^n(s-q)\prod_{i=0}^{n-1}[1+(f_i(s)-q)h(f_i(s))]. \tag{8.10}$$

Since $|f_i(s)-q|\leqq c^i\Delta$, the product $\prod_{i=0}^{n-1}$ converges uniformly for $|s-q|\leqq\Delta$, as $n\to\infty$, to a continuous function $G(s)$, with $G(q)=1$, and (8.10) then shows that $|f_n(s)-q|\leqq c_1(f'(q))^n|s-q|$, and hence $\left|1-\prod_{i=n}^{\infty}\right|\leqq c_2(f'(q))^n$. The reader can now complete the proof by putting $s=f_k(0)$ in (8.10), where k is a fixed integer sufficiently large so that $|q-f_k(0)|\leqq\Delta$ and so that $G(f_k(0))$ is positive. □

8.4. Local limit theorems when $m>1$. Čistyakov (1957) has given a result on the asymptotic form of the probabilities $P(Z_n=j)$ when j is in the neighborhood of the mean m^n, for the case $m>1$. For part of the proof, it appears to have been assumed that the sequence Z_n is imbeddable in a scheme with a continuous time parameter. (See Chapter V.) Accordingly we shall give the result in Chapter V.

8.5. Examples[1]. In only a few cases can K, the distribution of W, be found explicitly; we shall see in Chapter V how some of these cases arise. When $f(s)$ has the form (7.3) with $n=1$, and $m>1$, then $K(u)$ has a jump at $u=0$ of magnitude $q=s_0$ (see (7.2)), and for $u>0$, $K'(u)=(1-q)^2e^{-(1-q)u}$. If f has the form

$$f(s)=\frac{s}{[m-(m-1)s^k]^{1/k}},$$

where k is a positive integer (see (7.5)), then K is continuous at 0 and

$$K'(u)=k^{-1/k}\Gamma\left(\frac{1}{k}\right)u^{(1/k)-1}e^{-u/k}, \qquad u>0. \tag{8.11}$$

[1] Harris (1948).

9. Asymptotic results when $m<1$

Since $Z_n \to 0$ when $m<1$, the limiting distribution of Z_n is not interesting. However, we obtain a nontrivial limiting distribution if we consider the *conditional* distribution of Z_n, given that $Z_n \neq 0$.

Definition 9.1. If s is a complex number, we shall denote by $S_1(s)$ (or $S_2(s)$, etc.) the set of points s', with $|s'| \leqq 1$, that are interior to some circle centered at s. The radius of the circle will depend on the requirements of the problem, and when it is asserted that some relation $\mathscr{R}$ holds when $s' \in S_1(s)$, the meaning is that there exists an $S_1(s)$ such that $\mathscr{R}$ holds for $s' \in S_1(s)$.

The next result is due to YAGLOM (1947).

Theorem 9.1. *Suppose $m<1$ and $\mathscr{E} Z_1^2 < \infty$. Then for each $j=1, 2, \ldots,$*

$$\lim_{n\to\infty} P(Z_n = j | Z_n \neq 0) = b_j \tag{9.1}$$

exists, and $\sum_{j=1}^{\infty} b_j = 1$. Moreover, the generating function $g(s) = \sum b_j s^j$ satisfies the functional equation

$$g[f(s)] = m g(s) + 1 - m, \qquad |s| \leqq 1. \tag{9.2}$$

Proof. Under the hypotheses of the theorem we have

$$\left.\begin{array}{c} f(s) = 1 + m(s-1) + \frac{1}{2} f''(1)(s-1)^2 + o(s-1)^2, \\ s \in S_1(1). \quad \text{(See Definition 9.1.)} \end{array}\right\} \tag{9.3}$$

Arguing as in the proof of Theorem 8.4, we obtain the relation

$$f_n(s) = 1 + m^n B_1(s)(s-1)[1 + R_n(s)], \qquad s \in S_2(1), \tag{9.4}$$

where B_1 is continuous, $B_1(1) \neq 0$, and $|R_n(s)| \leqq d_1 m^n$. By putting $s = f_k(0)$ in (9.4), where k is sufficiently large so that $f_k(0) \in S_2(1)$, we find that

$$1 - f_n(0) \sim c_1 m^n, \qquad n \to \infty, \tag{9.5}$$

where c_1 is positive. This result was given by KOLMOGOROV (1938). From (9.4) and (9.5) we obtain

$$\left.\begin{aligned} \lim_{n\to\infty} \frac{f_n(s) - f_n(0)}{1 - f_n(0)} &= \lim_{n\to\infty} \frac{f_n(s) - 1}{1 - f_n(0)} + 1 \\ &= \left(\frac{1}{c_1}\right) B_1(s)(s-1) + 1 \\ &= g(s), \qquad \text{say}, \qquad s \in S_2(1). \end{aligned}\right\} \tag{9.6}$$

The functions $[f_n(s) - f_n(0)]/[1 - f_n(0)]$ are probability generating functions for each n and therefore are bounded in the unit circle uniformly in n. Hence (see TITCHMARSH (1939, p. 168)) these generating functions converge uniformly in any closed region interior to the unit circle, and the limit $g(s)$ must then be analytic whenever $|s|<1$. Since the derivatives of these functions at $s=0$ also converge to the derivatives of g at $s=0$ (TITCHMARSH (1939, pp. 95, 98)), we see that g is a power series whose coefficients $b_1, b_2, \ldots$, which are all nonnegative, are the limits of the conditional probabilities (9.1). From the relation between g and B in (9.6), we see that g is continuous at $s=1$ and that $g(1)=1$, and hence $\sum b_j=1$.

Now replace s by $f(s)$ in equation (9.6), taking s close enough to 1 so that both s and $f(s)$ are in $S_2(1)$. Using (9.5) we then obtain (9.2) for $s\in S_2(1)$, $f(s)\in S_2(1)$. Hence, by analytic continuation, equation (9.2) holds throughout the unit circle. □

Unfortunately an explicit expression is not available for $g'(1)$, although we can evaluate it numerically by calculating the iterates $f_n(0)$. The procedure, which incidentally proves that $g'(1)<\infty$, is as follows. Repeated application of (9.2) for $s=0$ gives

$$g[f_n(0)]-1=m^n[g(0)-1]=-m^n, \qquad n=1,2,\ldots. \tag{9.7}$$

Dividing both sides of (9.7) by $f_n(0)-1$, letting $n\to\infty$, and using (9.5), we obtain

$$g'(1)=\lim_{n\to\infty}\frac{g[f_n(0)]-1}{f_n(0)-1}=\lim_{n\to\infty}\frac{m^n}{1-f_n(0)}=\frac{1}{c_1}. \tag{9.8}$$

Remark on uniqueness. Since $g(1)=1$ and $g'(1)<\infty$, g has the form $g(s)=1+(s-1)\theta(s)$, where θ is continuous and $\theta(1)=g'(1)$. We leave it to the reader to show that (9.2) cannot have another solution of this sort.

From (5.3) we deduce that if $m<1$ and $f''(1)<\infty$, then $f_n''(1)=O(m^n)$. Hence the fractions in (9.6) have second derivatives that are bounded for $|s|\leqq 1$, uniformly in n. Thus $g''(1)<\infty$. We can then determine the value of $g''(1)$ by twofold differentiation of (9.2) at $s=1$. Presumably, we can determine higher derivatives of g at 1 (i.e., higher factorial moments) by a similar process.

The reader may wish to verify that if $f(s)$ is a fractional linear function with $f'(1)<1$, then g is also a fractional linear function, which can be calculated explicitly.

10. Asymptotic results when $m=1$

10.1. Form of the iterates f_n when $m=1$. The limiting distributions when $m\neq 1$ assume a great variety of forms. In fact, f is uniquely determined by φ in the case $m>1$, through the relation $f(s)=\varphi[m\varphi^{-1}(s)]$ (see (8.7)), and similarly f is determined by g in case $m<1$. This makes it rather surprising that when $m=1$, we have an exponential limiting distribution, regardless of the form of f provided $f'''(1)<\infty$. (However, see Sec. V.11.4.)

Our treatment will be based on Lemma 10.1, which follows in essence the treatment of FATOU (1919). The part of the lemma making use of the existence of the fourth moment for Z_1 is not needed in the present section, but will be required in Sec. 11.

In the remainder of Sec. 10, $O(x)$ *will designate a quantity bounded by* $C|x|$, *where the constant* C *does not depend on* n, s, *or other variables.*

Lemma 10.1. *Suppose* $m=1$ *and* $f'''(1)<\infty$. *Let* S *denote the points* s *that either* (a) *are interior to the unit circle or* (b) *lie on the segment of the unit circle* $-\theta_0\leqq \arg s\leqq\theta_0$, *excluding the point* $s=1$, *where* θ_0 *is a positive number to be specified in the proof. Then*

$$\frac{1}{1-f_n(s)}=\frac{1}{1-s}+\frac{nf''(1)}{2}+O(\log n),\qquad s\in S,\ n\to\infty. \tag{10.1}$$

If in addition $f^{\text{iv}}(1)<\infty$, *then the expression*

$$\frac{1}{1-f_n(s)}-\frac{1}{1-s}-\frac{nf''(1)}{2}-\left(\frac{(f''(1))^2}{4}-\frac{f'''(1)}{6}\right)\sum_{j=0}^{n-1}\frac{1}{\dfrac{1}{1-s}+\dfrac{1}{2}jf''(1)} \tag{10.2}$$

converges to a finite limit as $n\to\infty$, *uniformly for* s *in* S. *The limit is bounded for* $s\in S$.

Proof. We assume $|s|\leqq 1$ throughout. Since $f(e^{i\theta})$ is the characteristic function of a discontinuous distribution with more than one step, there is a $\theta_0>0$ such that $|f(e^{i\theta})|<1$ for $|\theta|\leqq\theta_0$, $\theta\neq 0$. (See LOÈVE (1960, p. 202).) We use this θ_0 in defining S. Now the Basic Assumptions of Sec. 3 imply that $|f(s)|<1$ if $|s|<1$. Hence if $s\in S$, none of the iterates $f_0(s), f_1(s), \ldots$ is equal to 1, and all except possibly $f_0(s)$ are less than 1 in absolute value. Moreover, since $f_n(0)\to 1$, we have $\lim_{n\to\infty} f_n(s)=1$ uniformly for $s\in S$.

Our hypotheses imply that

$$\left.\begin{aligned} 1-f(s)&=1-s-a(1-s)^2+b(1-s)^3-\gamma(s),\\ a&=\tfrac{1}{2}f''(1),\qquad b=\tfrac{1}{6}f'''(1),\end{aligned}\right\} \tag{10.3}$$

where $\gamma(s)=o(1-s)^3$ if $f'''(1)<\infty$, and $\gamma(s)=O(1-s)^4$ if $f^{\text{iv}}(1)<\infty$.

Using (10.3) we obtain, if $f'''(1)<\infty$,

$$\left.\begin{aligned}\frac{1}{1-f_j(s)}&=\frac{1}{1-f(f_{j-1}(s))}\\ &=\frac{1}{1-f_{j-1}(s)}+a+(a^2-b)\left(1-f_{j-1}(s)\right)\\ &\quad+\frac{\gamma(f_{j-1}(s))}{(1-f_{j-1}(s))^2}+\delta\left(f_{j-1}(s)\right),\qquad s\in S;\quad n=1,2,\ldots,\end{aligned}\right\} \tag{10.4}$$

where $\delta(s)=O(1-s)^2$. Summing both sides of (10.4) on j from 1 to n, we have

$$\left.\begin{aligned}\frac{1}{1-f_n(s)}&=\frac{1}{1-s}+na+(a^2-b)\sum_{j=0}^{n-1}\left(1-f_j(s)\right)\\&\quad+\sum_{j=0}^{n-1}\frac{\gamma(f_j(s))}{(1-f_j(s))^2}+\sum_{j=0}^{n-1}\delta\left(f_j(s)\right).\end{aligned}\right\}\tag{10.5}$$

Since $f_n(s)\to 1$ uniformly for s in S as $n\to\infty$, each of the last three sums on the right side of (10.5) is $o(n)$. Moreover, $Re[1/(1-s)]\geqq 0$ for $|s|\leqq 1$. Hence, taking the reciprocals of the two sides of (10.5), we see that $1-f_n(s)=O(1/n)$, and hence each of the last three sums in (10.5) is $O(\log n)$. Thus (10.1) is proved.

Next, using (10.1), we have (defining u_j by the context)

$$\left.\begin{aligned}u_j(s)&=1-f_j(s)-\frac{1}{\dfrac{1}{1-s}+ja}\\&=\frac{O(\log j)}{\left[\dfrac{1}{1-s}+ja+O(\log j)\right]\left[\dfrac{1}{1-s}+ja\right]},\end{aligned}\right\}\tag{10.6}$$

and since $Re[1/(1-s)]\geqq 0$, the right side of (10.6) is $O(\log j/j^2)$. Hence $\sum|u_j(s)|<\infty$ for $s\in S$. Returning to (10.5) and assuming now that $f^{\text{iv}}(1)<\infty$, we see that both the last two sums in (10.5) converge absolutely. Hence the expression

$$\frac{1}{1-f_n(s)}-\frac{1}{1-s}-na-(a^2-b)\sum_{j=0}^{n-1}\frac{1}{\dfrac{1}{1-s}+ja}\tag{10.7}$$

approaches a limit as $n\to\infty$, for each $s\in S$. Since the bounds on the u_j and on the terms of the last two sums in (10.5) are uniform for $s\in S$, the lemma is proved. □

10.2. The probability of extinction when n is large. It was shown by Kolmogorov (1938) that if $m=1$ and $f'''(1)<\infty$, then

$$P\{Z_n>0\}\sim\frac{2}{nf''(1)}.\tag{10.8}$$

This result can be obtained by putting $s=0$ in (10.1).

10.3. Distribution of Z_n when n is large. Since

$$\mathscr{E}Z_n=1=\mathscr{E}(Z_n|Z_n\neq 0)\,P(Z_n\neq 0),$$

we see from (10.8) that

$$\mathscr{E}(Z_n|Z_n\neq 0)\sim\frac{nf''(1)}{2},\qquad n\to\infty,\tag{10.9}$$

which suggests that we look for a conditional distribution not for Z_n but for $2Z_n/nf''(1)$. We then obtain the following result, due to YAGLOM (1947). The result for the special case when f has the Poisson form was given by FISHER (1930a).

Theorem 10.1. *Suppose* $m=1$ *and* $f'''(1)<\infty$. *Then*

$$\lim_{n\to\infty} P\left(\frac{2Z_n}{nf''(1)} > u \middle| Z_n \neq 0\right) = e^{-u}, \qquad u \geqq 0.$$

Proof. The conditional characteristic function of the random variable $2Z_n/nf''(1)$, given that $Z_n \neq 0$, is

$$\Psi_n(t) = \frac{f_n(e^{2it/nf''(1)}) - 1}{1 - f_n(0)} + 1, \qquad n=1, 2, \ldots; \quad -\infty < t < \infty. \tag{10.10}$$

Let I be the closed interval $[-t_0, t_0]$, *excluding the point* $t=0$, where t_0 is an arbitrary positive number. If n is sufficiently large, then $e^{2it/nf''(1)}$ belongs to S (see Lemma 10.1) for any t in I. Hence we can apply (10.1) to the right side of (10.10), obtaining

$$\Psi_n(t) = 1 - \frac{\frac{nf''(1)}{2}[1 + O(\log n/n)]}{\frac{1}{1 - e^{2it/nf''(1)}} + \frac{nf''(1)}{2} + O(\log n)}, \qquad t \in I. \tag{10.11}$$

Letting $n \to \infty$ in (10.11), we see that $\Psi_n(t) \to 1/(1-it)$, $t \in I$, this being also true, as can be seen directly, for $t=0$. Since t_0 is arbitrary, the functions Ψ_n must converge for each t to $1/(1-it)$, which is the characteristic function of the exponential distribution. □

ČISTYAKOV (1957) states that N. V. SMIRNOV has given a limit theorem for the individual probabilities in the case $m=1$.

10.4. The past history of a surviving family when $m=1$. Suppose a family for which $m=1$ happens to be surviving after a large number N of generations. Theorem 10.1 gives the conditional distribution of Z_N. However, we may be interested in the conditional distribution of the size of the family at time $n<N$, given that it has survived N generations. If n and $N-n$ are both large, there is again a universal limiting conditional distribution[1]. In fact, the conditional distribution of $2Z_n/nf''(1)$, given $Z_N>0$, as $n\to\infty$ and $N-n\to\infty$, approaches the distribution whose density function is ue^{-u}, $u>0$. We omit the proof.

11. Stationarity of Z_n

We have shown (Theorem 6.2) that Z_n either goes to 0 or goes to ∞ and thus is not the sort of process that can have a stationary probability distribution, except the trivial one $Z_n \equiv 0$. Nevertheless we shall see

[1] HARRIS (1951).

that if $p_0 > 0$ we can always find a stationary "distribution", or *stationary measure*, if we allow "probabilities" whose sum is infinite, and if we give special status to the absorbing state 0. After establishing the general results, we shall give in Sec. 12 an interesting application of stationary measures to genetics, due to R. A. FISHER (1930a, 1930b). In Chapter V we shall see how stationary measures can be determined in more explicit form in the case of continuous time.

11.1. Stationary probabilities. Let (P_{ij}) be the transition matrix for a Markov chain with states 0, 1, 2, ... By a set of *stationary probabilities* for (P_{ij}) we mean a set of numbers π_i, $i=0, 1, 2, \ldots$, satisfying

$$\pi_i = \sum_{j=0}^{\infty} \pi_j P_{ji}, \qquad i=0,1,2,\ldots; \ \pi_i \geqq 0; \ \sum_{i=0}^{\infty} \pi_i = 1. \tag{11.1}$$

The importance of stationary probabilities is explained in textbook discussions of Markov chains. We observe that if $X_0, X_1, \ldots$ are the variables of a Markov chain and if X_0 has the probability distribution $\mathrm{Prob}(X_0=i)=\pi_i$, $i=0, 1, \ldots$, then $X_1, X_2, \ldots$ all have this same distribution. Furthermore, under conditions found in texts, we have $\lim_{n\to\infty} \mathrm{Prob}(X_n=j|X_0=i)=\pi_j$ for any i and j.

If (P_{ij}) is the transition matrix for a Galton-Watson process, then it is obvious that $(1, 0, 0, \ldots)$ is a set of stationary probabilities. We can show that there is no other set of stationary probabilities by the following argument: (a) Suppose $p_0 > 0$. Then $P_{j0} > 0$, $j=0, 1, 2, \ldots$. From the equation $\pi_0 = \pi_0 P_{00} + \pi_1 P_{10} + \cdots$ and the fact that $P_{00}=1$, we see that $\pi_i = 0$ for $i > 0$. (b) Suppose $p_0 = 0$. Then $P_{ij}=0$ for $j < i$, and hence from (11.1) $\pi_i = \pi_1 P_{1i} + \cdots + \pi_i P_{ii}$, $i=1, 2, \ldots$. Let π_n be the first one of the numbers $\pi_1, \pi_2, \ldots$ that is different from 0. Then the above equation gives $\pi_n = P_{nn}\pi_n$. Now P_{nn} is the coefficient of s^n in the expansion of $(f(s))^n$ and it is a consequence of the Basic Assumptions of Sec. 3 that $P_{nn} < 1$. But this implies that $\pi_n = 0$, contrary to hypothesis. Hence $\pi_1, \pi_2, \ldots$ must all be 0.

11.2. Stationary measures. A stationary measure for a Markov chain is a set of nonnegative numbers π_i, not all 0, satisfying (11.1) except that the sum $\sum \pi_i$ need not be finite. For example, there is always a stationary measure if the chain is indecomposable and recurrent (DERMAN (1954, 1955)), and sometimes (but not always) if the chain is transient (HARRIS (1957a)). For further discussion of stationary measures and their uses see HARRIS and ROBBINS (1953), HARRIS (1956), and CHUNG (1960). As shown by DERMAN (1955), the π_i determine a stationary distribution for a set of particles in the states 0, 1, 2, ..., each wandering independently of the others according to the transition law of the chain. The number of particles in the state i has a Poisson distribution with mean proportional to π_i, and the numbers in different states at a given time are independent.

Strictly speaking, the Galton-Watson process has no stationary measure other than $(1, 0, 0, \ldots)$, even allowing $\sum \pi_i = \infty$. This follows from the argument in Sec. 11.1, which did not assume $\sum \pi_i < \infty$. The intuitive reason is that if $p_0 = 0$, the particles will all drift away to the right. (The trivial case $p_1 = 1$ is excluded.) On the other hand, if $p_0 > 0$, then an unlimited amount of mass will pile up in the state 0. However, this allows the possibility of a stationary measure for the states other than 0. The next result shows that there is one. For this result, it is not necessary to require the existence of any moments of Z_1, even the first. We also allow the case $f(s) = p_0 + (1 - p_0)s$, excluded by our Basic Assumptions.

11.3. Existence of a stationary measure for the Galton-Watson process.

Theorem 11.1.[1] *Let (P_{ij}) be the transition matrix for a Galton-Watson process for which $0 < p_0 < 1$. Then there exist nonnegative numbers $\pi_1, \pi_2, \ldots$, not all 0, satisfying the equations*

$$\pi_i = \sum_{j=1}^{\infty} \pi_j P_{ji}, \qquad i = 1, 2, \ldots, \tag{11.2}$$

and $\sum \pi_i = \infty$. We shall refer to the numbers $\pi_1, \pi_2, \ldots$ as a stationary measure for the Galton-Watson process. The generating function

$$\pi(s) = \sum_{j=1}^{\infty} \pi_j s^j \tag{11.3}$$

is analytic for $|s| < q$, where q is the probability of extinction, and satisfies the functional equation (Abel's equation)

$$\pi[f(s)] = 1 + \pi(s), \qquad |s| < q, \tag{11.4}$$

provided the numbers π_i are multiplied by a constant factor so that $\pi[f(0)] = 1$.

Remark 1. We can also interpret the set $(\infty, \pi_1, \pi_2, \ldots)$ as a stationary measure on the states $0, 1, 2, \ldots$, provided we interpret $0 \cdot \infty$ as 0.

Remark 2. Some generating functions f correspond to just one stationary measure (except for a constant multiplicative factor), and some correspond to more than one. See Sec. 11.4 and page 29.

Although, as stated above, Theorem 11.1 is true with no requirements on the moments of Z_1, we shall give a proof, of a purely analytic nature, that requires some assumptions on the moments. The result can be proved without these assumptions by a probabilistic method that we shall only discuss briefly below, since it is rather lengthy.

---

[1] This result and its relation to FISHER's work were announced in HARRIS (1958).

(a) Proof of Theorem 11.1 *when* $m<1$, $\mathscr{E}Z_1^2<\infty$. Let $g(s)$ be the function of Sec. 9, satisfying (9.2). The function

$$\pi(s)=\frac{\log[1-g(s)]}{\log m}=\frac{-g-\frac{1}{2}g^2-\frac{1}{3}g^3-\cdots}{\log m} \tag{11.5}$$

is obviously a power series with nonnegative coefficients, satisfying (11.4), and analytic for $|s|<1$, with $\pi(0)=0$. Comparing coefficients of powers of s on the two sides of (11.4), we see that the coefficients satisfy (11.2). Putting $s=1$ in (11.4) we see that $\pi(1)=1+\pi(1)$. Hence $\sum\pi_i=\infty$.

(b) Proof of Theorem 11.1 *when* $m>1$. (Second moment not required.) This is very similar to (a). Using an argument similar to that which established Theorem 9.1 and (9.2), we see that there is a function $g^*(s)=b_1^*s+b_2^*s^2+\cdots$, with nonnegative coefficients, satisfying $g^*(q)=1$, and

$$g^*(s)=\lim_{n\to\infty}\frac{f_n(s)-f_n(0)}{q-f_n(0)},$$

$$g^*[f(s)]=f'(q)\,g^*(s)+1-f'(q), \qquad |s|\leqq q.$$

We may then take

$$\pi(s)=\frac{\log[1-g^*(s)]}{\log f'(q)}, \qquad |s|<q.$$

From (11.4), $\pi(q)=\infty$. Hence $\sum\pi_i=\infty$.

(c) Proof of Theorem 11.1 *when* $m=1$, $f^{\text{iv}}(1)<\infty$. Here we follow Fatou (1919) except for special arguments affected by the positivity of the coefficients. By elementary arguments (see, e.g., Courant (1937, Vol. 1, p. 381)), we find that if a is defined by (10.3) and $0\leqq s<1$, then the expression

$$\sum_{j=0}^{n-1}\frac{1}{\dfrac{1}{1-s}+ja}-\frac{\log[1+na(1-s)]}{a} \tag{11.6}$$

is bounded uniformly in n and s, and approaches a limit as $n\to\infty$ for each s, $0\leqq s<1$. From (11.6) and (10.2) we see that if we put

$$R_n(s)=\frac{1}{1-f_n(s)}-\frac{1}{1-f_n(0)},$$

then

$$R_n(s)=\frac{s}{1-s}+\frac{a^2-b}{a}\log\left(\frac{1+na(1-s)}{1+na}\right)+\delta_n(s), \tag{11.7}$$

where the functions δ_n are bounded uniformly in n and s, $0\leqq s<1$, and approach a bounded limit as $n\to\infty$. From (11.7) and the definition of the functions R_n we see that they are power series with nonnegative coefficients that are bounded uniformly in n in the region $|s|\leqq r<1$, where r is as close to 1 as we please. (The bound will depend on r.)

Hence (TITCHMARSH (1939, p. 168)) the functions R_n converge for $|s|<1$ to a limiting function $R(s)$, analytic for $|s|<1$. Moreover, we see from (11.7) that

$$R(s) - \frac{s}{1-s} - \frac{a^2-b}{a}\log(1-s) \tag{11.8}$$

is bounded for $0 \leqq s < 1$.

From (10.2) we see that

$$\lim\left(\frac{1}{1-f_{n+1}(0)} - \frac{1}{1-f_n(0)}\right) = \frac{f''(1)}{2}.$$

Hence

$$\left.\begin{aligned} R[f(s)] &= \lim\left\{\frac{1}{1-f_{n+1}(s)} - \frac{1}{1-f_{n+1}(0)} + \frac{1}{1-f_{n+1}(0)} - \frac{1}{1-f_n(0)}\right\} \\ &= R(s) + \tfrac{1}{2}f''(1). \end{aligned}\right\} \tag{11.9}$$

Hence we see that $2R(s)/f''(1)$ is a solution of equation (11.4), and the coefficients in its expansion are a solution of (11.2).

(d) Proof of Theorem 11.1 *in the general case.* We can prove that (11.2) has a nonnegative solution, without assuming the existence of moments, and without considering the cases $m<1$, $m=1$, and $m>1$ separately, by making use of a probabilistic method essentially the same as the method used to prove the existence of stationary measures for certain transient chains (HARRIS (1957a, Theorem 2)). A brief indication of the argument follows.

Let $f(s)$ have the form $p_0 + p_{r_1}s^{r_1} + p_{r_2}s^{r_2} + \cdots$, $p_{r_i} > 0$, $p_0 > 0$, where $0 < r_1 < r_2 < \cdots$. Let T be the set of positive integers (or states) of the form $n_1 r_1 + n_2 r_2 + \cdots$, where $n_1, n_2, \ldots$ are nonnegative integers. If i and j belong to T, then $P_{ij}^n > 0$ for infinitely many values of n, where P_{ij}^n is the n-step transition probability from i to j. However, if $j \notin T$, $j \neq 0$, then P_{ij}^n is 0 for every i and every $n > 0$.

Let ℓ be any state in T; ℓ will remain fixed in the rest of this discussion. Now consider the quantities

$$\pi_i^{(k)} = \sum_{n=1}^{\infty} P_{ki}^n \Big/ \sum_{n=1}^{\infty} P_{k\ell}^n, \qquad k \in T;\ i = 1, 2, \ldots.$$

It can be shown that there is a sequence $k_1 < k_2 < \cdots$ of integers, each belonging to T, such that $\lim_{k_r\to\infty} \pi_i^{(k_r)} = \pi_i$ exists and is positive and finite for $i \in T$ and 0 for $i \notin T$, $i \neq 0$, and that the π_i are a stationary measure. The main work in the proof is in verifying that a condition is satisfied analogous to the sufficient condition of Theorem 2 of HARRIS (1957a)[1].

[1] The author takes this opportunity to make the following minor corrections to HARRIS (1957a). Page 938, line 20, "p_{ij}" should be "P_{ij}". Page 940, line 7, replace "$k>j$" by "$k \geqq j > i$"; line 9, replace "As before, $L_{ki} = L_{ki}(0)$" by "L_{ki}"; line 18, replace "P_{0i}^n" by "${}_0P_{0i}^n$".

11.4. The question of uniqueness. It has long been known that (11.4) has more than one solution (ABEL (1881, posthumous)). If $\omega(s)$ is defined for all s and has period 1, and if π is a solution of (11.4), then $\pi(s)+\omega(\pi(s))$ is also a solution, and, at least in a suitable region, every solution has this form. J. F. C. KINGMAN, answering negatively a conjecture of the author, has given an example, with $m>1$, where (11.4) has distinct power series solutions with nonnegative coefficients, leading to distinct stationary measures. Sometimes there is only one (see the end of Sec. 11.6). Concerning (11.4), see FATOU (1919, 1920).

11.5. The form of the π_i when i is large, for the case $m=1$. The behavior of the function $R(s)$ indicated in (11.8) suggests that if we take $\pi(s)$ as the function $2R(s)/f''(1)$ when $m=1$, then the π_i have, for large i, the form

$$\pi_i = \frac{2}{f''(1)} - \left[1 - \frac{2}{3}\,\frac{f'''(1)}{(f''(1))^2}\right]\frac{1}{i} + o\left(\frac{1}{i}\right). \tag{11.10}$$

However, the relation (11.10) cannot always be true. For example, if the p_i differ from 0 only when i is even, then the π_i have the same property. This leaves open the question whether (11.10) holds except in such "periodic" cases.

We can at least assert something about the averages, since it follows from (11.8) and a well-known result of HARDY and LITTLEWOOD that

$$\lim_{n\to\infty}\frac{1}{n}\sum_{i=1}^{n}\pi_i = \frac{2}{f''(1)} \tag{11.11}$$

(TITCHMARSH (1939, p. 226)). Without the normalization $\pi(f(0))=1$, the right side of (11.11) must be multiplied by $\pi(f(0))$.

In fact, (11.11) holds for *any* stationary measure satisfying (11.2) in case $m=1$ and $f^{\text{iv}}(1)<\infty$. Since we have not proved uniqueness, we shall at least prove this.

Theorem 11.2. *Suppose $p_0>0$, $m=1$, and $f^{\text{iv}}(1)<\infty$. Let $\tau_1, \tau_2, \ldots$ be a stationary measure. Then the series $\tau(s)=\sum\tau_i s^i$ converges for $|s|<1$. If we normalize so that $\tau(f(0))=1$, then τ satisfies (11.4) for $|s|<1$ and the τ_i satisfy (11.11), with π_i replaced by τ_i.*

Proof. Let $f(s)=p_0+p_1 s+\cdots$ as usual, and let p_r be the first of the numbers $p_1, p_2, \ldots$ that is positive. Then the transition probability P_{jr} has the form $j p_r p_0^{j-1}$, $j=1, 2, \ldots$. Since $\tau_r=\sum\tau_j P_{jr}<\infty$, we see that $\sum j\tau_j p_0^{j-1}<\infty$. Hence the series $\tau(s)=\sum\tau_j s^j$ is convergent at least for $|s|<p_0$. Multiplying both sides of the equality $\tau_i=\sum_j\tau_j P_{ji}$ by s^i, summing from $i=1$ to ∞, and changing the order of summation on the right, we thus obtain

$$\tau(s)=\sum_{j=1}^{\infty}\tau_j\left(f(s)^j-f(0)^j\right)<\infty, \qquad 0\leqq s<p_0. \tag{11.12}$$

Since $f(s)^j - f(0)^j \geqq (f(s) - f(0)) f(s)^{j-1}$ for $0 \leqq s < p_0$, we see from (11.12) that $\sum \tau_j (f(s))^j$ is finite when $0 \leqq s < p_0$, and hence the radius of convergence of $\tau(s)$ is at least $f(p_0)$, which is greater than p_0. Repeating the argument given above, we see that the equation in (11.12) holds for $0 \leqq s < f(p_0)$, and hence that the radius of convergence of τ is at least $f(f(p_0)) = f_2(p_0)$. Similarly the radius is at least $f_n(p_0)$, $n=3, 4, \ldots$, and hence is at least 1. If we normalize so that $\sum \tau_i (p_0)^i = 1$, we can then assert from (11.12) that

$$\tau(f(s)) = \tau(s) + 1, \qquad |s| < 1. \tag{11.13}$$

Letting $s \to 1$ in (11.13) we see that $\sum \tau_i = \infty$.

Next, using an argument deriving in essence from ABEL, replace $f(s)$ by $\pi^{-1}[\pi(f(s))]$ on the left side of (11.13) and replace s by $\pi^{-1}(\pi(s))$ on the right side, $0 \leqq s < 1$, where π is the function defined just above (11.10). Replacing $\pi[f(s)]$ by $\pi(s)+1$ and then letting $\pi(s) = z$, we have $\tau[\pi^{-1}(z+1)] = \tau[\pi^{-1}(z)] + 1$, which shows that $\tau[\pi^{-1}(z)] - z$, which we denote by $\omega(z)$, has period 1 for $0 \leqq z < \infty$. Hence $\tau(s) = \pi(s) + \omega[\pi(s)]$, $0 \leqq s < 1$. Notice that we have used no properties of π except that it satisfies (11.4) and increases continuously from 0 to ∞ as s increases from 0 to 1.

The function ω is analytic at every point of the positive real axis. Since it is periodic, it is bounded. Hence $\tau(s)$ must have the same asymptotic behavior as π when $s \to 1$, and hence the theorem of HARDY and LITTLEWOOD can be applied to the τ_i. □

R. A. FISHER (1930a) considered the case where f has the Poisson form e^{s-1} and arrived at the expression $\pi_i \sim 1 - 1/6i$, which agrees with (11.10) except for a multiplicative factor. We shall consider FISHER's application in Sec. 12. HALDANE (1939) extended some of FISHER's work to a general f, although in neither case was the existence of a stationary measure actually demonstrated.

11.6. Example: fractional linear generating function. When $f(s)$ has the fractional linear form (7.3) or (7.4), the following sequences represent stationary measures, normalized so that $\pi[f(0)] = 1$:

$$\pi_i = \frac{1-c}{c}, \qquad i = 1, 2, \ldots; \qquad m = 1; \tag{11.14}$$

$$\pi_i = \frac{1}{\log(1/m)} \frac{1 - (1/s_0)^i}{i}, \qquad i = 1, 2, \ldots; \qquad m < 1; \tag{11.15}$$

$$\pi_i = \frac{1}{\log m} \frac{(1/s_0)^i - 1}{i}, \qquad i = 1, 2, \ldots; \qquad m > 1. \tag{11.16}$$

(Let us recall that $s_0 > 1$ when $m < 1$, and that $s_0 = q < 1$ when $m > 1$.)

It is easy to verify directly that (11.14) gives a stationary measure when $m=1$, since in this case we obtain $\pi(s)=\left(\frac{1-c}{c}\right)\left(\frac{s}{1-s}\right)$, and direct substitution shows that $\pi[f(s)]-\pi(s)=$constant. In case $m<1$, consider the rational function

$$F(s)=\frac{(s_0-f(s))(1-s)}{(1-f(s))(s_0-s)},$$

where f is defined by (7.3) with $n=1$. This function is bounded and is hence constant, but is obviously not zero. Since $\pi(s)$, calculated from (11.15), is proportional to $\log\left(\frac{s_0-s}{s_0(1-s)}\right)$, we see that $\pi[f(s)]-\pi(s)$ is proportional to $F(s)$ and hence is constant. A similar argument holds when $m>1$.

The forms (11.14), (11.15), and (11.16) can be deduced most readily by considering certain processes with a continuous time parameter. We shall do so in Chapter V.

Since the uniqueness property for stationary measures fails to hold in general, it is of interest that the measure (11.14) is unique, except for a multiplicative constant. For suppose $\tau_1, \tau_2, \ldots$ is another stationary measure for the case $m=1$, and put $\tau(s)=\sum\tau_i s^i$. As indicated above,

$$\tau(s)=\pi(s)+\omega[\pi(s)], \tag{11.17}$$

where ω is periodic and analytic and $\pi(s)=(1-c)s/c(1-s)$. If we differentiate (11.17) with respect to s and multiply by π^2/π' we obtain

$$\pi^2+\pi^2\omega'(\pi)=\frac{\tau'\pi^2}{\pi'}=\frac{(1-c)s^2\tau'}{c}, \quad 0\leqq s<1. \tag{11.18}$$

The right side of (11.18) is an increasing function of s, while the derivative of the left side is

$$\pi(s)\pi'(s)[2+2\omega'(\pi(s))+\pi(s)\omega''(\pi(s))]. \tag{11.19}$$

Since ω' is bounded, since $\pi(1)=\infty$, and since ω'' must take negative values unless ω is constant, the expression (11.19) must take negative values unless ω is constant. Hence we obtain a contradiction unless ω is constant. This constant must obviously be 0.

12. An application of stationary measures

In this section we shall interpret in terms of stationary measures part of the treatment of gene fixation by R. A. Fisher (1930a, Chapter IV). The reader is warned that neither the mathematics nor the genetics given below is rigorous. The reader should also consult Moran (1962, Chapter V) for further explanation, and Haldane (1939).

Consider a population of N animals, where N is large and remains approximately the same from generation to generation. In the population are many genetic factors α, β, γ, ..., where α controls some property such as length of tail, β controls perhaps hair color, etc. Each animal has two genes for each factor, each gene having one of two alternative forms or *alleles*; for example, α has the alleles a and A, β has b and B, etc. Thus, in relation to each factor an animal may be one of three genotypes; for the α-factor, the types are aa, aA, and AA.

For convenience, we shall refer to the alleles $a, b, c, \ldots$ as *small* alleles. Let x_n be the number of small alleles of the factor α in the population in the n-th generation. Thus each aa animal contributes 2 to x_n, each aA 1, and each AA 0. Then $0 \leqq x_n \leqq 2N$. If x_n reaches either 0 or $2N$, then either the a's or the A's will have all disappeared, and in the absence of mutations (which we here assume) will not return. We then say that the factor α has been *fixed*.

Let us suppose that $x_1, x_2, \ldots$ is a Markov process, although it is only if x_n/N or $(2N - x_n)/N$ is small that this seems at all plausible. Let (Q_{ij}) be the transition matrix for this process, and let us suppose that the same matrix would apply for any of the other factors $\beta, \gamma, \ldots$. Then if $\pi_i(n)$ is the expected number of factors in the n-th generation represented by i small alleles in the population, we have

$$\pi_i(n+1) = \sum_{j=1}^{2N-1} \pi_j(n)\, Q_{ji}, \qquad i = 1, 2, \ldots, 2N-1; \tag{12.1}$$

we can omit 0 and $2N$ from the range of i and j in (12.1), since they are absorbing states.

The matrix (Q_{ij}), where i and j range only from 1 to $2N-1$, will have a positive eigenvalue λ, $0 < \lambda < 1$, corresponding to a positive eigenvector $(\pi_1^*, \pi_2^*, \ldots, \pi_{2N-1}^*)$; thus

$$\lambda \pi_i^* = \sum_{j=1}^{2N-1} \pi_j^*\, Q_{ji}, \qquad i = 1, 2, \ldots, 2N-1. \tag{12.2}$$

Now if i and j are small compared to N, it is plausible that Q_{ij} is close to P_{ij}, where (P_{ij}) is the transition matrix for a branching process with generating function $f(s) = p_0 + p_1 s + \cdots$. This is because if x_n/N is small, then under the assumptions of random mating, the only probable mating involving the a-allele is $aA \times AA$, and since there are only a few a-alleles, they should not interfere with one another. We suppose $f'(1) = 1$, to accord with a population of stationary size.

If N is large, we should expect λ to be close to 1, since the rate of fixation should be slow compared to the total number of factors. Roughly speaking, this is because Q_{i0} and $Q_{i,2N}$ are small unless i is close to 0 or $2N$. Hence it is plausible that the numbers π_i^* satisfy approximately

$$\pi_i^* = \sum_{j=1} \pi_j^*\, P_{ji}, \tag{12.3}$$

when i/N is small. We have omitted the upper limit of summation in (12.3). Presumably, only the initial terms are important, so we may think of the upper limit as infinite. That is, *the* π_i^* *are approximately a stationary measure for a branching process.* In other words, *the relative frequencies of factors having* 1, 2, 3, ..., *i small alleles are (in a situation of steady decline) proportional to the first i components* $\pi_1, \ldots, \pi_i$ *of a stationary measure, if* i/N *is small.*

We can make an additional deduction (following FISHER) if we are willing to assume that π_i^* is close to π_i even if i/N is not small. There is some plausibility to this, but we do not discuss the plausibility here. The point is discussed by MORAN (1962). Let us suppose we have a situation of steady decline, so that in a given generation the number of factors with i small alleles is π_i^*. Then $\sum_{i=1}^{2N-1} \pi_i^* \sim \sum_{i=1}^{2N-1} \pi_i$ is the total number of unfixed factors. From (11.11) we have

$$\left(\frac{1}{2N-1}\right) \sum_{i=1}^{2N-1} \pi_i \sim \frac{2}{f''(1)} \pi(f(0)).$$

But $\pi(f(0)) = \pi_1 p_0 + \pi_2 p_0^2 + \cdots$ is the number of factors for which the small allele vanishes in the next generation and, from symmetry, $2\pi(f(0))$ is the number for which the small allele either vanishes or becomes $2N$. Hence, *the proportion of factors becoming fixed in each generation is approximately* $f''(1)/2N$ if N is large. (FISHER obtained the result for the Poisson case in which $f''(1)=1$, and HALDANE for the case in which f is arbitrary.)

There are other ways of treating the problem, which has been discussed by many geneticists, and for further references the reader should consult MORAN (1962). See also FELLER (1951) and KARLIN and McGREGOR (1962) for mathematical treatments of certain related problems.

13. Further results on the Galton-Watson process and related topics

In this section we give a number of further results concerning the Galton-Watson process and related topics, referring to published papers for the proofs. Recall that $Z_0=1$.

13.1. Joint generating function of the various generations[1]. Let $g_n(s_1, s_2, \ldots, s_n)$ be the joint generating function of $Z_1, Z_2, \ldots, Z_n$. Then

$$g_n(s_1, \ldots, s_n) = f(s_1 f(s_2 f(\ldots s_{n-1} f(s_n) \ldots))). \tag{13.1}$$

It can be deduced from (13.1) that the conditional distribution of the vector $(Z_2, Z_3, \ldots, Z_n)$, given $Z_1=k$, is that of the sum of k independent random vectors, each distributed like the vector $(Z_1, Z_2, \ldots, Z_{n-1})$.

[1] GOOD (1955). This result is a special case of his.

13.2. Distribution of $Z_0+Z_1+\cdots+Z_n$ and of $Z=Z_0+Z_1+\cdots$. Let $F_n(s)$ be the generating function of $Z_0+\cdots+Z_n$, $n=0, 1, \ldots$, and let $F(s)=\sum_{i=1}^{\infty} P(Z=i)\,s^i$. Then $F(1)=1$ or $F(1)<1$, depending on whether or not Z is finite with probability 1. The generating functions $F_1, F_2, \ldots$ are connected by the recurrence relation

$$F_{n+1}(s)=s f[F_n(s)], \qquad n=0, 1, \ldots, \tag{13.2}$$

and the function F satisfies the relation[1]

$$F(s)=s f[F(s)]. \tag{13.3}$$

The coefficients in the expansion of F can be determined by using Lagrange's expansion.

We can use equation (13.3) to investigate the form of the probabilities $P(Z=i)$ when i is large. In this connection we have the following result.

Theorem 13.1.[2] *Suppose that $p_0>0$ and that there is a point $a>0$ in the interior of the circle of convergence of f for which $f'(a)=f(a)/a$. (This is true, for example, if $1<m\leqq\infty$ or if $f(s)$ is entire or if $f'(\varrho)=\infty$, where ϱ is the radius of convergence of f. The point $(a, f(a))$ is then the point where the graph of f, for positive real s, is tangent to a line through the origin.) Let $\alpha=a/f(a)$ and let d be the largest integer such that $p_r\neq 0$ implies that r is a multiple of d, $r=1, 2, \ldots$. If $r-1$ is not divisible by d, then we have $P(Z=r)=0$, while if $r-1$ is divisible by d, then*

$$P(Z=r)=d\left(\frac{a}{2\pi\alpha f''(a)}\right)^{\frac{1}{2}}\alpha^{-r}r^{-\frac{3}{2}}+O(\alpha^{-r}r^{-\frac{5}{2}}), \qquad r\to\infty. \tag{13.4}$$

Notice that $\alpha\geqq 1$, the equality holding if and only if $m=1$.

13.3. The time to extinction. In case $m\leqq 1$, let N be the smallest integer n such that $Z_n=0$. Then $P(N=n)=f_n(0)-f_{n-1}(0)$, $n=1, 2, \ldots$. In general it is not easy to determine these probabilities explicitly, or even to determine the moments of N. The problem is discussed briefly in Harris (1948).

13.4. Estimation of parameters. The question of estimating the parameters p_r from observations on a sample sequence $Z_1, Z_2, \ldots$ has not been treated systematically. The ratio $(Z_1+\cdots+Z_n)/(Z_0+\cdots+Z_{n-1})$ is a maximum likelihood estimate of m. Moreover, if $m>1$, if $\mathscr{E}Z_1^2<\infty$, and if ε is any positive number, then[3]

$$\lim_{n\to\infty} P\left(\left|\frac{Z_1+\cdots+Z_n}{Z_0+\cdots+Z_{n-1}}-m\right|>\varepsilon \,\middle|\, Z_n\neq 0\right)=0.$$

[1] Hawkins and Ulam (1944), Good (1949), Otter (1949). If the initial object is not counted, the equation becomes $F(s)=f[sF(s)]$.

[2] Otter (1949).

[3] Harris (1948).

13.5. Variable generating function. Suppose that an object in the n-th generation has the generating function $f^{(n)}(s)$ for the number of its children. Then the generating function for the number of objects in the n-th generation is $f^{(0)}(f^{(1)} \ldots (f^{(n-1)}(s)) \ldots)$. Formula (13.1) can be generalized in a similar way[1].

13.6. Trees, etc. We have considered $Z_0, Z_1, \ldots$ as an integer-valued Markov process representing the total number of objects in successive generations. Such a representation does not distinguish between different family trees, as long as the numbers in the various generations are the same. Thus consider the three families in Fig. 3, where the children of an object are the dots (vertices) at the end of the segments going to the right of it.

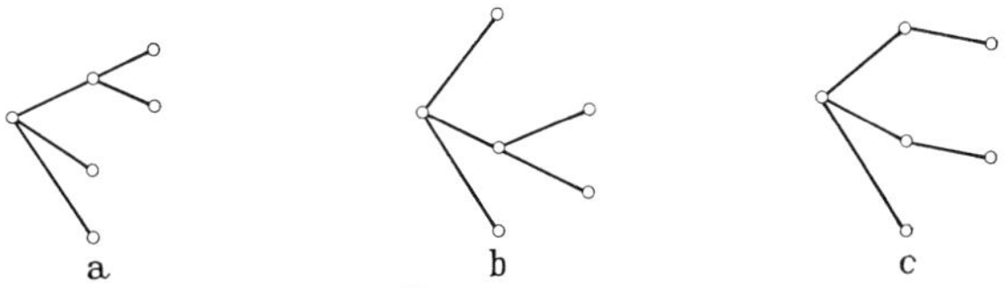

Fig. 3

In all three cases we have $Z_1=3$, $Z_2=2$. However, if we stipulate, regarding two dots linked to the same vertex, that the higher one corresponds to the older child, then (a), (b), and (c) represent three different family structures.

In some problems in chemistry it is required to enumerate the trees having a given number of vertices[2]. However, in these applications the two trees (a) and (b) would not be distinguished; thus the enumeration problem is not the same as the one usually considered for family trees.

In Chapter VI we shall discuss the problem of defining a branching process by means of an explicitly given family tree.

There is an intimate connection between trees and random walks[3], and there are also connections between branching processes and the formation of queues[4].

13.7. Percolation processes. The mathematical treatment of branching processes is simple because the reproduction of an object is assumed to be independent of past history and of the presence of other objects. Once the assumption of independence is dropped, there is no simple way of classifying the resulting processes. They may be Markov (or non-Markov) processes of quite general types.

[1] Good (1955).
[2] Pólya (1937); Otter (1948).
[3] Good (1951); Harris (1952).
[4] Kendall (1951).

Percolation processes may be considered mathematical models for the spread of an infection among objects arranged in a lattice structure. The most important difference between such a spread and a branching process is that the objects surrounding a newly infected object may already have been infected. Thus the further spread of the infection depends on the past history of the infection. Percolation processes may also be considered models for the penetration or "percolation" of a fluid into a porous medium through channels that are subject to random blocking or damming.

As defined by BROADBENT and HAMMERSLEY (1957), and studied further by HAMMERSLEY (1957a, 1957b, 1959), a percolation process consists of a set of *atoms* (or vertices) of which certain pairs are connected by *bonds* (or links). The connection may be in both directions or only one. Each bond has a certain probability of being dammed (i.e., impervious to the transport of fluid), independently of other bonds. One can then study questions such as the probability distribution of the number of atoms connected to a given atom by chains of undammed bonds, or the probability that a given atom is connected to infinitely many other atoms.

Percolation processes have some relation to mathematical theories that have been proposed for random nerve networks[1] and to theories of random clusters of points on a lattice[2], although the studies on these subjects have considered different kinds of problems from those mentioned above.

Since the subject is rather complicated, we refer the reader to the references cited above. A special result for percolation processes on lattices of squares in the plane was given by HARRIS (1960a).

Chapter II

Processes with a finite number of types

1. Introduction

A first step in generalizing the simple Galton-Watson process is the consideration of processes involving several types of objects. For example, in cosmic-ray cascades electrons produce photons, and photons produce electrons. In the reproduction of certain bacteria the usual form may produce a mutant form that behaves differently. Or, objects may be characterized by a continuous variable, such as position or age, but as an approximation the range of the variable may be divided

[1] E.g., ECCLES (1952), pp. 251ff.

[2] E.g., KRISHNA IYER (1955).

into a finite number of portions, each portion corresponding to one type. This device was suggested by SCOTT and UHLENBECK (1942) in their treatment of cosmic rays, where the continuous variable is energy, and was used by BARTLETT (1946) and LESLIE (1948) in dealing with human populations, where the continuous variable is age.

From the point of view of the general theory of Markov processes, it makes little difference whether one or several types are considered, since it is still a question of Markov processes with a denumerable number of states. However, from the point of view of the special theorems for branching processes that were given in Chapter I, some new problems arise; for example, we must consider the ratios of the numbers of the different types; and the classification of the different types, according to the ability of one to produce another, becomes complicated.

A special case of a Galton-Watson process with two types was treated by BARTLETT (1946). The first general formulation and treatment of processes with several types appears to have been given by KOLMOGOROV and DMITRIEV (1947), who treated chiefly the case of a continuous time parameter. Early results for the case of a discrete time parameter — the case that will be treated in the present chapter — were obtained by KOLMOGOROV and SEVAST'YANOV (1947), EVERETT and ULAM (1948a, b, c, d), and SEVAST'YANOV (1948). References to further results will be given later in this chapter.

A thorough exposition of the theory of these processes was given by SEVAST'YANOV (1951), covering the various complications that arise when some types cannot produce others. For the most part we shall avoid these complications in the present chapter. In some cases they can be treated by simple extensions of the results presented here.

2. Definition of the multitype Galton-Watson process

Notation. We shall denote vectors and matrices by boldface letters. The components of a vector will be indicated by the corresponding lightface letter indexed in some cases by superscripts and in some cases by subscripts. A matrix or vector whose elements are zeros will be denoted simply by 0. A vector premultiplying (postmultiplying) a matrix will be considered a row (column) vector; otherwise we do not distinguish between row and column vectors. All vectors (matrices) will be of order $k\,(k\times k)$, where k is a fixed positive integer, representing the number of different types.

Definitions 2.1. Let T denote the set of all k-dimensional vectors whose components are nonnegative integers. Let $\boldsymbol{e}_i$, $1\leqq i\leqq k$, denote the vector whose i-th component is 1 and whose other components are 0.

Definition 2.2. The *multitype (or vector) Galton-Watson process* is a temporally homogeneous vector Markov process $\boldsymbol{Z}_0, \boldsymbol{Z}_1, \boldsymbol{Z}_2, \ldots$, whose states are vectors in T. We shall always assume that $\boldsymbol{Z}_0$ is nonrandom. We interpret Z_n^i, the i-th component of $\boldsymbol{Z}_n$, as the number of objects of type i in the n-th generation.

The transition law for the process is as follows. If $\boldsymbol{Z}_0 = \boldsymbol{e}_i$, then $\boldsymbol{Z}_1$ will have the generating function

$$f^i(s_1, \ldots, s_k) = \sum_{r_1, \ldots, r_k = 0}^{\infty} p^i(r_1, \ldots, r_k)\, s_1^{r_1} \ldots s_k^{r_k}, \qquad |s_1|, \ldots, |s_k| \leqq 1, \tag{2.1}$$

where $p^i(r_1, \ldots, r_k)$ is the probability that an object of type i has r_1 children of type 1, ..., r_k of type k. In general, if $\boldsymbol{Z}_n = (r_1, \ldots, r_k) \in T$, then $\boldsymbol{Z}_{n+1}$ is the sum of $r_1 + \cdots + r_k$ independent random vectors, r_1 having the generating function f^1, r_2 having the generating function $f^2, \ldots, r_k$ having the generating function f^k. If $\boldsymbol{Z}_n = 0$, then $\boldsymbol{Z}_{n+1} = 0$.

Probabilities and expected values for the process will be designated by P and $\mathscr{E}$. When necessary to specify the starting vector, say $\boldsymbol{Z}_0 = \boldsymbol{z}$, we shall use the notation of conditional probabilities, e.g., $P(\cdot|\boldsymbol{Z}_0 = \boldsymbol{z})$.

Definition 2.3. The generating function of $\boldsymbol{Z}_n$, when $\boldsymbol{Z}_0 = \boldsymbol{e}_i$, will be denoted by $f_n^i(s_1, \ldots, s_k) = f_n^i(\mathbf{s})$, $n = 0, 1, \ldots$; $i = 1, \ldots, k$. Then f_1^i is the function f^i introduced in (2.1). The vector $(f^1(\mathbf{s}), \ldots, f^k(\mathbf{s}))$ will sometimes be written as $\boldsymbol{f}(\mathbf{s})$.

3. The basic result for generating functions

We now state the generalization of the basic theorem of Watson given in the preceding chapter.

Theorem 3.1.[1] *The generating functions f_n^i are functional iterates, being connected by the relations*

$$\left.\begin{aligned} f_{n+1}^i(\mathbf{s}) &= f^i[f_n^1(\mathbf{s}), \ldots, f_n^k(\mathbf{s})], \qquad n = 0, 1, \ldots; \\ f_0^i(\mathbf{s}) &= s_i, \qquad i = 1, 2, \ldots, k. \end{aligned}\right\} \tag{3.1}$$

The proof of Theorem 3.1 is a straightforward generalization of that of Theorem I.4.1, and is omitted.

Similarly we have, in vector form,

$$\boldsymbol{f}_{n+N}(\mathbf{s}) = \boldsymbol{f}_n[\boldsymbol{f}_N(\mathbf{s})], \qquad n, N = 0, 1, 2, \ldots. \tag{3.2}$$

4. First and second moments[2]; basic assumption

Definition 4.1. Let $\boldsymbol{M} = (m_{ij})$ be the matrix of first moments,

$$m_{ij} = \mathscr{E}(Z_1^j | \boldsymbol{Z}_0 = \boldsymbol{e}_i) = \frac{\partial f^i(1, \ldots, 1)}{\partial s_j}, \qquad i, j = 1, \ldots, k.$$

[1] Kolmogorov and Dmitriev (1947).

[2] General expressions for the first and second moments were first given by Everett and Ulam (1948a).

Basic assumption. We assume throughout the remainder of the chapter that all the first moments m_{ij} are finite and that they are not all 0.

By a straightforward generalization to vectors of the argument leading to formula (I. 8.2), we obtain the relations

$$\mathscr{E}(\boldsymbol{Z}_{n+N}|\boldsymbol{Z}_N)=\boldsymbol{Z}_N\boldsymbol{M}^n, \qquad n, N=0, 1, 2, \ldots . \tag{4.1}$$

Remark. The reader can show that if $\boldsymbol{Z}_N$ is given, then the conditional distribution of $\boldsymbol{Z}_{n+N}$ is that of a sum of $Z_N^1+\cdots+Z_N^k$ independent random vectors, of which $\boldsymbol{Z}_N^i$ have the generating function f_n^i, $i=1, \ldots, k$.

Definitions 4.2. If $\boldsymbol{X}$ is a random vector whose components have finite second moments, then the *variance* (here used synonymously with *covariance matrix*), $\operatorname{Var}\boldsymbol{X}$, is the matrix $(\mathscr{E}X^iX^j-\mathscr{E}X^i\mathscr{E}X^j)$. Let $\boldsymbol{V}_i$ denote $\operatorname{Var}\boldsymbol{Z}_1$, calculated for $\boldsymbol{Z}_0=\boldsymbol{e}_i$, and let $\boldsymbol{C}_n$ denote the matrix $(\mathscr{E}Z_n^iZ_n^j)$, calculated for whatever initial vector $\boldsymbol{Z}_0$ is being used.

Assumption 4.1. The variances $\boldsymbol{V}_i$ are finite, $i=1, \ldots, k$.

If Assumption 4.1 holds, the reader can verify, by considering the conditional expectations $\mathscr{E}(Z_{n+1}^iZ_{n+1}^j|Z_n^1, \ldots, Z_n^k)$, that we obtain

$$\boldsymbol{C}_{n+1}=\boldsymbol{M}'\boldsymbol{C}_n\boldsymbol{M}+\sum_{i=1}^{k}\boldsymbol{V}_i\mathscr{E}(Z_n^i), \qquad n=0, 1, \ldots, \tag{4.2}$$

where $\boldsymbol{M}'$ is the transpose of $\boldsymbol{M}$. Applying (4.2) repeatedly with $n=0, 1, \ldots$, we obtain

$$\boldsymbol{C}_n=\boldsymbol{M}'^n\boldsymbol{C}_0\boldsymbol{M}^n+\sum_{j=1}^{n}\boldsymbol{M}'^{n-j}\left(\sum_{i=1}^{k}\boldsymbol{V}_i\mathscr{E}Z_{j-1}^i\right)\boldsymbol{M}^{n-j}, \qquad n=1, 2, \ldots . \tag{4.3}$$

From (4.1) and (4.3) we see that the behavior of the mean and variance of $\boldsymbol{Z}_n$ for large n depends on the behavior of $\boldsymbol{M}^n$, which we now examine.

5. Positivity properties

Definitions 5.1. We shall call a vector or matrix *positive, nonnegative*, or 0 if all its components have those properties. If $\boldsymbol{u}$ and $\boldsymbol{v}$ are vectors or matrices, then $\boldsymbol{u}>\boldsymbol{v}$ ($\boldsymbol{u}\geqq\boldsymbol{v}$) means that $\boldsymbol{u}-\boldsymbol{v}$ is positive (nonnegative). We use absolute value signs enclosing a vector or matrix to denote the sum of the absolute values of the elements; thus $|\boldsymbol{Z}_n|=\sum_i|Z_n^i|$.

We shall use the following basic results from the theory of matrices[1].

Theorem 5.1. *Let $\boldsymbol{M}$ be a nonnegative matrix of order k, such that $\boldsymbol{M}^N$ is positive for some positive integer N. Then $\boldsymbol{M}$ has a positive characteristic*

[1] Perron (1907a, b), Frobenius (1908, 1909, 1912), and the general theory of matrices. See also Birkhoff and Varga (1958) for a discussion of more general results. For a textbook treatment, see Chapter 13, Volume 2, of Gantmacher (1959).

root ϱ that is simple and greater in absolute value than any other characteristic root; ϱ corresponds to positive right and left eigenvectors $\boldsymbol{\mu}=(\mu^i)$ and $\boldsymbol{\nu}=(\nu^i)$, $i=1, 2, \ldots, k$, which are the only nonnegative eigenvectors. Moreover, we have

$$\boldsymbol{M}^n=\varrho^n \boldsymbol{M}_1+\boldsymbol{M}_2^n, \qquad n=1, 2, \ldots, \tag{5.1}$$

where $\boldsymbol{M}_1=(\mu^i \nu^j)$, with the normalization $\sum \mu^i \nu^i=1$. Hence $\boldsymbol{M}_1\boldsymbol{M}_1=\boldsymbol{M}_1$. Furthermore, (a) $\boldsymbol{M}_1\boldsymbol{M}_2=\boldsymbol{M}_2\boldsymbol{M}_1=0$; (b) $|\boldsymbol{M}_2^n|=O(\alpha^n)$ *for some* $\alpha, 0<\alpha<\varrho$; (c) *for each* r, $1\leqq r\leqq k$, *we have* $\min_{1\leqq i\leqq r}\sum_{j=1}^{r} m_{ij}\leqq\varrho$; (d) *if j is a positive integer then ϱ^j corresponds to $\boldsymbol{M}^j$ just as ϱ does to $\boldsymbol{M}$.*

Definition 5.2. A multitype Galton-Watson process is called *positively regular* if $\boldsymbol{M}^N$ is positive for some positive integer N.

6. Transience of the nonzero states

The state $\boldsymbol{z}$ is said to be *transient* if $P(\boldsymbol{Z}_n=\boldsymbol{z}$ for some $n=1, 2, \ldots \mid \boldsymbol{Z}_0=\boldsymbol{z})<1$, or equivalently if $P(\boldsymbol{Z}_n=\boldsymbol{z}$ infinitely often $\mid \boldsymbol{Z}_0=\boldsymbol{z})=0$; the latter probability is known to be either 0 or 1 in any Markov chain and hence will be known to be 0 as soon as we prove it to be less than 1. (See FELLER (1957, Chapter 15).)

We shall show that, barring one complication, all states in a positively regular process, other than the state 0, are transient; *that is, no state except* 0 *can occur infinitely often.*

Definition 6.1. Let S be the set of types i such that

$$P(\boldsymbol{Z}_n=0 \mid \boldsymbol{Z}_0=\boldsymbol{e}_i)=0, \qquad n=1, 2, \ldots.$$

We suppose that the types in S, if it is not empty, are numbered $1, 2, \ldots, r$.

If i is not in S, then obviously $P(\boldsymbol{Z}_n=0 \mid \boldsymbol{Z}_0=\boldsymbol{e}_i)$ is positive for all sufficiently large n.

We shall need the following lemma. (Positive regularity is not required in the lemma.)

Lemma 6.1. *If S is not empty, then*

$$P(Z_{n+1}^1+\cdots+Z_{n+1}^r \geqq Z_n^1+\cdots+Z_n^r)=1, \qquad n=0, 1, \ldots.$$

Proof. Suppose $\boldsymbol{z}\in T$ and $z^1+\cdots+z^r=0$. From the definition of S and the remark following formula (4.1), it follows that

$$P(\boldsymbol{Z}_n=0 \mid \boldsymbol{Z}_0=\boldsymbol{z})>0$$

for all sufficiently large n. Now if $i\in S$, then

$$\left.\begin{aligned} 0&=P(\boldsymbol{Z}_{n+1}=0 \mid \boldsymbol{Z}_0=\boldsymbol{e}_i)\\ &\geqq P(\boldsymbol{Z}_{n+1}=0 \mid \boldsymbol{Z}_1=\boldsymbol{z})\, P(\boldsymbol{Z}_1=\boldsymbol{z} \mid \boldsymbol{Z}_0=\boldsymbol{e}_i). \end{aligned}\right\} \tag{6.1}$$

Since $P(\boldsymbol{Z}_{n+1}=0|\boldsymbol{Z}_1=\boldsymbol{z})$ is positive for n sufficiently large, it follows that $P(\boldsymbol{Z}_1=\boldsymbol{z}|\boldsymbol{Z}_0=\boldsymbol{e}_i)=0$, $i\in S$. Hence

$$P(Z_1^1+\cdots+Z_1^r\geqq 1|\boldsymbol{Z}_0=\boldsymbol{e}_i)=1, \qquad i\in S. \tag{6.2}$$

From (6.2) and the definition of the process, we then see that

$$P(Z_1^1+\cdots+Z_1^r\geqq Z_0^1+\cdots+Z_0^r)=1.$$

Lemma 6.1 is now a consequence of the temporal homogeneity of the process. □

Definition 6.2. A multitype Galton-Watson process will be called *singular* if the generating functions $f^1(s_1,\ldots,s_k)$, $f^2(s_1,\ldots,s_k),\ldots$, $f^k(s_1,\ldots,s_k)$ are all linear in $s_1,\ldots,s_k$, with no constant terms; i.e., each object has exactly one child.

Theorem 6.1. *Suppose the process is positively regular and not singular, and suppose $\boldsymbol{z}\in T$, $\boldsymbol{z}\neq 0$. Then $P(\boldsymbol{Z}_n=\boldsymbol{z}$ infinitely often$)=0$.* (See Def. 2.1 for T.)

Proof. We distinguish two cases. The vector $\boldsymbol{z}$ will remain fixed throughout.

Case 1. *S is empty.* Then for each i the probability $P(\boldsymbol{Z}_n=0|\boldsymbol{Z}_0=\boldsymbol{e}_i)$ will be positive for some n, and hence for all sufficiently large n. From the definition of the process it follows that $P(\boldsymbol{Z}_n=0|\boldsymbol{Z}_0=\boldsymbol{z})$ is positive for some n. Hence $\boldsymbol{z}$ is transient, since the state $\boldsymbol{z}$ cannot occur after 0 occurs.

Case 2. *S is not empty.* We first carry out the proof on the assumption that $\boldsymbol{M}$ is positive.

As before, let $1, 2, \ldots, r\leqq k$ be the integers corresponding to types in S. For any vector $\boldsymbol{w}$ let $\zeta(\boldsymbol{w})$ be the sum of the first r components of $\boldsymbol{w}$. We shall show that for some integer n_0 we have

$$P\big(\zeta(\boldsymbol{Z}_{n_0})>\zeta(\boldsymbol{z})\,|\boldsymbol{Z}_0=\boldsymbol{z}\big)>0. \tag{6.3}$$

Equation (6.3), together with Lemma 6.1, implies that $\boldsymbol{z}$ is a transient state for the following reason: If $n>n_0$, Lemma 6.1 implies that the conditional probability that $\boldsymbol{Z}_n=\boldsymbol{z}$, given $\zeta(\boldsymbol{Z}_{n_0})>\zeta(\boldsymbol{z})$, is 0.

To proceed with the proof, if $r=k$, then at least one of the probabilities $P\big(\zeta(\boldsymbol{Z}_1)>1\,|\boldsymbol{Z}_0=\boldsymbol{e}_i\big)$, $i=1,\ldots,k$, is positive; otherwise, from Lemma 6.1, the process would be singular. Let j be a value of i for which this probability is positive. If $\boldsymbol{z}_1$ is a vector in T such that $z_1^j\geqq 1$, then from the definition of the process we see that $P\big(\zeta(\boldsymbol{Z}_2)>\zeta(\boldsymbol{z}_1)\,|\boldsymbol{Z}_1=\boldsymbol{z}_1\big)>0$. From the positivity of $\boldsymbol{M}$, there is a positive probability that $Z_1^j\geqq 1$. Moreover, Lemma 6.1 implies that $\zeta(\boldsymbol{Z}_1)\geqq\zeta(\boldsymbol{Z}_0)$ with probability 1. Hence (6.3) holds with $n_0=2$.

Now suppose $1 \leqq r < k$. If we are given $\boldsymbol{Z}_1 = \boldsymbol{z}_1$, where $\boldsymbol{z}_1$ is an arbitrary vector in T, then $\boldsymbol{Z}_2$ is a sum of independent vectors $\boldsymbol{Z}_2' + \boldsymbol{Z}_2''$, the children of the first r and last $k-r$ components of $\boldsymbol{z}_1$, respectively. Now $\zeta(\boldsymbol{Z}_2') \geqq \zeta(\boldsymbol{z}_1)$ with probability 1, from Lemma 6.1, and similarly $\boldsymbol{Z}_1$ takes only values $\boldsymbol{z}_1$ such that $\zeta(\boldsymbol{z}_1) \geqq \zeta(\boldsymbol{z})$. From the positivity of $\boldsymbol{M}$ it follows that Z_1^{r+1} has a positive probability of being positive and hence $\zeta(\boldsymbol{Z}_2'')$ has a positive probability of being positive. Since $\zeta(\boldsymbol{Z}_2) = \zeta(\boldsymbol{Z}_2') + \zeta(\boldsymbol{Z}_2'')$, (6.3) holds with $n_0 = 2$. This completes the proof when $\boldsymbol{M}$ is positive.

Finally, suppose that $\boldsymbol{M}^N$ is positive for some integer N. From the remark following formula (4.1), we see that the random vectors $\boldsymbol{Z}_0, \boldsymbol{Z}_N, \boldsymbol{Z}_{2N}, \ldots$ are a Galton-Watson process (we call it the "N-process") with moment matrix $\boldsymbol{M}^N$. The N-process is nonsingular. This can be seen as follows. If the N-process were singular, then we would have $f_N^i(0, 0, \ldots, 0) = 0$, $i = 1, \ldots, k$. This in turn implies $f^i(0, 0, \ldots, 0) = 0$, $i = 1, \ldots, k$, since $f_n^i(0, 0, \ldots, 0)$ is nondecreasing as n increases. Then, since the original process is nonsingular, one of the f^i must be nonlinear. Suppose it is f^1. Then $f_2^1 = f^1(f^1, \ldots, f^k)$ must be nonlinear, since the f^i have no constant terms, and by induction f_n^1 is nonlinear for each n. This is a contradiction.

We are still considering the case in which S is not empty; moreover, the N-process has exactly the same set S as the original one. Hence, arguing for the N-process as we did before for the original process, we see that (6.3) must hold with n_0 equal to some multiple of N. But this implies the truth of the theorem for the original process. □

Remark 1. Theorem 6.1 is true even if some of the first moments m_{ij} are infinite, provided we define the process to be positively regular if some power of the matrix (m_{ij}') is positive, where $m_{ij}' = m_{ij}$ if $m_{ij} < \infty$ and $m_{ij}' = 1$ if $m_{ij} = \infty$. This is obvious, since the argument in Theorem 6.1 did not depend on the finiteness of the m_{ij}.

Remark 2. The following example shows that recurrent (nontransient) states other than 0 exist in some nonsingular processes. Take $f^1(s_1, s_2) = s_1 s_2$, $f^2(s_1, s_2) = 1$. Then the state $(1, 1)$ is not transient; in fact, $P(\boldsymbol{Z}_n = (1, 1) \mid \boldsymbol{Z}_0 = (1, 1)) = 1$.

Theorem 6.1 implies that if $\boldsymbol{z} \neq 0$, then (under the conditions of the theorem) $\lim_{n\to\infty} P(\boldsymbol{Z}_n = \boldsymbol{z}) = 0$ for any initial vector $\boldsymbol{Z}_0$. This was proved by EVERETT and ULAM (1948a).

7. Extinction probability

As we should expect, the eigenvalue ϱ plays a role similar to that of the expectation m of Chapter I in determining whether extinction occurs. (See Theorem 5.1 and Def. 5.1.)

Definitions 7.1. Let q^i be the *extinction probability* if initially there is one object of type i, $i=1, 2, \dots, k$. That is, $q^i = P(\boldsymbol{Z}_n = 0$ for some $n | \boldsymbol{Z}_0 = \boldsymbol{e}_i)$. The vector $(q^1, \dots, q^k)$ is denoted by $\boldsymbol{q}$. Let $\mathbf{1}$ denote the vector $(1, 1, \dots, 1)$.

Theorem 7.1.[1] *Suppose the process is positively regular and not singular. If* $\varrho \leqq 1$, *then* $\boldsymbol{q} = \mathbf{1}$. *If* $\varrho > 1$, *then* $0 \leqq \boldsymbol{q} < \mathbf{1}$, *and* $\boldsymbol{q}$ *satisfies the equation*

$$\boldsymbol{q} = \boldsymbol{f}(\boldsymbol{q}). \tag{7.1}$$

Remark 1. We leave it to the reader to formulate the analogous result if some of the m_{ij} are infinite.

Remark 2. If the process is singular, then $\boldsymbol{M}$ is a Markov matrix, $\varrho = 1$, and it is clear that $\boldsymbol{q} = 0$.

Proof of Theorem 7.1. From Theorem 6.1 it follows that if N is any positive number then $P(0 < |\boldsymbol{Z}_n| < N$ for infinitely many values of $n) = 0$. Hence $P(|\boldsymbol{Z}_n| \to 0) + P(|\boldsymbol{Z}_n| \to \infty) = 1$. Now from (4.1), Theorem 5.1, and the assumption $\varrho \leqq 1$ it follows that $\mathscr{E}|\boldsymbol{Z}_n|$ is a bounded function of n. Hence $P(|\boldsymbol{Z}_n| \to \infty) = 0$ and thus $P(|\boldsymbol{Z}_n| \to 0) = 1$, proving the theorem in case $\varrho \leqq 1$.

Next suppose $\varrho > 1$. We have

$$\boldsymbol{f}_{n+N}(0) = \boldsymbol{f}_n[\boldsymbol{f}_N(0)], \qquad n, N = 0, 1, \dots. \tag{7.2}$$

Moreover, just as in the simpler case of Chapter I, we see that $f_n^i(0)$ is nondecreasing in n and approaches q^i, $i = 1, \dots, k$, as $n \to \infty$. Putting $n=1$ in (7.2) and letting $N \to \infty$, we see that equation (7.1) is satisfied.

Furthermore, the assumption of positive regularity implies that $f_n^i(0) < 1$, $i = 1, \dots, k$, $n = 1, 2, \dots$. For suppose $f_N^i(0) = 1$; then $\boldsymbol{M}^n$ would have zeros in the i-th row for all $n \geqq N$, which is impossible.

Next, suppose that $\mathbf{s}$ is nonnegative and that $|1 - s_i| \leqq 1$, $i = 1, \dots, k$. Using Taylor's theorem with the remainder, we have for any positive integer n

$$\boldsymbol{f}_n(\mathbf{1} - \mathbf{s}) = \mathbf{1} - \boldsymbol{M}^n \mathbf{s} + o(|\mathbf{s}|), \qquad |\mathbf{s}| \to 0, \tag{7.3}$$

where the magnitude of $o(|\mathbf{s}|)$ may depend on n. From Theorem 5.1 it follows that there is a positive integer n_0 such that for any nonnegative vector $\mathbf{s}$, $\mathbf{s} \neq 0$, we have $|\boldsymbol{M}^{n_0} \mathbf{s}| > 2|\mathbf{s}|$. By (7.3) we can then choose an $\varepsilon > 0$ such that if $\mathbf{s} \geqq 0$, $\mathbf{s} \neq 0$, and $|\mathbf{s}| < \varepsilon$, then

$$|\mathbf{1} - \boldsymbol{f}_{n_0}(\mathbf{1} - \mathbf{s})| > |\mathbf{s}|. \tag{7.4}$$

---

[1] This theorem is actually a special case of a more general result due to Sevast'yanov (1948), given in Sec. 10. The present theorem was also given by Everett and Ulam (1948a). The proof given here is somewhat different from those cited.

We next show that $\boldsymbol{q} \neq \mathbf{1}$. For if $\boldsymbol{q}$ were $\mathbf{1}$, we could pick N sufficiently large so that $|\mathbf{1}-\boldsymbol{f}_N(0)| < \varepsilon$. Then from (7.4), we would have

$$|\mathbf{1}-\boldsymbol{f}_{n_0+N}(0)| = |\mathbf{1}-\boldsymbol{f}_{n_0}[\boldsymbol{f}_N(0)]| > |\mathbf{1}-\boldsymbol{f}_N(0)|,$$

which contradicts the fact that $|\boldsymbol{q}-\boldsymbol{f}_N(0)| \to 0$ monotonically as $N \to \infty$. Hence $\boldsymbol{q} \neq \mathbf{1}$.

We complete the proof by showing that under the hypotheses of the theorem it is impossible that some of the q^i are 1 and some <1. Suppose, e.g., that $q^1 = \cdots = q^t = 1$, where $t<k$, and $q^{t+1}<1, \ldots, q^k<1$. The first t equations of (7.1) become

$$1 = f^i(1, \ldots, 1, q^{t+1}, \ldots, q^k), \qquad i=1, \ldots, t. \tag{7.5}$$

Equations (7.5) imply that the functions $f^i(s_1, \ldots, s_k)$, $i=1, \ldots, t$, are independent of $s_{t+1}, \ldots, s_k$; that is, $m_{ij}=0$ for $i=1, \ldots, t$ and $j=t+1, \ldots, k$. But this is impossible, since it implies that no positive integral power of $\boldsymbol{M}$ is positive. ☐

The next result is of practical importance in that it insures that we can obtain a solution of (7.1) by using successive approximations, using *any* initial vector in the unit cube other than $\mathbf{1}$. As a byproduct, we assure the uniqueness of the solution among such vectors.

Theorem 7.2.[1] *Under the assumptions of Theorem 7.1, if $\boldsymbol{q}_1$ is any vector in the unit cube other than $\mathbf{1}$, then $\lim_{n\to\infty} \boldsymbol{f}_n(\boldsymbol{q}_1) = \boldsymbol{q}$.*

Corollary 1. *The only solutions of equation (7.1) in the unit cube are $\boldsymbol{q}$ and $\mathbf{1}$.*

Corollary 2. *Let $\boldsymbol{q}_1$ and $\boldsymbol{q}_2$ be vectors in the unit cube, $\boldsymbol{q}_1$ being different from $\mathbf{1}$. If $0 \leqq \boldsymbol{q}_1 \leqq \boldsymbol{f}(\boldsymbol{q}_1) \leqq \mathbf{1}$, then $\boldsymbol{q}_1 \leqq \boldsymbol{q}$. If $0 \leqq \boldsymbol{f}(\boldsymbol{q}_2) \leqq \boldsymbol{q}_2 \leqq \mathbf{1}$, then $\boldsymbol{q}_2 \geqq \boldsymbol{q}$.*

Proof of Theorem 7.2. First suppose $0 \leqq q_1^i < 1$, $i=1, 2, \ldots, k$. If N is a positive integer, then

$$\left.\begin{aligned} f_n^i(\boldsymbol{q}_1) = {} & P(\boldsymbol{Z}_n = 0 \mid \boldsymbol{Z}_0 = \boldsymbol{e}_i) \\ & + \sum_{0<|\boldsymbol{z}|\leqq N} P(\boldsymbol{Z}_n = \boldsymbol{z} \mid \boldsymbol{Z}_0 = \boldsymbol{e}_i)\,(q_1^1)^{z^1} \ldots (q_1^k)^{z^k} \\ & + \sum_{|\boldsymbol{z}|>N} P(\boldsymbol{Z}_n = \boldsymbol{z} \mid \boldsymbol{Z}_0 = \boldsymbol{e}_i)\,(q_1^1)^{z^1} \ldots (q_1^k)^{z^k}. \end{aligned}\right\} \tag{7.6}$$

The last sum on the right in (7.6) is bounded by $(\max_i q_1^i)^N P(|\boldsymbol{Z}_n| > N)$; the preceding sum goes to 0 as $n \to \infty$, since $|\boldsymbol{Z}_n|$ goes to 0 or ∞ (see the proof of Theorem 7.1). Moreover, the first term on the right approaches q^i. Since N is arbitrary, the theorem is proved whenever $\max_i q_1^i < 1$.

[1] EVERETT and ULAM (1948a). The uniqueness property is contained in the results of SEVAST'YANOV (1951).

If some but not all the q_1^i are 1, let n be an integer such that $\boldsymbol{M}^n$ is positive. Then for each $i=1, \ldots, k$, the function $f_n^i(s_1, \ldots, s_k)$ must involve all the variables $s_1, \ldots, s_k$ explicitly. Hence we must have $f_n^i(q_1^1, \ldots, q_1^k)<1$, $i=1, \ldots, k$. Since whenever $N \geqq n$ the relations $f_N^i(\boldsymbol{q}_1)=f_{N-n}^i[\boldsymbol{f}_n(\boldsymbol{q}_1)]$ hold, $i=1, \ldots, k$, the argument of the preceding paragraph shows that $\lim_{N\to\infty} f_N^i(\boldsymbol{q}_1)=q^i$, $i=1, \ldots, k$. □

Corollary 1 now follows readily, since if $\boldsymbol{q}_1 \neq \boldsymbol{1}$ is a solution of (7.1) in the unit cube, then $\boldsymbol{q}_1=\boldsymbol{f}_n(\boldsymbol{q}_1)=\lim_{n\to\infty} \boldsymbol{f}_n(\boldsymbol{q}_1)$, but we have just seen that $\boldsymbol{f}_n(\boldsymbol{q}_1) \to \boldsymbol{q}$ if $\boldsymbol{q}_1 \neq \boldsymbol{1}$.

We leave the proof of Corollary 2 to the reader. It is a convenient way of setting bounds on $\boldsymbol{q}$, since if we can find a pair of such vectors $\boldsymbol{q}_1$ and $\boldsymbol{q}_2$, then $\boldsymbol{q}_1 \leqq \boldsymbol{q} \leqq \boldsymbol{q}_2$.

8. A numerical example

Consider a process with three types, designated by integers 1, 2, and 3. Let the generating functions have the form

$$\left.\begin{aligned} f^1 &= \tfrac{4}{7} s_1^2 + \tfrac{3}{7} s_2, \\ f^2 &= \tfrac{1}{35} + \tfrac{16}{35} s_1 s_3 + \tfrac{18}{35} s_2^2, \\ f^3 &= \tfrac{3}{7} s_2 + \tfrac{4}{7} s_3^2. \end{aligned}\right\} \tag{8.1}$$

The expectation matrix $\boldsymbol{M}$ has the form

$$\boldsymbol{M}=\begin{pmatrix} \frac{8}{7} & \frac{3}{7} & 0 \\ \frac{16}{35} & \frac{36}{35} & \frac{16}{35} \\ 0 & \frac{3}{7} & \frac{8}{7} \end{pmatrix}.$$

It can be verified that $\boldsymbol{M}^2$ is positive and that the eigenvalue ϱ is $\frac{12}{7}$ corresponding to the positive left eigenvector (4, 5, 4). (In Sec. 11 we shall consider a class of matrices of which this is one, where ϱ can be found explicitly.)

Starting with the trial values $s_1=s_3=0.03$, $s_2=0.015$ — which were suggested by the course of several previous iterations of (8.1), beginning with the values $s_1=s_2=s_3=\frac{1}{2}$ — we arrive at the bounds

$$\left.\begin{aligned} 0.0125 < q^1 = q^3 < 0.0126, \\ 0.0290 < q^2 < 0.0291. \end{aligned}\right\} \tag{8.2}$$

We can be sure of the bounds in (8.2) because by substitution in (8.1) we obtain

$$\left.\begin{aligned} f^1(0.0126, 0.0291, 0.0126) = f^3 < 0.0126, \\ f^2(0.0126, 0.0291, 0.0126) < 0.0291, \end{aligned}\right\} \tag{8.3}$$

and the inequalities are reversed if we replace 0.0126 by 0.0125 and 0.0291 by 0.0290 on both sides of (8.3). (See Corollary 2 to Theorem 7.2.)

9. Asymptotic results for large n

9.1. Results when $\rho<1$. In this case, provided the process is positively regular with finite second moments, JIŘINA (1957) has obtained the following analogue of the theorem of YAGLOM (Theorem I.9.1). We omit the proof.

Theorem 9.1. *Suppose $\varrho<1$, the process is positively regular, and the second moments $\mathscr{E}(Z_1^i Z_1^j | Z_0 = e_r)$ are finite, $i, j, r=1, \ldots, k$. Then, if $Z_0=e_i$, the conditional probability distribution of Z_n, given $Z_n \neq 0$, approaches a limiting probability distribution, $i=1, \ldots, k$. If $g^i(s_1, \ldots, s_k)$ is the limiting generating function when $Z_0=e_i$, then*

$$g^i[f^1(\mathbf{s}), \ldots, f^k(\mathbf{s})] = \varrho g^i(\mathbf{s}) + 1 - \varrho, \qquad i=1, \ldots, k. \tag{9.1}$$

The first moments of the limiting distributions are given by $\partial g^i(1, \ldots, 1)/\partial s_j = \mu^i v^j / K_i$, where $\boldsymbol{\mu}$ and $\boldsymbol{v}$ are the positive right and left eigenvectors of M, normalized so that $\sum \mu^i v^i = 1$, and

$$K_i = \lim_{n\to\infty} [1 - f_n^i(0, \ldots, 0)]/\varrho^n, \qquad i=1, \ldots, k.$$

9.2. The case $\rho=1$. T. W. MULLIKIN (1963) has shown that an analogue of the theorem of YAGLOM in Sec. I.10 holds. That is, the conditional distribution of Z_n/n, given $Z_n \neq 0$, approaches the distribution of a vector having the fixed direction $\boldsymbol{v}$, while any one of the components has an exponential distribution. For continuous time YAGLOM'S result has been generalized by SEVAST'YANOV (1959) and ČISTYAKOV (1961).

9.3. Results when $\rho>1$. In this case the vector random variables Z_n/ϱ^n converge with probability 1 to a vector random variable W. The length of W is truly random (except in special cases to be noted below), but its direction, in case $W \neq 0$, is fixed and in fact is the direction of the positive left eigenvector $\boldsymbol{v}$ of the matrix M. For example, if we are dealing with a mathematical model for a biological population, where the k types represent k age groups, then the total population size, divided by ϱ^n, converges to a random variable, but the relative proportions of the various age groups approach fixed limits.

Theorem 9.2. *Under the conditions of Theorem 9.1, except that $\varrho>1$, the random vectors Z_n/ϱ^n converge with probability 1 to a random vector W. With probability 1 the direction of W, if $W \neq 0$, coincides with that of $\boldsymbol{v}$, the positive left eigenvector of M.*

EVERETT and ULAM (1948c) proved that the direction of Z_n/ϱ^n converges to $\boldsymbol{v}$ if Z_n does not go to 0. The rest of the theorem was given by HARRIS (1951), where a proof may be found. The proof depends on the second moments and may be sketched very briefly as follows. (If X is a vector $(X_1, \ldots, X_k)$, the matrix $(X_i X_j)$ will be denoted by $X'X$.)

Using (4.3) and Theorem 5.1, we have

$$\varrho^{-2n} \boldsymbol{C}_n = \boldsymbol{C} + O(\alpha_1^n), \qquad 0 < \alpha_1 < 1, \tag{9.2}$$

where

$$\boldsymbol{C} = \boldsymbol{M}_1' \boldsymbol{C}_0 \boldsymbol{M}_1 + \varrho^{-2} \boldsymbol{M}_1' \left(\sum_{i=1}^{k} \delta^i \boldsymbol{V}_i \right) \boldsymbol{M}_1 . \tag{9.3}$$

Here δ^i is the i-th component of the vector $\boldsymbol{Z}_0 (\boldsymbol{I} - \varrho^{-2} \boldsymbol{M})^{-1}$ and $\boldsymbol{I}$ is the identity matrix. Note that

$$\boldsymbol{C}\boldsymbol{M}^p = \boldsymbol{M}'^p \boldsymbol{C} = \varrho^p \boldsymbol{C}, \qquad p = 0, 1, \ldots . \tag{9.4}$$

Putting $\boldsymbol{W}_n = \varrho^{-n} \boldsymbol{Z}_n$, we deduce from (4.1) and (9.2) — (9.4) that $\mathscr{E} \boldsymbol{W}_n' \boldsymbol{W}_{n+p}$ and $\mathscr{E} \boldsymbol{W}_{n+p}' \boldsymbol{W}_n$ each differs from $\boldsymbol{C}$ by a matrix that is $O(\alpha_2^n)$, $0 < \alpha_2 < 1$, independently of the value of $p = 0, 1, 2, \ldots$. It follows from this that $\mathscr{E}(\boldsymbol{W}_{n+p}' - \boldsymbol{W}_n')(\boldsymbol{W}_{n+p} - \boldsymbol{W}_n)$ is $O(\alpha_2^n)$. From this we deduce, much as in Sec. I.8, that the sequence $\{\boldsymbol{W}_n\}$ converges with probability 1 to a vector random variable $\boldsymbol{W}$, with $\mathscr{E}\boldsymbol{W} = \left(\sum_{i=1}^{k} Z_0^i \mu^i \right) \boldsymbol{\nu}$ and $\mathscr{E} \boldsymbol{W}' \boldsymbol{W} = \boldsymbol{C}$. From the definition of $\boldsymbol{C}$ we see that it must have the form $c(\nu^i \nu^j)$, where c is a scalar, and hence $\mathscr{E}(\nu^i W^j - \nu^j W^i)^2 = 0$, $i, j = 1, 2, \ldots, k$. That is, *if* $\boldsymbol{W} \neq 0$, *then* $\boldsymbol{W}$ *has the same direction as* $\boldsymbol{\nu}$.

The variance of $\boldsymbol{W}$ is the second term on the right side of (9.3). The reader can verify that if any one element of any of the matrices $\boldsymbol{V}_1, \ldots, \boldsymbol{V}_k$ is positive, and if $\boldsymbol{Z}_0 \neq 0$, then each element of the matrix $\operatorname{Var} \boldsymbol{W}$ is positive. Hence except in the trivial cases where all the $\boldsymbol{V}_i$ are 0 or where $\boldsymbol{Z}_0 = 0$, the vector $\boldsymbol{W}$ has a nonzero variance.

The moment-generating functions $\varphi^i(s) = \mathscr{E}(e^{-s|\boldsymbol{W}|} | \boldsymbol{Z}_0 = \boldsymbol{e}_i)$ satisfy a set of equations analogous to equation (I.8.7), namely[1],

$$\left. \begin{aligned} &\varphi^i(\varrho s) = f^i(\varphi^1(s), \ldots, \varphi^k(s)), \qquad Re(s) \geqq 0, \quad i = 1, 2, \ldots, k, \\ &\varphi^i(0) = 1, \qquad \frac{d\varphi^i(0+)}{ds} = -\mu^i(\nu^1 + \cdots + \nu^k). \end{aligned} \right\} \tag{9.5}$$

10. Processes that are not positively regular

In dealing with processes that are not positively regular, complications arise similar to those encountered in the theory of Markov chains with decomposable matrices. A detailed treatment of the non-positively-regular cases has been given by SEVAST'YANOV (1951).

Non-positively-regular processes can often be studied conveniently by applying results for positively regular processes. For example, consider the process with two types, with generating functions $f^1(s_1, s_2) = \frac{1}{2} + \frac{1}{2} s_1^2 s_2^2$, $f^2(s_1, s_2) = \frac{1}{2} + \frac{1}{2} s_2^2$. The matrix $\boldsymbol{M}$ has the form

[1] See HARRIS (1951).

$\boldsymbol{M}=\begin{pmatrix}1 & 1\\ 0 & 1\end{pmatrix}$, with powers $\boldsymbol{M}^n=\begin{pmatrix}1 & n\\ 0 & 1\end{pmatrix}$, $n=1, 2, \ldots$. Thus,

$$\lim_{n\to\infty} \mathscr{E}(|\boldsymbol{Z}_n|\,|\boldsymbol{Z}_0=\boldsymbol{e}_1)=\infty.$$

Nevertheless we have $q^1=q^2=1$. The fact that $q^2=1$ is obvious, since objects of type 2 produce only objects of type 2, and an object of type 2 thus initiates a process with one type and expected number 1 in every generation. Moreover, an initial object of type 1 gives rise to a finite total number of objects of type 1 in all generations. With each such object there is produced exactly one object of type 2, which then has a finite number of descendants. Thus $q^1=1$ also. In order to make such considerations rigorous, it is convenient to be able to apply probabilistic arguments based on family trees, which we shall not consider until Chapter VI.

We shall give two results for the general case, due to SEVAST'YANOV (1948) and KOLMOGOROV and SEVAST'YANOV (1947). The proofs, which are quite complicated, are not given here, but may be found in the paper of SEVAST'YANOV (1951). We must first give some definitions.

Definitions 10.1. Let $m_{ij}^{(n)}$ be the element in the i-th row and j-th column of $\boldsymbol{M}^n$. The types i and j *communicate* if there is a positive integer n such that $m_{ij}^{(n)}>0$, and a positive integer n' such that $m_{ji}^{(n')}>0$; (i and j may be the same). A type that communicates with some other type obviously communicates with itself; a type that communicates neither with itself nor with any other type is called *singular*. The types thus fall uniquely into singular types and mutually exclusive *classes*; we define a class as a set of types, each pair of which communicate, that is not contained in any larger set having this property. A class may consist of only one type that communicates with itself. A *final class* C is a class having the property that any object whose type is in C has probability 1 of producing in the next generation exactly one object whose type is in C (objects whose types are not in C may also be produced).

Theorem 10.1.[1] *In order that the probability of extinction be* 1 *for every* $\boldsymbol{Z}_0\in T$, *it is necessary and sufficient that* (a) $\varrho\leqq 1$ *and* (b) *there are no final classes.*

Definitions 10.2. A *final group* is a final class whose types cannot produce any types outside the class. If the sequence $\boldsymbol{Z}_1, \boldsymbol{Z}_2, \ldots$ is such that for some (random) integer N all nonzero components of $\boldsymbol{Z}_N$ correspond to types in final groups, the process is said to *reach completion* in this instance.

[1] SEVAST'YANOV (1948). When this result was quoted by HARRIS (1951), the word "groups" was erroneously substituted for "classes".

Next suppose that $H_1, \ldots, H_d$ are the final groups for a process, if there are any. Let $q^i(r_1, \ldots, r_d)$ be the probability that the process reaches completion with r_1 objects in $H_1, \ldots, r_d$ objects in H_d, if $\boldsymbol{Z}_0 = \boldsymbol{e}_i$. Define

$$\left.\begin{aligned} &\psi^i(s_1, \ldots, s_d) = \sum_{r_1, \ldots, r_d = 0}^{\infty} q^i(r_1, \ldots, r_d)\, s_1^{r_1} \ldots s_d^{r_d}, \\ &|s_j| \leqq 1, \; j = 1, \ldots, d; \qquad i = 1, 2, \ldots, k. \end{aligned}\right\} \tag{10.1}$$

In Theorem 10.2 below, we shall assume that $f^i(0) = 0$, $i = 1, \ldots, k$. However, this is not a real restriction, since we can always add a $(k+1)$-th type, assuming that each object of type i, $i = 1, \ldots, k$ produces one object of type $k+1$ in addition to the other objects produced, and that each object of type $k+1$ produces just one object of type $k+1$ and no others. Thus a new final group is added.

Theorem 10.2.[1] *Suppose $f^i(0) = 0$, $i = 1, \ldots, k$. Then the functions ψ^i are uniquely determined by the equations*

$$\left.\begin{aligned} &\psi^i(s_1, \ldots, s_d) = f^i(\psi^1, \ldots, \psi^k), \qquad i \notin H_1 \cup \cdots \cup H_d, \\ &\psi^i(s_1, \ldots, s_d) = s_j, \quad i \in H_j; \quad j = 1, 2, \ldots, d; \quad |s_1|, \ldots, |s_d| \leqq 1. \end{aligned}\right\} \tag{10.2}$$

10.1. The total number of objects of various types. Let there be k types, as previously, and let $P_i(s_1, \ldots, s_k)$ be the generating function (we may have $P_i(\mathbf{1}) < 1$) for the total numbers of the various types in all generations, starting with one object of type i. (The initial object is counted.) Then the P_i satisfy the functional equations[2]

$$P_i = s_i f^i(P_1, \ldots, P_k), \qquad i = 1, \ldots, k. \tag{10.3}$$

The coefficients in the expansions of the functions P_i can be obtained by a generalization of LAGRANGE's expansion due to GOOD (1960), provided each type of object has a positive probability of having no children. Equation (10.3) is valid whether or not the process is positively regular.

11. An example from genetics[3]

The gene, at one time considered the basic unit of the inheritance process, seems in some instances actually to be composed of a number

[1] KOLMOGOROV and SEVAST'YANOV (1947). The proof is due to SEVAST'YANOV (1951). The result is quoted in HARRIS (1951, p. 311) where, however, six lines above the statement of the theorem, "theorem 7" should read "theorem 6".

[2] These equations were given by EVERETT and ULAM (1948b, p. 16) in a somewhat different form, for the case $\varrho < 1$, and they apply equally when $\varrho \geqq 1$. The present form was given independently by GOOD (1955).

[3] The material in this section was suggested by the treatment of KIMURA (1957) except that, as explained below, we consider an increasing rather than a stationary population. In KIMURA's discussion the chromosome on which the gene is situated has a multiple structure, which is shared by each gene on the chromosome. For our purpose it will suffice to talk about a gene at one location on the chromosome.

of subunits. When a cell containing the gene prepares to reproduce by splitting, each subunit doubles, and each of the two new cells receives a gene with half the subunits, i.e., with the same number as the original gene.

One or more subunits in a gene may be a mutant form. It is assumed that when doubling occurs, mutant or nonmutant subunits produce respectively mutant or nonmutant subunits. Moreover, the subunits are assumed to be divided between the two new genes randomly, mutants and nonmutants alike.

Consider a branching process where the multiplying objects are genes with k subunits, the type of a gene being the number of mutant subunits. According to the hypotheses given above, if an object is of type i, $0 \leqq i \leqq k$, then either of its children has probability

$$\binom{2i}{j}\binom{2k-2i}{k-j}\Big/\binom{2k}{k} \tag{11.1}$$

of being an object of type j, $0 \leqq j \leqq k$. (A binomial coefficient $\binom{a}{b}$ is considered 0 if $a < b$.) The types $j=1, \ldots, k-1$ will be called *mixed*; types 0 and k will be called *pure*. Type k is sometimes of special interest in that a gene, all of whose subunits are mutant, may cause the death of its possessor, while type 0 is of interest in that it will produce no more of the mutant form.

It is customary, in genetical discussions, to treat the case of a stationary population, where just one of the children of a cell survives; from a different point of view, we may suppose that we are following a single line of descent. Thus in the treatment of KIMURA (1957) a line of descent is treated as a Markov process whose state at time n (i.e., in the n-th generation) is the number of mutant subunits in the cell under consideration in that generation. The probability that the state is j in the $(n+1)$-th generation if it is i in the n-th generation is given by (11.1). The state will eventually reach 0 or k with probability 1; that is, the line will become pure. KIMURA has studied the distribution of the number of generations required for this to happen.

Rather than dealing with a single line of descent, let us use a branching process as a model for a situation in which the environment allows continued doubling of the population. In this case, as we shall see, the mixed types have a positive probability of surviving indefinitely (more accurately, of surviving until limitations of the environment make it impossible for the population to continue doubling). It is sufficient to consider only the mixed types $1, 2, \ldots, k-1$, regarding a gene of type 0 or k as dead.

The element m_{ij} of the expectation matrix $\boldsymbol{M}$ is now twice the value in formula (11.1), $i, j=1, \ldots, k-1$. Each element that is on, just above, or just below the main diagonal of $\boldsymbol{M}$ is positive, and it is easy to verify from this that we have the positively regular case. The eigenvalues of $\boldsymbol{M}$ are given by[1]

$$\varrho_j = 2^{j+2}\binom{2k-j-1}{k-j-1}\Big/\binom{2k}{k}, \qquad j=1,\ldots,k-1, \tag{11.2}$$

and from (11.2) we see that $\varrho_1 > \varrho_2 > \cdots > \varrho_{k-1}$. Hence

$$\varrho_1 = 4(k-1)/(2k-1)$$

is the eigenvalue ϱ of Theorem 5.1, and $1 < \varrho_1 < 2$ if $k \geqq 2$. From Theorem 7.1 it follows that the extinction probabilities q^i are <1 and are determined by equation (7.1).

If we take $k=4$, we obtain the numerical example of Sec. 8, the probability of extinction of the mixed types being quite small in this example. The author does not know what should be considered a reasonable value for k.

12. Remarks

12.1. Martingales. Let $\boldsymbol{\mu}$ be a nonnegative right eigenvector of $\boldsymbol{M}$, corresponding to the eigenvalue ϱ. Then the random variables $\xi_n = \sum_{i=1}^{k} Z_n^i \mu^i/\varrho^n$, $n=0, 1, \ldots$, are a martingale. In Chapter VI we shall see that this relation (generalized) has an interesting interpretation for processes where the type of an object is its age.

12.2. The expectation process. We make the familiar observation that if $\boldsymbol{M}$ has the positive right eigenvector $\boldsymbol{\mu}$ corresponding to the eigenvalue ϱ, the matrix $(p_{ij}) = (m_{ij}\,\mu^j/\varrho\mu^i)$ is a Markov transition matrix, corresponding to a Markov process that we shall call the *expectation process*. The elements of the n-th power of (p_{ij}) are related to the elements of $\boldsymbol{M}^n$ by the equations $p_{ij}^{(n)} = m_{ij}^{(n)}\mu^j/\varrho^n\mu^i$. We shall find this remark useful in the more general context of Chapter III.

12.3. Fractional linear generating functions. Suppose that $k=2$ (a similar remark applies when k is larger), and that f^1 and f^2 have the fractional linear forms

$$f^1(s_1, s_2) = \frac{a+bs_1+cs_2}{d+es_1+fs_2}, \qquad f^2(s_1, s_2) = \frac{g+hs_1+is_2}{d+es_1+fs_2}. \tag{12.1}$$

(Note that the denominators are the same.) Then it is known that the iterates f_n^1 and f_n^2, $n=1, 2, \ldots$ are all fractional linear. Thus at least one family of iterates can be calculated explicitly.

---

[1] These were obtained by KIMURA (1957), using a method due to FELLER (1951). It is necessary to multiply KIMURA's values by 2, since our population doubles in each generation, and then to discard the two eigenvalues equal to 2, which disappear when we drop the types 0 and k.

Chapter III

The general branching process

1. Introduction

In the branching processes of most interest in the physical and biological sciences an object is characterized by a parameter x, which describes its position, age, energy, or a combination of these or other factors. Thus in the mathematical models commonly used to describe biological populations the parameter x represents age; in the case of nuclear fission x may represent the position and velocity of a neutron; in the case of cosmic-ray showers x may represent the energy of an electron or other particle.

In this chapter we shall give a systematic treatment of branching processes where the "type" of an object is described by a point x in a Euclidean space. The objects in a single generation will be described by a random finite set of points $x_1, x_2, \ldots$ representing their types, i.e., a *random point-distribution.* Thus our processes are examples of the *point stochastic processes* studied by WOLD (1949) and BARTLETT (1954, 1955).

The first five sections of this chapter will provide some preliminary results concerning random point-distributions. For the most part these are known results about random set functions, put together in such a way as to be most directly applicable to branching processes. In Sec. 6 we define the general branching process as a generalized Markov process and then proceed to study its properties. Most of the results we obtain are analogues of those for a finite number of types, but the example of the nucleon cascade (see Sec. 8.1 and the end of Sec. 10, as well as Sec. 16) shows that fundamentally different situations can arise. We shall treat the case of a discrete time parameter except for a brief account in Sec. 17 of some results for continuous time. Chapters V—VII will deal with various special cases in continuous time.

J. E. MOYAL (1961a, 1961b, 1962) has recently given an independent treatment of many of the problems treated in this chapter, as well as some more general ones; the author was privileged to see this work before its publication. MOYAL considers the case in which the type of an object is a point in a not necessarily Euclidean space. This may be of interest if each object is some very complicated entity; e.g., an object may itself be a population. Much of Secs. 2—9 below, and parts of the remaining sections, are parallel to MOYAL's work, although the approaches differ.

Additional treatments of various branching processes with general states and discrete or continuous time have been given by BARTLETT

(1960) on models for population growth and spread, SEVAST'YANOV (1958, 1961) on branching processes for diffusing objects, and NEYMAN and SCOTT (1957a, 1957b) on the spread of populations. Various papers on special cases will also be cited in Chapters VI and VII.

2. Point-distributions and set functions

We shall be concerned with sets of objects, each object being described by a quantity x, which we shall suppose is a point in a d-dimensional Euclidean space R. We shall refer to x as the *type* of the object. In many cases x will be naturally restricted to some part of R, which we shall denote by X. Thus if x is age, then R is the real line and X is the nonnegative part of R. The space R and the set[1] X (which may be all of R) will be supposed fixed in the subsequent discussion, and it is always understood that x is in X.

We can describe a set of objects by specifying their x-values, allowing for the possibility of several objects of the same type.

Definition 2.1. A *point-distribution* ω on X, $\omega=(x_1, n_1; x_2, n_2; \ldots; x_k, n_k)$, is a finite set of distinct points $x_1, x_2, \ldots, x_k$ in X, with a positive integer "weight" n_i attached to x_i, $i=1, \ldots, k$; k may be any nonnegative integer, including 0, which corresponds to the null point-distribution. The order in which the pairs x_i, n_i appear is immaterial[2].

We shall interpret ω as a set of n_1 objects of type $x_1, \ldots, n_k$ of type x_k, and the null point-distribution corresponds to the absence of any objects.

Definition 2.2. Let Ω_X denote the set of all point-distributions. (We omit the phrase "on X" when no confusion arises.)

2.1. Set functions. A point-distribution ω determines in a natural way a set function $\tilde{\omega}$, defined for each subset A of X; that is, if A is a subset of X then $\tilde{\omega}(A)$ is the total number of objects with type in A,

$$\tilde{\omega}(A)=\sum_{x_i\in A} n_i. \tag{2.1}$$

If ω is null, we put $\tilde{\omega}(A)=0$.

If ω is any point-distribution, then the corresponding set function $\tilde{\omega}$ is easily verified to have the following properties whenever $A, A_1, A_2, \ldots$ are sets in X:

(a) $\tilde{\omega}(A)$ is a nonnegative integer;

---

[1] All the sets in R to be introduced will, without further notice, be assumed to be Borel sets. All functions on R that are introduced are assumed to be Borel measurable.

[2] MOYAL (1962) has treated also the case of distinguishable points; for example, if the x's are ages, then x_1 might be the age of the oldest, x_2 the next oldest, etc.

(b) if $A_1, A_2, \ldots, A_n$ are disjoint sets, then

$$\tilde{\omega}(A_1 \cup \cdots \cup A_n) = \tilde{\omega}(A_1) + \cdots + \tilde{\omega}(A_n);$$

(c) if $A_1 \supset A_2 \supset \cdots$ and $\bigcap_i A_i = 0$, then $\tilde{\omega}(A_i)$ is 0 for all sufficiently large i.

Conversely, suppose that $\tilde{\omega}$ is a set function satisfying (a), (b), and (c). (We shall see, in Sec. 3, that we need only assume that (a), (b), and (c) are satisfied when the sets $A, A_1, A_2, \ldots$ belong to a certain denumerable family of sets.) Then there is a unique point-distribution ω that generates $\tilde{\omega}$ according to (2.1). The proof of this proposition is left to the reader.

Thus there is a one-to-one correspondence between point-distributions and set functions satisfying (a), (b), *and* (c). *In view of this correspondence we shall use the same symbol* ω *to denote a point-distribution and the set function that it generates. We shall find it convenient to think sometimes in terms of set functions, sometimes in terms of point-distributions.*

Moyal (1961 b) has discussed the correspondence between set functions and point-distributions when x is not in a Euclidean space. The discussion is more complicated than in the Euclidean case.

Examples can be given of set functions that are defined for every subset of a Euclidean space and that satisfy (a) and (b) above, but not (c). Such functions are complicated. (See, e.g., Sikorski (1960, p. 18).) There are quite simple functions, however, that satisfy (a) and (b) but not (c), if we require that they be defined only for certain classes of sets, e.g., unions of a finite number of intervals. Since we are not concerned with such functions, we leave the construction of one to the interested reader.

3. Probabilities for point-distributions

The set of all point-distributions do not comprise a finite-dimensional space, since there is no upper limit to the integer k that appears in Def. 2.1. Thus we have to do with probabilities in an infinite-dimensional space.

If we are to define probabilities for point-distributions we must at least be able to define such probabilities as

$$\mathrm{Prob}\{\omega(A_1) = r_1, \omega(A_2) = r_2, \ldots, \omega(A_n) = r_n\}, \tag{3.1}$$

where $r_1, \ldots, r_n$ are nonnegative integers. Accordingly, we assume that we have a set of functions $p(A_1, \ldots, A_n; r_1, \ldots, r_n)$ defined for each positive integer n, each set of n nonnegative integers $r_1, \ldots, r_n$, and each set of n sets $A_1, \ldots, A_n$. The form of the functions p will be suggested by the problem at hand and they will be interpreted as the probabilities indicated in (3.1).

The functions p must satisfy certain relations of consistency required in the case of any stochastic process and must also express the properties (a), (b), and (c) of point-distributions given in Sec. 2.1. We are thus led to Conditions 3.1 and the explanation following them.

Conditions 3.1. (n is any positive integer; $A_1, A_2, \ldots$ are subsets of X; and $r_1, \ldots, r_n$ are any nonnegative integers.)

(1) $p(A_1, A_2, \ldots, A_n; r_1, r_2, \ldots, r_n)$ is a probability distribution on n-tuples of nonnegative integers $r_1, r_2, \ldots, r_n$. If the A's and r's undergo the same permutation, the value of p is not changed. Thus

$$p(A_1, A_2; r_1, r_2) = p(A_2, A_1; r_2, r_1).$$

(2) The functions p are "consistent". For example,

$$\sum_{r_2=0}^{\infty} p(A_1, A_2; r_1, r_2) = p(A_1, r_1).$$

(3) If $A_1, A_2, \ldots, A_n$ are disjoint sets and $A = A_1 \cup A_2 \cup \cdots \cup A_n$, then $p(A, A_1, \ldots, A_n; r, r_1, \ldots, r_n) = 0$ unless $r = r_1 + \cdots + r_n$, and

$$p(A, A_1, \ldots, A_n; r_1 + \cdots + r_n, r_1, \ldots, r_n) = p(A_1, \ldots, A_n; r_1, \ldots, r_n).$$

(4) If $A_1 \supset A_2 \supset A_3 \supset \cdots$ and $\bigcap_i A_i = 0$, then $\lim_{i \to \infty} p(A_i; 0) = 1$.

Explanation. We are dealing here with a random function ω, which is a function not of a time parameter t, as is more usually the case, but of a set A. Conditions (1) and (2) are simply conditions required of any family of distributions in order that they should define a physically meaningful random function (see Kolmogorov (1933, p. 27)). Condition (3) expresses the set-function nature of ω and means that

$$\text{Prob}\{\omega(A) = \omega(A_1) + \cdots + \omega(A_n)\} = 1;$$

it corresponds to (b) of Sec. 2.1. Condition (4) means that

$$\lim_{i \to \infty} \text{Prob}\{\omega(A_i) = 0\} = 1.$$

This condition is suggested by (c) of Sec. 2.1.

Remark. It is sometimes convenient to be able to define the functions p by prescribing their values only when the sets $A_1, A_2, \ldots, A_n$ are disjoint. Suppose we have a set of functions p_0 $(A_1, \ldots, A_n; r_1, \ldots, r_n)$ defined whenever the A's are disjoint. We can regard the functions p_0 as defining a joint distribution for random variables $\omega(A_1), \ldots, \omega(A_n)$ whenever the A_i are disjoint. Suppose that Conditions 3.1 (1) and 3.1 (2) are satisfied whenever the A_i are disjoint, and suppose Condition 3.1 (4) is satisfied. Suppose further that if $A_1, A_2, \ldots, A_n$ are any disjoint sets, each being a union of a finite number of disjoint sets, $A_i = A_{i1} \cup A_{i2} \cup \ldots$, then the joint distribution of $\omega(A_1), \omega(A_2), \ldots, \omega(A_n)$ is the same as the joint distribution of $\sum_j \omega(A_{1j}), \sum_j \omega(A_{2j}), \ldots, \sum_j \omega(A_{nj})$. (For example, if A, B, and C are disjoint sets, we require $p_0(A, B \cup C; r_1, r_2) = \sum_{r_3+r_4=r_2} p_0(A, B, C; r_1, r_3, r_4)$.) Then the functions p_0 can be extended in a unique way to functions p that satisfy Conditions 3.1 and agree with p_0 when the A's are disjoint. The proof will now be sketched.

Let $A_1, A_2, \ldots$ be a finite collection of sets, not necessarily disjoint. We can always express them in the form $A_i = \sum_j \alpha_{ij} B_j$, where the B_i are some finite collection of *disjoint* sets and the α_{ij} are 0 or 1; the notation means a union where B_j is included if and only if $\alpha_{ij} = 1$. We can then define the joint distribution of $\omega(A_1), \omega(A_2), \ldots$ to be that of $\sum_j \alpha_{1j} \omega(B_j), \sum_j \alpha_{2j} \omega(B_j), \ldots$. Suppose further that the B_j can be expressed in terms of a finite collection of disjoint sets $C_1, C_2, \ldots$ as $B_j = \sum_k \beta_{jk} C_k$. The expression for the A_i in terms of the C_k is then $A_i = \sum_k \gamma_{ik} C_k$, where $\gamma_{ik} = \sum_j \alpha_{ij} \beta_{jk}$ and $\gamma_{ik} = 0$ or 1. Our hypothesis implies that the joint distribution of $\sum_k \gamma_{1k} \omega(C_k), \sum_k \gamma_{2k} \omega(C_k), \ldots$ is the same as that of $\sum_j \alpha_{1j} \omega(B_j), \sum_j \alpha_{2j} \omega(B_j), \ldots$, since $\omega(B_1), \omega(B_2), \ldots$ are jointly distributed like $\sum_k \beta_{1k} \omega(C_k), \sum_k \beta_{2k} \omega(C_k), \ldots$. Now if we also express the A_i in terms of different disjoint sets B'_j, we can find disjoint sets C_k such that both the B_j and the B'_j have expressions in terms of the C_k,

$$B_j = \sum_k \beta_{jk} C_k, \qquad B'_j = \sum_k \beta'_{jk} C_k.$$

Since the expression for the A_i in terms of the C_k is unique, the above argument shows that our definition is independent of the particular sets B_j that are used. It can then be verified that the extended function p so defined satisfies Conditions 3.1 (1)—(4).

Example. Let X be the interval (0, 1) on the real line and let λ be a positive number. If $A_1, \ldots, A_n$ are disjoint sets on (0, 1), then, putting q_i = one-dimensional measure of A_i, define

$$p(A_1, \ldots, A_n; r_1, \ldots, r_n) = e^{-\lambda(q_1 + \cdots + q_n)} \frac{(\lambda q_1)^{r_1}}{r_1!} \frac{(\lambda q_2)^{r_2}}{r_2!} \cdots \frac{(\lambda q_n)^{r_n}}{r_n!}. \tag{3.2}$$

We leave it to the reader to verify that Conditions 3.1 are satisfied. (Use the preceding remark.)

The rest of Sec. 3 describes just how the functions p determine a probability measure P on the point-distributions.

3.1. Rational intervals, basic sets, cylinder sets. The measure-theoretic treatment is simplified if we can base everything on a denumerable number of sets A. Accordingly, we define a denumerable family of basic sets that generate the Borel sets in X.

We introduce a set of Cartesian coordinates $(\xi_1, \xi_2, \ldots, \xi_d)$ in R.

Definition 3.1. A *rational interval* is the intersection of X with a point set of the form $a_i < \xi_i \leqq b_i$, $i = 1, \ldots, d$, where the a_i and b_i are rational. We also allow $a_i = -\infty$ or $b_i = \infty$, or both; if $b_i = \infty$, we replace $\leqq$ by $<$.

Definition 3.2. A *basic set* is a finite union of rational intervals, or the empty set.

Notice that if A and B are basic sets, then $A - B$, $A \cup B$, and $A \cap B$ are basic sets; also every basic set is the union of a finite number of disjoint rational intervals.

Let ω be a set function satisfying (a), (b), and (c) of Sec. 2.1 whenever the sets $A, A_1, \ldots$ are basic sets. Then (see HALMOS (1950, p. 54)) ω can be extended in a unique manner to a set function satisfying (b) and (c) whenever $A_1, A_2, \ldots$ are Borel sets. Since for any Borel set A we can find a basic set A' such that $|\omega(A) - \omega(A')|$ is arbitrarily small, the extended ω must also satisfy (a) for any Borel set A. The extended ω is generated by a unique point-distribution, which we identify with ω.

We recall from Def. 2.2 that Ω_X is the set of all point-distributions on X.

Definitions 3.3. If $A_1, \ldots, A_n$ are any basic sets, then a set of point-distributions determined by conditions on $\omega(A_1), \ldots, \omega(A_n)$ is a *cylinder set* in Ω_X. Let C be the family of cylinder sets and let C^* be the Borel extension of C. The sets in C^* will be the measurable sets in Ω_X.

3.2. Definition of a probability measure on the point-distributions.

Theorem 3.1. *Let functions $p(A_1, \ldots, A_n; r_1, \ldots, r_n)$ be given, defined whenever $A_1, \ldots, A_n$ are basic sets and satisfying Conditions 3.1 when the sets involved are basic sets. Then there is a unique probability measure P defined on C^* satisfying*

$$\left.\begin{array}{l} P\{\omega(A_1) = r_1, \ldots, \omega(A_n) = r_n\} = p(A_1, \ldots, A_n; r_1, \ldots, r_n), \\ \qquad r_1, \ldots, r_n = 0, 1, \ldots, \end{array}\right\} \quad (3.3)$$

whenever the A's are basic sets.

Theorem 3.1 is proved in Appendix 1 to this chapter.

In the subsequent treatment, any set in Ω_X that is introduced is assumed to belong to C^*. The sets that arise in the treatment will all belong to C^*. (See Appendix 2.)

4. Random integrals

Suppose ω is the point-distribution $(x_1, n_1; x_2, n_2; \ldots; x_k, n_k)$. If $h(x)$ is a function of x, then we define the integral $\int_X h\, d\omega$ to be 0 if ω is null, and otherwise

$$\int_X h\, d\omega = \int_X h(x)\, d\omega(x) = n_1 h(x_1) + n_2 h(x_2) + \cdots + n_k h(x_k). \qquad (4.1)$$

We shall sometimes omit the range of integration and write simply $\int h\, d\omega$.

Since ω is random, according to the probability measure P determined in Sec. 3, the expression (4.1), which is a function of ω, defines a random variable. (See Appendix 2 concerning the measurability of the integral. We remind the reader that h, and any other functions introduced, are assumed to be Borel measurable.)

Note that if $h(x)$ is the function defined to be 1 when x belongs to A and 0 otherwise, then $\int_X h(x)\, d\omega(x) = \omega(A)$.

5. Moment-generating functionals

The usefulness of the moment-generating function in treating scalar and vector random variables suggests finding an analogous device for dealing with random point-distributions. We shall define a *moment-generating functional*, which we shall sometimes abbreviate by MGF (as opposed to mgf for moment-generating function).

The MGF is a modification of the *characteristic functional* introduced by LeCam (1947) for random point functions and by Bochner (1947, 1955) for random set functions. Bartlett and Kendall (1951) demonstrated the importance of such functionals in applications; see also Bartlett (1955), where a number of applications are discussed. Bochner (1955) applies the term *generating functional* to what we shall call the MGF. Bochner gives a general treatment of characteristic functionals but the treatment to be given here seems best suited for our needs[1].

In Sec. 3 we defined a probability measure P on the set Ω_X of point-distributions. We shall use the symbol $\mathscr{E}$ to denote an expectation calculated with P.

If $s(x)$ is a nonnegative function defined on X, then $e^{-\int s\, d\omega}$ is a random variable that from (4.1) is positive and not greater than 1. Hence the functional Φ defined by

$$\Phi(s) = \mathscr{E} e^{-\int s\, d\omega} = \int_{\Omega_X} \{e^{-\int s\, d\omega}\}\, dP \qquad (5.1)$$

[1] Added in proof: According to Prohorov (1961), the characteristic functional was employed by Kolmogorov (1935) for distributions in a Banach space.

is positive and does not exceed 1 when $s \geqq 0$. We call Φ the *moment-generating functional* (MGF) associated with the probability measure P, and we shall say that Φ is the MGF of a random point-distribution.

If X is a finite set, consisting of say k distinct points, then a function s on X is simply a vector with k components; a point-distribution ω is then simply a random vector $(\omega_1, \ldots, \omega_k)$ with nonnegative integer components. Hence in this case $\int s\, d\omega = s_1\omega_1 + \cdots + s_k\omega_k$, and Φ coincides with the ordinary mgf.

Before giving another example of an MGF, we introduce the following notation.

Definition 5.1. If A is a set of points in X, the *indicator function* of A, denoted by s_A, is defined by $s_A(x) = 1$ if x belongs to A, $s_A(x) = 0$ otherwise.

Suppose s_A is the indicator function of a set A and t is a nonnegative number. Then $\Phi(t s_A) = \mathscr{E} e^{-t\int s_A\, d\omega} = \mathscr{E} e^{-t\omega(A)}$. Hence $\Phi(t s_A)$, considered as a function of t, is the ordinary moment-generating function of the random variable $\omega(A)$. Thus we can obtain moments of $\omega(A)$ by differentiating $\Phi(t s_A)$ at $t = 0$.

Example of an MGF. Suppose P is the probability measure determined by the functions p in the example of Sec. 3, formula (3.2). First suppose that $s(x) = c_1 s_1(x) + \cdots + c_n s_n(x)$, where $c_1, c_2, \ldots$ are positive constants and $s_1, s_2, \ldots$ are the indicator functions of disjoint sets $A_1, A_2, \ldots$. We again let q_i be the measure of A_i and set q_0 equal to $1 - q_1 - q_2 - \cdots - q_n$. Then $\int s\, d\omega = c_1\omega(A_1) + \cdots + c_n\omega(A_n)$, and from (3.2) we find

$$\left.\begin{aligned} \mathscr{E} e^{-\int s\, d\omega} &= e^{-\lambda(q_1 + \cdots + q_n)} \sum_{r_1, r_2, \ldots} \frac{(\lambda q_1)^{r_1}}{r_1!} \frac{(\lambda q_2)^{r_2}}{r_2!} \cdots e^{-c_1 r_1 - c_2 r_2 - \cdots} \\ &= \exp\{-\lambda(q_1 + \cdots + q_n) + \lambda q_1 e^{-c_1} + \lambda q_2 e^{-c_2} + \cdots\} \\ &= \exp\{-\lambda + \lambda(q_0 + q_1 e^{-c_1} + q_2 e^{-c_2} + \cdots)\} \\ &= \exp\{-\lambda + \lambda \int_0^1 e^{-s(x)}\, dx\} = \Phi(s)\,. \end{aligned}\right\} \quad (5.2)$$

Thus Φ is defined by the last line of (5.2) whenever s has the special form above. It will follow from Theorem 5.1 (c) that Φ has this form for all s.

5.1. Properties of the MGF of a random point-distribution. We continue to use the abbreviations MGF and mgf for moment-generating functional (moment-generating function).

We shall next list several properties possessed by the MGF of a random point-distribution.

Theorem 5.1. *Suppose P is a probability measure on the point-distributions, defined as in Sec. 3 through a set of functions p satisfying*

Conditions 3.1. *Then the MGF* Φ *defined by* (5.1) *has the following properties:*

(a) *If* s *is a nonnegative function on* X, *then* $0<\Phi(s)\leqq 1$.

(b) *If* $A_1, \ldots, A_n$ *are sets in* X *and* $s_{A_1}, \ldots, s_{A_n}$ *are their indicator functions* (*Def.* 5.1), *then the function of nonnegative numbers* $t_1, \ldots, t_n$ *defined by*

$$\varphi(A_1, \ldots, A_n; t_1, \ldots, t_n) = \Phi(t_1 s_{A_1} + \cdots + t_n s_{A_n}) \tag{5.3}$$

is the mgf of an n-dimensional random vector whose components are nonnegative integers. This random vector has the probability function $p(A_1, \ldots, A_n; r_1, \ldots, r_n)$ *of Sec.* 3.

(c) *If* $s, s_1, s_2, \ldots$ *are nonnegative functions such that* $\lim_{n\to\infty} s_n(x) = s(x)$ *for each* x *in* X, *then* $\lim_{n\to\infty} \Phi(s_n) = \Phi(s)$. (*That is,* Φ *is continuous.*)

Proof. Statement (a) is obvious from (5.1). Next, from the definitions of φ and Φ, we have

$$\varphi(A_1, \ldots, A_n; t_1, \ldots, t_n) = \mathscr{E} e^{-t_1\omega(A_1) - \cdots - t_n\omega(A_n)}.$$

From (3.3), this expectation is given by

$$\sum_{r_1, \ldots, r_n = 0}^{\infty} e^{-t_1 r_1 - \cdots - t_n r_n} p(A_1, \ldots, A_n; r_1, \ldots, r_n),$$

which proves (b). Finally, suppose $s_n(x) \to s(x)$ for each x. Then from the definition of a random integral we see that $\int s_n d\omega \to \int s\, d\omega$ for each ω. Since the random variables $e^{-\int s_n d\omega}$ are bounded by 1, the Lebesgue bounded convergence theorem (HALMOS (1950, p. 110)) implies that $\mathscr{E} e^{-\int s_n d\omega} \to \mathscr{E} e^{-\int s\, d\omega}$. □

It may appear that we did not use Condition 3.1 (4). However, without this we could not have constructed a probability measure on the point-distributions, as we did in Sec. 3.2.

From the definition of φ in (5.3) we see that it satisfies permutation and consistency conditions analogous to those in Conditions 3.1 (1) and (2); e.g., $\varphi(A_1, A_2;\ t_1, t_2) = \varphi(A_2, A_1; t_2, t_1)$, and $\varphi(A_1, A_2; t_1,\ 0) = \varphi(A_1; t_1)$. Moreover, from the way in which φ is defined in (5.3) we see that if $A_1, A_2, \ldots, A_n$ are disjoint sets whose union is A, then $\varphi(A, A_1, A_2, \ldots, A_n; t, t_1, t_2, \ldots, t_n) = \varphi(A_1, A_2, \ldots, A_n; t+t_1, t+t_2, \ldots, t+t_n)$. This is equivalent to Condition 3.1 (3).

Suppose now that we are given some functional $\Phi(s)$. When can we say that Φ is the MGF of a random point-distribution? The answer is given by the following theorem, whose proof is left to the reader. (Use the remark in Sec. 3.)

Theorem 5.2. *Let* $\Phi(s)$ *be a functional defined whenever* s *is a function of the form* $\sum_{i=1}^{n} t_i s_{A_i}$, *where the* t_i *are nonnegative and* s_{A_i} *is the indicator* *of the set* A_i. *Suppose that whenever* $A_1, A_2, \ldots, A_n$ *are disjoint sets, then* $\Phi(t_1 s_{A_1} + t_2 s_{A_2} + \cdots + t_n s_{A_n})$ *is the mgf of an n-dimensional random vector whose components are nonnegative integers. We denote the probability function of this vector by* $p_0(A_1, \ldots, A_n; r_1, \ldots, r_n)$. *Suppose further that if* $A_1 \supset A_2 \supset \ldots$ *and* $\cap A_i = 0$ *then* $\Phi(t s_{A_i}) \to 1$ *for each* $t \geqq 0$. *Then the functions* p_0 *can be extended in a unique manner to functions* p, *defined whether or not the sets* A_i *are disjoint. The functions* p *satisfy Conditions* 3.1 *and hence define a random point-distribution whose MGF has the same value as* Φ *for any function* s *of the above form.*

We can then define Φ for an arbitrary Borel-measurable function s by a limiting process.

At first sight it might appear that additional conditions on Φ should be given in Theorem 5.2 to insure that Condition 3.1 (3) will hold, but this is not so.

We can use Theorem 5.2 to obtain the following result, whose detailed proof is again left to the reader.

Theorem 5.3. *Let* $\Phi_1, \Phi_2, \ldots, \Phi_n$ *be MGF's of random point-distributions. Then the functional given by the product* $\Phi(s) = \Phi_1(s)\,\Phi_2(s) \ldots \Phi_n(s)$ *is the MGF of a random point-distribution.*

If we define the sum of a finite number of point-distributions as the point-distribution corresponding to the sum of their corresponding set functions, then we can think of Φ as the MGF of a sum of independent random point-distributions whose MGF's are $\Phi_1, \Phi_2, \ldots, \Phi_n$.

5.2. Alternative formulation. Suppose, following Moyal (1961b), that we choose to specify probabilities on the point-distributions by a sequence $f_n(x_1, \ldots, x_n)\, dx_1 \ldots dx_n$, representing the probability that there are n objects of which the first is in the interval $(x_1, x_1 + dx_1)$, the second in $(x_2, x_2 + dx_2)$, etc. (For simplicity we consider here the special case in which X is one dimensional, the probability distributions have densities, and we can distinguish a "first object", "second object", etc. For example, the first might be the largest, oldest, etc. Moyal treats the general case.) Let f_0 be the probability of no objects. Then the MGF $\Phi(s)$ is given by

$$f_0 + \sum_{n=1}^{\infty} \int \cdots \int f_n(x_1, \ldots, x_n)\, e^{-s(x_1) - \cdots - s(x_n)}\, dx_1 \ldots dx_n. \tag{5.4}$$

It can be verified that Φ as defined by (5.4) satisfies the criteria of Theorem 5.2.

6. Definition of the general branching process

The development of a family will be described by a sequence of point-distributions $Z_0, Z_1, \ldots,$ where Z_n represents the objects in the

n-th generation (each object being described perhaps by the value of x that it has at its time of birth), or alternatively Z_n might represent the objects present at the n-th of a sequence of points in time. We shall denote by $Z_n(A)$ the number of objects with types in A.

We shall suppose that (a) the quantity x is a complete enough description of an object so that if we know the point-distribution Z_n, then knowledge of previous generations adds nothing to our ability to predict the future; and (b) the ability of an object to procreate is unaffected by the presence of other objects.

The supposition (a) corresponds to the mathematical assumption that the sequence $Z_0, Z_1, \ldots$ is a generalized Markov process whose states are not numbers but point-distributions. The supposition (b) corresponds to the special assumption that makes our Markov process a branching process. We shall suppose that the law relating one generation to the next does not change with n, although the basic formulas of Secs. 7 and 9 can easily be modified to fit the contrary case.

We prescribe a generalized Markov process, just as in the simple case, by means of an initial distribution for Z_0 and a transition probability function

$$P^{(1)}(\omega, B) = \text{Prob}\,[Z_{n+1} \in B \mid Z_n = \omega]. \tag{6.1}$$

Then, just as in the simple case, we have the recurrence relation

$$P^{(m+n)}(\omega, B) = \int_{\Omega_X} P^{(n)}(\omega', B)\, d_{\omega'} P^{(m)}(\omega, \omega'), \tag{6.2}$$

where $P^{(n)}(\omega, B)$ is the n-step transition function. This is almost all that we shall require of the theory of generalized Markov processes, some of which is described in Doob (1953).

6.1. Definition of the transition function. Suppose that for each x in X there is defined a family p_x of probabilities $p_x(A_1; r_1)$, $p_x(A_1, A_2; r_1, r_2), \ldots,$ where $p_x(A_1, \ldots, A_n; r_1, \ldots, r_n)$ is interpreted as the probability that an object of type x has r_1 children with types in the set $A_1, \ldots,$ and r_n children with types in the set A_n. The form of the functions p_x will be determined by the particular problem; we shall consider some examples later. We shall assume that for each x the functions p_x satisfy Conditions 3.1. The only additional assumption is that if the arguments $A_1, \ldots, A_n$, $r_1, \ldots, r_n$ are fixed, then p_x is a Borel-measurable function of x.

By Theorem 3.1, p_x determines for each x a probability measure $P_x^{(1)}$ on the set Ω_X of point-distributions, and $P_x^{(1)}$ determines in turn an MGF $\Phi_x^{(1)}$. Next, if ω is a point-distribution, $\omega = (x_1, n_1; \ldots; x_k, n_k)$ (see Def. 2.1), we put

$$\Phi(\omega, s) = [\Phi_{x_1}^{(1)}(s)]^{n_1} \ldots [\Phi_{x_k}^{(1)}(s)]^{n_k}. \tag{6.3}$$

If ω is null, we put $\Phi(\omega, s)=1$, all s. The functional Φ defined by (6.3) is, for each ω, the MGF of a point-distribution (Theorem 5.3), and is interpreted as the MGF for the children of n_1 objects of type $x_1, \ldots$, and n_k objects of type x_k. For each ω the MGF Φ corresponds to a probability measure $P^{(1)}(\omega, B)$ on sets B of point-distributions. We take $P^{(1)}$ to be the transition probability function for our process. (See the first sentence of Appendix 2 to this chapter.)

Having defined $P^{(1)}$, we can in principle define $P^{(2)}, P^{(3)}, \ldots$ by means of (6.2), although the explicit calculation would in general be difficult.

We shall always suppose that the initial point-distribution Z_0 is non-random. The results we obtain can easily be modified in the contrary case.

6.2. Notation. We shall use the symbols P_x and $\mathscr{E}_x$ to denote probabilities and expectations for our process, if initially there is one object of type x, i.e., if $Z_0=(x, 1)$. We shall write $\int s(y)\,dZ_n(y)=\int s\,dZ_n$ for the random integral $\int s\,d\omega$ in case the measure on the point-distributions ω is the one defined for Z_n in our Markov process. We denote the MGF of Z_n, if initially there is one object of type x, by $\Phi_x^{(n)}(s)=\mathscr{E}_x e^{-\int s\,dZ_n}$. Then $\Phi_x^{(0)}(s)=e^{-s(x)}$.

If g is a measurable function on Ω_X, then the conditional expectation $\mathscr{E}[g(Z_{n+m})\,|\,Z_n]$ is evaluated by taking $\mathscr{E}[g(Z_{n+m})\,|\,Z_n=\omega]=\int_{\Omega_X} g(\omega')\,d_{\omega'}P^{(m)}(\omega, \omega')$.

7. Recurrence relation for the moment-generating functionals

In the recurrence relation $f_{n+1}(s)=f[f_n(s)]$ of Chapter I, replace s by e^{-s}, and put $\psi_n(s)=f_n(e^{-s})$ and $\psi(s)=f(e^{-s})$. We then obtain for the mgf's ψ_n the relation $\psi_{n+1}(s)=\psi(-\log\psi_n(s))$, which we now wish to generalize.

When it is necessary to distinguish a function on X from its value at a particular point, we shall use a dot; thus $f(\cdot)$ is the function whose value at x is $f(x)$, etc.

Theorem 7.1.[1] *If $\omega=(x_1, n_1; \ldots; x_k, n_k)$, then*

$$\left.\begin{aligned}\mathscr{E}\{e^{-\int s\,dZ_{n+m}}\,|\,Z_n=\omega\} &= [\Phi_{x_1}^{(m)}(s)]^{n_1}\cdots[\Phi_{x_k}^{(m)}(s)]^{n_k}\\ &= e^{\int_X \log\Phi_x^{(m)}(s)\,d\omega(x)}, \qquad n, m=0, 1, \ldots\end{aligned}\right\}\tag{7.1}$$

Furthermore,

$$\Phi_x^{(m+n)}(s)=\Phi_x^{(n)}(-\log\Phi_{\cdot}^{(m)}(s)), \qquad n, m=0, 1, \ldots. \tag{7.2}$$

---

[1] Moyal (1962); formula (7.2), with $n=1$, appears in Harris (1959a, abstract).

Proof. Since our Markov process is temporally homogeneous, we have, using (6.3),

$$\left.\begin{aligned}\mathscr{E}\{e^{-\int s\,dZ_{n+1}}|Z_n=\omega\} &= \int\{e^{-\int s\,d\omega'}\}d_{\omega'}P^{(1)}(\omega,\omega')\\ &= \Phi(\omega,s) = [\Phi^{(1)}_{x_1}(s)]^{n_1}\dots[\Phi^{(1)}_{x_k}(s)]^{n_k}.\end{aligned}\right\}\tag{7.3}$$

This proves (7.1) for $m=1$. Taking expectations $\mathscr{E}_x$ of both sides of (7.3), we obtain

$$\left.\begin{aligned}\Phi^{(n+1)}_x(s) &= \mathscr{E}_x[\Phi^{(1)}_{x_1}(s)]^{n_1}\dots[\Phi^{(1)}_{x_k}(s)]^{n_k}\\ &= \mathscr{E}_x\, e^{\int\log\Phi^{(1)}_y(s)\,dZ_n(y)} = \Phi^{(n)}_x\left(-\log\Phi^{(1)}_{\cdot}(s)\right),\end{aligned}\right\}\tag{7.4}$$

which proves (7.2) when $m=1$. From (7.4) we obtain, e.g.,

$$\begin{aligned}\Phi^{(3)}_x(s) &= \Phi^{(2)}_x\left(-\log\Phi^{(1)}_{\cdot}(s)\right) = \Phi^{(1)}_x[-\log\Phi^{(1)}_{\cdot}(-\log\Phi^{(1)}_{\cdot}(s))]\\ &= \Phi^{(1)}_x[-\log\Phi^{(2)}_{\cdot}(s)].\end{aligned}$$

Proceeding in this way, taking n successively equal to $3, 4, \dots$, we obtain (7.2) for all $m, n=0, 1, \dots$.

Next we have

$$\begin{aligned}&\mathscr{E}\{e^{-\int s\,dZ_2}|Z_0=\omega\} = \mathscr{E}[\mathscr{E}\{e^{-\int s\,dZ_2}|Z_1\}|Z_0=\omega]\\ &= \mathscr{E}[\Phi(Z_1,s)|Z_0=\omega] = \int e^{\int\log\Phi^{(1)}_x(s)\,d\omega'(x)}d_{\omega'}P^{(1)}(\omega,\omega')\\ &= \Phi\left(\omega,\,-\log\Phi^{(1)}_{\cdot}(s)\right) = e^{\int\log\Phi^{(1)}_x(-\log\Phi^{(1)}_{\cdot}(s))\,d\omega(x)}\\ &= e^{\int\log\Phi^{(2)}_x(s)\,d\omega(x)}.\end{aligned}$$

Because of temporal homogeneity, this proves (7.1) for any n when $m=2$, and a similar procedure proves (7.1) for any m and n. □

8. Examples

8.1. The nucleon cascade and related processes. In an important class of branching processes a particle is transformed into two or more new particles that share some element of the original one, for example, its energy or mass. The most familiar example is the *nucleon cascade* occurring in cosmic-ray showers. (The word *nucleon* means either a proton or a neutron.) A similar problem was treated by Kolmogorov (1941) in connection with the grinding of particles. This work has been generalized by Filippov (1961). See also Sec. 16 below.

For a detailed discussion of nucleon cascades we refer the reader to the paper by Messel in the collection of Wilson (1954). Here we formulate the model used by Messel, except that we consider only the successive generations, not the development as a function of depth or time. We shall take up the latter sort of treatment in connection with another type of cosmic-ray shower in Chapter VII.

A very energetic nucleon (the *primary*) enters the atmosphere from outer space. Eventually it will suffer a collision in which it detaches another nucleon (the *secondary*) from an atomic nucleus. The secondary, whose energy was previously neglible compared to that of the primary, now has an appreciable portion of the energy of the primary, which likewise loses energy by other mechanisms whose nature need not concern us. Both the primary and the secondary are now able to detach further nucleons, and a branching process proceeds until the energies of the nucleons become too small to maintain it.

Let E be the energy of the primary before its first collision. We assume that after the collision the primary has the energy $U_1 E$ and the secondary has $U_2 E$, $0 \leqq U_1, U_2$; $U_1 + U_2 \leqq 1$, where U_1 and U_2 have the continuous probability density $f(u_1, u_2)$, which is symmetric in u_1 and u_2 and does not depend on E. The primary and the secondary form the first generation, and so on.

We define a branching process where x is the energy of a nucleon and X is the set of positive numbers. From (5.4) the appropriate form for $\Phi_x^{(1)}(s)$, the MGF for the nucleons produced in one generation by a nucleon of energy x, is

$$\Phi_x^{(1)}(s) = \iint\limits_{\substack{0 < u_1 + u_2 < 1 \\ 0 < u_1, u_2}} e^{-s(x u_1) - s(x u_2)} f(u_1, u_2)\, d u_1\, d u_2 . \tag{8.1}$$

Putting $n=1$ and replacing m by n in formula (7.2), we obtain

$$\Phi_x^{(n+1)}(s) = \iint \Phi_{x u_1}^{(n)}(s)\, \Phi_{x u_2}^{(n)}(s)\, f(u_1, u_2)\, d u_1\, d u_2, \qquad n = 0, 1, \ldots . \tag{8.2}$$

From (8.1) and (8.2) we see by induction that $\Phi_x^{(n)}(s) = \Phi_1^{(n)}(s^{(x)})$, where $s^{(x)}(y) = s(xy)$, $n = 1, 2, \ldots$.

Let $\varphi^{(n)}(t, y)$ be the mgf of the number of nucleons in the n-th generation with energy $\leqq y$, starting with one of energy 1; that is, $\varphi^{(n)}(t, y) = \Phi_1^{(n)}(t s_y)$, where $s_y(u) = 1$ if $0 < u \leqq y$ and 0 otherwise. Then $\varphi^{(0)}(t, y) = 1$ if $y < 1$ and $\varphi^{(0)}(t, y) = e^{-t}$ if $y \geqq 1$. From (8.2) we then have the recurrence relation

$$\left.\begin{aligned} &\varphi^{(n+1)}(t, y) = \iint \varphi^{(n)}\left(t, \frac{y}{u_1}\right) \varphi^{(n)}\left(t, \frac{y}{u_2}\right) f(u_1, u_2)\, d u_1\, d u_2, \\ &\qquad n = 0, 1, \ldots; t \geqq 0, y > 0 . \end{aligned}\right\} \tag{8.3}$$

8.2. A one-dimensional neutron model[1]. The following one dimensional model for neutron multiplication has many features of more complicated models; we shall discuss a variation of it in Sec. 10 of Chapter IV. Such a model appears to have been first used by Fermi

[1] See also Bellman, Kalaba, and Wing (1958); Harris (1960 b).

(1936, pp. 23 ff.), who considered collisions and absorption of neutrons without fission; see also AMALDI (1959, pp. 557–562).

Neutrons can move backward and forward in a "rod" of length L and can be lost at the ends. We take the space X to be the interval $(0, L)$, and x represents the position of a neutron at birth. A neutron born at x goes right or left with probability $\frac{1}{2}$ each. In any interval of length dx in the rod it has a probability $\alpha\, dx$ of being transformed into two neutrons, each having probability $\frac{1}{2}$ of going right or left, independently of the other. A neutron leaving the end of the rod does not return. Since $e^{-\alpha x}$ is the probability of traveling a distance x in the rod without a collision, the appropriate MGF for the first generation is then

$$\Phi_x^{(1)}(s) = \tfrac{1}{2}\left(e^{-\alpha x} + e^{-\alpha(L-x)}\right) + \tfrac{1}{2}\alpha \int_0^L e^{-\alpha|y-x|}\, e^{-2s(y)}\, dy.$$

9. First moments[1]

If s_A is the indicator of the set A in X, then $\Phi_x^{(1)}(t s_A)$ is the mgf of the random variable $Z_1(A)$, if there is one initial object of type x, and hence

$$\mathscr{E}_x Z_1(A) = -\left.\frac{\partial \Phi_x^{(1)}(t s_A)}{\partial t}\right|_{t=0}. \tag{9.1}$$

Definition 9.1. Let $M(x, A) = \mathscr{E}_x Z_1(A)$.

If $A_1, A_2, \ldots$ are disjoint sets, then

$$Z_1(A_1 \cup A_2 \cup \cdots) = Z_1(A_1) + Z_1(A_2) + \cdots.$$

Taking expectations $\mathscr{E}_x$ of both sides of this equality, we see that $M(x, A_1 \cup A_2 \cup \cdots) = M(x, A_1) + M(x, A_2) + \cdots$. That is, M is, for each fixed x, a *countably additive measure*.

In the rest of this chapter we make the following assumption about M.

Condition 9.1. The first moment $M(x, X)$ is a bounded function of x.

We must now define the iterates of M, which, as we shall see, will correspond to the first moments of the successive generations. *When symbols such as $dM(x, y)$ are used, the differential will always be with respect to the second variable.*

Definition 9.2. Let $M_1(x, A) = M(x, A)$, and let

$$M_{n+1}(x, A) = \int_X M_n(y, A)\, dM(x, y) = \int_X M(y, A)\, dM_n(x, y), \qquad n = 1, 2, \ldots.$$

We also put $M_0(x, A) = 1$ if x is in A and 0 otherwise.

[1] The material of this section was given essentially by MOYAL (1961a, 1961b, 1962).

Remark. Condition 9.1 insures that $M_n(x, X)$ is, for each $n=1, 2, \ldots$, a bounded function of x, and in fact, if $M \leqq c$, then

$$M_n(x, X) = \int M(y, X)\, dM_{n-1}(x, y) \leqq c M_{n-1}(x, X) \leqq \cdots \leqq c^n. \tag{9.2}$$

Moreover, it follows by induction that M_n is for each x a countably additive measure. See also the first sentence of Appendix 2 to this chapter.

9.1. Expectations of random integrals. We shall sometimes write $\int h\, dZ_n$ for the random integral defined in Sec. 4 if the point-distribution ω corresponds to the n-th generation in a branching process. If s_A is the indicator function of the set A, then $\mathscr{E}_x \int s_A\, dZ_1 = \mathscr{E}_x Z_1(A) = M(x, A)$. Hence the relation

$$\mathscr{E}_x \int_X s(y)\, dZ_1(y) = \int_X s(y)\, dM(x, y) \tag{9.3}$$

holds whenever s is an indicator function. It likewise holds when s is a finite linear combination of indicator functions. Hence, by the usual limiting arguments of the theory of integration, it follows that (9.3) is true whenever the integral on the right side is defined.

9.2. First moment of Z_n. The next step is to identify the iterates of M with the successive first moments of $Z_1, Z_2, \ldots$. In formula (7.1) put $m=1$ and $s = t s_A$, $t \geqq 0$, where s_A is the indicator function of A. Then (7.1) becomes

$$\mathscr{E}\{e^{-tZ_{n+1}(A)} \mid Z_n = \omega\} = [\Phi_{x_1}^{(1)}(t s_A)]^{n_1} \cdots [\Phi_{x_k}^{(1)}(t s_A)]^{n_k}, \tag{9.4}$$

where $\omega = (x_1, n_1; \ldots; x_k, n_k)$. The left side of (9.4) is $\Phi(\omega, t s_A)$ and the t-derivative of the left side at $t=0$ is $-\mathscr{E}\{Z_{n+1}(A) \mid Z_n = \omega\}$. Equating derivatives of the two sides of (9.4) we obtain

$$\mathscr{E}\{Z_{n+1}(A) \mid Z_n = \omega\} = \sum_i n_i M(x_i, A) = \int M(y, A)\, d\omega(y). \tag{9.5}$$

Now put $n=1$ in (9.5) and take expectations of both sides. We obtain, using (9.3) and Def. 9.2,

$$\mathscr{E}_x Z_2(A) = \int M(y, A)\, dM(x, y) = M_2(x, A). \tag{9.6}$$

Proceeding by induction on n we thus obtain the following result.

Theorem 9.1. *The functions M_n of Definitions 9.1 and 9.2 are the first moments: $\mathscr{E}_x Z_n(A) = M_n(x, A)$, $n=1, 2, \ldots$. Furthermore,*

$$\mathscr{E}\{Z_{n+m}(A) \mid Z_n = \omega\} = \int M_m(y, A)\, d\omega(y), \qquad m, n = 0, 1, \ldots. \tag{9.7}$$

Example 1. *The nucleon cascade* (Sec. 8.1).

Let $g(u)$ be the marginal probability density for the energy of the primary nucleon after a collision, if its prior energy is 1; i.e., $g(u)=\int_0^{1-u} f(u, u_2)\, du_2$, $0<u<1$; $g=0$ otherwise.

Then from (8.1) and (9.1), $M(x, A)=(2/x)\int_A g(y/x)\, dy$.

Example 2. *The one-dimensional neutron model* (Sec. 8.2).

In this case $M(x, A)=\int_A \alpha e^{-\alpha|y-x|}\, dy$, provided $0<x<L$ and A is in the interval $(0, L)$.

10. Existence of eigenfunctions for $\boldsymbol{M}$

For studying the asymptotic behavior of a general branching process, it is important to know whether the expectation operator M has eigenfunctions and eigenvalues analogous to those of the matrix $\boldsymbol{M}$ of Chapter II. Various generalizations are known of the results of FROBENIUS and PERRON in the preceding chapter. Of these, the earliest seems to be that of JENTZSCH (1912) for integral operators with a positive kernel. More recently, numerous results have been given by KREIN and RUTMAN (1948, 1950)[1], and we refer also to the papers of BIRKHOFF (1957) and KARLIN (1959).

Here we shall give results sufficient for many applications, although they are not the most general known.

At this point we must introduce a "volume" element dV in X; i.e., V will be a measure on X. If X has coordinates $(x_1, \ldots, x_d)$ and is finite in extent, then in many cases the choice $dV=dx_1\, dx_2 \ldots dx_d$ will be suitable, e.g., in the example of Sec. 8.2. More generally we may have $dV=v(x_1, \ldots, x_d)\, dx_1 \ldots dx_d$, where v is a positive function on X. In the case of a finite number of types we may suppose that V assigns measure 1 to each type.

Definition 10.1. Let V be a measure on X with $0<V(X)<\infty$.

Remark. If we make a statement such as "the function f is positive", where f is a function of x, we shall mean positive *everywhere* on X; we shall *not* mean "positive except for a set of points x of V-measure zero", unless this is explicitly stated. A similar remark applies to equalities.

Definition 10.2. A function f on X is *uniformly positive* if there is a positive constant c such that $f(x)\geqq c$ for each x. A similar statement applies for a function $f(x, y)$ of the pair (x, y).

---

[1] For corrections to the work of KREIN and RUTMAN, see SILVERMAN and TI YEN (1959).

Definition 10.3. We say that M_n has a *density* if there is a nonnegative function $m_n(x, y)$ such that for each x in X

$$M_n(x, A) = \int_A m_n(x, y)\, dV(y). \tag{10.1}$$

If M_n has a density, then M_{n+1} likewise has a density given by

$$m_{n+1}(x, y) = \int m_n(z, y)\, dM(x, z). \tag{10.2}$$

On the other hand, M_{n+1} may have a density even though M_n does not.

10.1. Eigenfunctions and eigenvalues. Suppose M has a density m. We shall say that the function μ, not identically 0, is a *right eigenfunction* of M, corresponding to the eigenvalue ϱ, if for each x we have

$$\varrho \mu(x) = \int m(x, y)\, \mu(y)\, dV(y). \tag{10.3}$$

Similarly ν is a *left eigenfunction* if it is not identically 0 and if for each y we have

$$\varrho \nu(y) = \int \nu(x)\, m(x, y)\, dV(x). \tag{10.4}$$

In case M does not have a density, we could define appropriate analogues of (10.3) and (10.4).

We shall next state a condition, by no means the most general possible, which implies an analogue for M of the matrix results of Chapter II.

Condition 10.1. $M(x, X)$ is a bounded function of x. M has a density m and there is an integer n_0 such that the density $m_{n_0}(x, y)$ is a uniformly positive bounded function, $0 < a \leqq m_{n_0}(x, y) \leqq b < \infty$.

It is not difficult to treat the case in which M does not have a density provided M_{n_0} does.

We do not exclude the possibility that m itself is unbounded. In three-dimensional neutron models (Chapter IV), m is unbounded but m_4 is bounded.

From (9.2) and Condition 10.1 it follows that $M(x, X)$ is uniformly positive. It can then be seen from (10.2) that m_n is, for each $n \geqq n_0$, bounded above and below by positive constants that may depend on n.

Theorem 10.1. *If Condition 10.1 is satisfied then M has a positive eigenvalue ϱ, larger in magnitude than any other eigenvalue, and corresponding to right and left eigenfunctions μ and ν that are bounded and uniformly positive. Moreover, μ and ν are the only nonnegative right and left eigenfunctions (bounded or not). Furthermore, if we normalize μ and ν so that $\int \mu(x)\nu(x)\, dV(x) = 1$, which will henceforth be assumed, then*

$$m_n(x, y) = \varrho^n \mu(x)\, \nu(y)\, [1 + O(\Delta^n)], \qquad 0 < \Delta < 1,\ n \to \infty, \tag{10.5}$$

where the bound Δ can be taken independently of x and y.

A proof of Theorem 10.1 is given in Appendix 3 to this chapter.

Examples. In the one-dimensional neutron model of Sec. 8.2 the conditions of Theorem 10.1 are satisfied if we take $dV=dx$. Following KAC (1945), we find that $\varrho=2/(1+\beta^2)$, where, if $L\neq\pi/2\alpha$, β is the smallest positive number satisfying the equation $\tan(\alpha L\beta)=-2\beta/(1-\beta^2)$. If $L=\pi/2\alpha$, then $\varrho=\beta=1$. The eigenfunctions μ and ν are both proportional (normalizing factors would be required) to the function $\sin(\beta\alpha y)+\beta\cos(\beta\alpha y)$, $0\leqq y\leqq L$.

The situation is quite different for the nucleon cascade. Let us restrict X to be the interval $(0, 1)$ and take $dV=dx$. Then $m(x, y)=(2/x)g(y/x)$. For each $\gamma>0$ the function $\mu(y)=y^\gamma$ is a right eigenfunction corresponding to the eigenvalue $\varrho_\gamma=2\int_0^1 u^\gamma g(u)\,du$. Since there are infinitely many right eigenfunctions, Conditions 10.1 cannot hold if $dV=dx$. Actually, m_n in this case does not have the form (10.5), so that no choice of V would make Conditions 10.1 hold.

11. Transience of Z_n

In Theorem 11.2 below we establish an analogue of Theorem II.6.1 regarding the transience of Z_n. We need a preliminary result on the behavior of the extinction probabilities. (See Sec. 6.2.)

Definitions 11.1. Let $q_n(x)=P_x\{Z_n(X)=0\}$, $n=0, 1, \ldots$. Let $q(x)=\lim_{n\to\infty} q_n(x)=P_x\{Z_n(X)=0 \text{ for some } n\}$.

The double definition of q is justified just as in (I.6.1), and q_n must be a nondecreasing function of n.

Theorem 11.1. *Suppose that Condition* 10.1 *holds, and that there is a positive integer N such that $q_N(x)>0$ for each x in X. Then $q_{n_0+N}(x)$ is a uniformly positive function of x (Def.* 10.2).

Proof. We shall give the proof for the case in which $n_0=N=1$, since the general argument is almost the same. In (7.2) take $m=n=1$, take $s=ts_X$, where $s_X\equiv 1$, and let $t\to\infty$. Since $\Phi_x^{(n)}(ts_X)\to P_x\{Z_n(X)=0\}$, we obtain

$$\left.\begin{aligned} q_2(x)&=\Phi_x^{(1)}(-\log q_1)=\mathscr{E}_x\, e^{\int \log q_1(y)\,dZ_1(y)}\\ &=\mathscr{E}_x\{q_1(x_1)^{n_1} q_1(x_2)^{n_2}\ldots q_1(x_k)^{n_k}\},\end{aligned}\right\} \tag{11.1}$$

where Z_1 is the point-distribution $(x_1, n_1; \ldots; x_k, n_k)$, as in Sec. 2, and the quantity $\{\,\}$ is taken as 1 if $Z_1(X)=0$.

For each $\varepsilon>0$ let A_ε be the set of points x such that $q_1(x)<\varepsilon$. From Condition 10.1, with $n_0=1$, and the fact that q_1 is positive, we see that we can find a sufficiently small positive ε, with $\varepsilon<1$, so that for each x we have $M(x, A_\varepsilon)<\frac{1}{2}$, and since $M(x, A_\varepsilon)\geqq 1-P_x\{Z_1(A_\varepsilon)=0\}$, this implies

$$P_x\{Z_1(A_\varepsilon)=0\}\geqq\tfrac{1}{2}. \tag{11.2}$$

For such an ε put $A_\varepsilon = A$. Then from (11.1),

$$\left.\begin{aligned} q_2(x) &= \int \{q_1(x_1)^{n_1} \dots q_1(x_k)^{n_k}\} \, dP_x \\ &\geqq \int_{Z_1(A)=0} \{\} \, dP_x \geqq \int_{Z_1(A)=0} \varepsilon^{Z_1(X-A)} \, dP_x . \end{aligned}\right\} \tag{11.3}$$

From the inequality between the arithmetic and geometric means (essentially HARDY, LITTLEWOOD, and PÓLYA (1952, p. 137)) the last term in (11.3) is bounded below by

$$P_x(B) \, \varepsilon^{\int_B Z_1(X-A) \, dP_x / P_x(B)}, \tag{11.4}$$

where B is the event $\{Z_1(A) = 0\}$. Now

$$\int_B Z_1(X-A) \, dP_x \leqq \int_{(\text{all } Z_1)} Z_1(X-A) \, dP_x = M(x, X-A),$$

and $P_x(B) \geqq \frac{1}{2}$ from (11.2). Hence from (11.3) and (11.4) we have $q_2(x) \geqq \frac{1}{2} \varepsilon^{2M(x, X-A)}$. Since M is bounded, the result is proved. ☐

Theorem 11.2. *Suppose that the conditions of Theorem 11.1 hold. Then for each x in X and each $K > 0$ we have*

$$P_x\{0 < Z_n(X) \leqq K \text{ infinitely often}\} = 0. \tag{11.5}$$

Proof. From Theorem 11.1 we know that q_n is uniformly positive for some n. We shall suppose $q_1(x) > c > 0$, since only slight changes are needed in the general case.

Put $s = t s_X$ as in the proof of Theorem 11.1. Putting $m = 1$ in (7.1) and letting $t \to \infty$ we have

$$P\{Z_{n+1}(X) = 0 \mid Z_n = \omega\} = q_1(x_1)^{n_1} \dots q_1(x_k)^{n_k}. \tag{11.6}$$

If $0 < \omega(X) \leqq K$, the right side of (11.6) is $\geqq c^K$. Thus whenever $0 < Z_n(X) \leqq K$, the probability is $\geqq c^K$ that $Z_{n+1}(X) = 0$. This makes it intuitively obvious that if the event $\{0 < Z_n(X) \leqq K\}$ occurs often enough, then $Z_n(X) = 0$ must occur eventually, and hence $\{0 < Z_n(X) \leqq K\}$ cannot occur subsequently (DOEBLIN (1940, p. 72)). A more complete proof follows. (This general proposition about Markov processes is also proved by CHUNG (1951, Proposition 8).)

Let $\Omega_0 = \{\omega : 0 < \omega(X) \leqq K\}$. If B is a subset of Ω_0 let $Q(\omega, B)$ be the probability, if $Z_0 = \omega$, that at least one of the point-distributions $Z_1, Z_2, \dots$ belongs to Ω_0 and that the first such belongs to B. Let

$$R(\omega) = P\{Z_n \in \Omega_0 \text{ infinitely often} \mid Z_0 = \omega\}.$$

Then

$$R(\omega) = \int_{\Omega_0} R(\omega') \, d_{\omega'} \, Q(\omega, \omega'), \qquad \omega \in \Omega_X. \tag{11.7}$$

Put $R_1 = \sup_{\omega \in \Omega_0} R(\omega)$. Then from (11.7)

$$R_1 \leqq R_1 \sup_{\omega \in \Omega_0} Q(\omega, \Omega_0) \leqq R_1 (1 - c^K). \tag{11.8}$$

Hence $R_1=0$, and from (11.7) it follows that $R(\omega)=0$ for each ω in Ω_X. This proves the theorem. □

12. The case $\rho \leqq 1$

Theorem 12.1.[1] *Suppose Condition* 10.1 *holds, the eigenvalue ϱ is $\leqq 1$, and there is a positive integer N such that for each x*

$$q_N(x) > 0. \tag{12.1}$$

Then $q(x) \equiv 1$. (See Def. 11.1.)

Proof. From Theorem 10.1 we see that $\mathscr{E}_x Z_n(X)$ is a bounded function of n and x when $\varrho \leqq 1$, and accordingly $P_x\{Z_n(X) \to \infty\}=0$ for each x. Using Theorem 11.2 we see then that $P_x\{Z_n(X) \to 0\}=1$. □

12.1. Limit theorems when $\rho \leqq 1$. If ϱ is <1, then presumably an analogue exists to the theorem of Yaglom, Sec. I.9, and its generalization for a finite number of types given in Sec. II.9.1, although such a result has not yet been proved. T. W. Mullikin (1963) has shown that if $\varrho=1$, then there is a limiting exponential distribution, generalizing the results of Sec. II.9.2, provided certain positivity and compactness conditions hold.

If $q(x) \equiv 1$, then the total number of objects is finite, and we may want to study its probability distribution. If $f(x, t)$ is the generating function of the total number of objects in all generations (including the initial object) if there is initially one object of type x, then the methods we have used previously lead to the functional equation

$$f(x, t) = t\,\Phi_x^{(1)}[-\log f(\cdot, t)].$$

Here t is a complex variable. In the next chapter we shall give a result due to Sevast'yanov (1958) enabling us to study f in certain cases.

13. Second moments

In the rest of Chapter III we shall impose the following condition, which, as we shall see, insures the finiteness of all the second moments to be considered.

Condition 13.1. The second moment $\mathscr{E}_x(Z_1(X))^2$ is a bounded function of x.

Definitions 13.1. If A and B are subsets of X, define

$$\left.\begin{aligned} M_n^{(2)}(x, A, B) &= \mathscr{E}_x\{Z_n(A) Z_n(B)\}, \qquad n=0, 1, \ldots; \\ v(x, A, B) &= M_1^{(2)}(x, A, B) - M(x, A)\, M(x, B). \end{aligned}\right\} \tag{13.1}$$

[1] Given by Harris (1959a) for the case $N=1$. Moyal (1962) gave results that are weaker than Theorem 12.1 but apply to cases where X is not necessarily Euclidean. Sevast'yanov (1958) gave a related result for certain processes.

Before proceeding, we shall introduce some terminology that is not quite standard but is convenient here.

Definitions 13.2. A function $F(A, B)$, where A and B are subsets of X, will be called a *bivariate measure* if (1) F is finite and nonnegative; (2) $F(A_1 \cup A_2 \cup \ldots, B) = F(A_1, B) + F(A_2, B) + \cdots$ whenever the A's are disjoint, and similarly for $F(A, B_1 \cup B_2 \cup \ldots)$. A function F will be called a *signed bivariate measure* if $F = F_1 - F_2$, where F_1 and F_2 are bivariate measures.

Since $\mathscr{E}_x\{Z_1(A_1 \cup A_2 \cup \ldots) Z_1(B)\} = \sum_i \mathscr{E}_x[Z(A_i) Z(B)]$ if the A's are disjoint, and similarly for A and $B_1, B_2, \ldots$, we see that $M_1^{(2)}(x, A, B)$ is, for each x, a bivariate measure. Since $M(x, A) M(x, B)$ is likewise a bivariate measure, $v(x, A, B)$ is a signed bivariate measure.

A bivariate measure F determines a measure on $X \times X$, the product space of pairs (ζ, η) where ζ and η are points of X, since $F(A, B)$ may be interpreted as the measure of the "rectangle" $A \times B$ of points (ζ, η) such that $\zeta \in A$ and $\eta \in B$. Hence we may define integrals $\iint f(\zeta, \eta)\, d_{\zeta,\eta} F(\zeta, \eta)$; if F is a signed bivariate measure, the integral may be defined in an obvious manner as a difference.

13.1. Expectations of random double integrals. Suppose $f(\zeta, \eta)$ is a function defined on $X \times X$. By analogy with Sec. 4, we define a random double integral by

$$\iint_{XX} f(\zeta, \eta)\, dZ_1(\zeta)\, dZ_1(\eta) = \sum_{i,j} n_i n_j f(x_i, x_j),$$

where Z_1 is the point-distribution $(x_1, n_1; x_2, n_2; \ldots)$. By an argument similar to that of Sec. 9.1, we then conclude that

$$\mathscr{E}_x \iint f(\zeta, \eta)\, dZ_1(\zeta)\, dZ_1(\eta) = \iint f(\zeta, \eta)\, d_{\zeta,\eta} M_1^{(2)}(x, \zeta, \eta), \tag{13.2}$$

whenever the integral on the right side of (13.2) is defined.

13.2. Recurrence relation for the second moments. In order to relate $M_{n+1}^{(2)}$ to $M_n^{(2)}$ we first define a transformation T that transforms any signed bivariate measure F into another one defined by

$$TF(A, B) = \iint_{XX} M(\zeta, A) M(\eta, B)\, d_{\zeta,\eta} F(\zeta, \eta).$$

The desired recurrence relation is then

$$\left.\begin{aligned} M_{n+1}^{(2)}(x, A, B) = TM_n^{(2)}(x, A, B) + \int_X v(y, A, B)\, d_y M_n(x, y), \\ n = 0, 1, \ldots . \end{aligned}\right\} \tag{13.3}$$

We can establish (13.3) in essentially the same way as (II.4.2). We see by induction from (13.3) that $M_n^{(2)}(x, X, X)$ is, for each n, a bounded function of x.

We observe that the iterates of T have the simple form

$$T^n F(A, B) = \int\int M_n(\zeta, A) M_n(\eta, B)\, d_{\zeta,\eta} F(\zeta, \eta). \tag{13.4}$$

13.3. Asymptotic form of the second moment when $\rho > 1$. If $\varrho > 1$ and if Conditions 10.1 and 13.1 hold, then we may evaluate $M_n^{(2)}$ by repeated application of (13.3), using Theorem 10.1 to estimate T^n. The procedure is essentially the same as in the similar step for the matrix case of Chapter II. We then obtain the asymptotic formula

$$\left.\begin{aligned} M_n^{(2)}(x, A, B) = \varrho^{2n}\Big[U(x) \int_A \nu(x)\, dV(x) \int_B \nu(x)\, dV(x) + O(\Delta_1^n)\Big], \\ n \to \infty \end{aligned}\right\} \tag{13.5}$$

where $0 < \Delta_1 < 1$, Δ_1 is independent of x, A, and B, and

$$\left.\begin{aligned} U(x) &= (\mu(x))^2 \\ &+ \sum_{k=1}^{\infty} \varrho^{-2k} \int_X \Big\{\int\int_{XX} \mu(y_1)\mu(y_2)\, d_{y_1 y_2} v(\zeta, y_1, y_2)\Big\} d_\zeta M_{k-1}(x, \zeta). \end{aligned}\right\} \tag{13.6}$$

The notation is that of Secs. 9 and 10 and formula (13.1).

13.4. Second-order product densities. Just as there is sometimes a density $m(x, y)$ such that $\mathscr{E}_x(Z_1(A)) = \int_A m(x, y)\, dV(y)$, so there may be a second-order density $m^{(2)}(x, y, y')$ such that $\mathscr{E}_x(Z_1(A))^2 = \int_A m^{(2)}(x, y, y)\, dV(y) + \int\int_{AA} m^{(2)}(x, y, y')\, dV(y)\, dV(y')$. The function $m^{(2)}$ is typically discontinuous at the "line" $y = y'$, and in fact $m^{(2)}(x, y, y)$ will ordinarily be identical with $m(x, y)$ if Z_1 cannot have points of weight greater than 1.

We shall not make use of the second-order density. Product densities seem to have been introduced as tools for treating random processes independently by BHABHA (1950) and RAMAKRISHNAN (1950).

14. Convergence of Z_n/ρ^n when $\rho > 1$

The asymptotic behavior of Z_n when $\varrho > 1$ is essentially the same as in the matrix case and is proved in a similar manner. Note that the result, which is given without proof in Theorem 14.1, corresponds to one initial object of type x.

Theorem 14.1.[1] *Suppose that Conditions* 10.1 *and* 13.1 *hold, and* $\varrho > 1$. *For each subset A of X put $W_n(A) = Z_n(A)/\varrho^n$, $n = 0, 1, \ldots$. Then for each x we have*

$$P_x\{\lim_{n\to\infty} W_n(A) = W(A)\} = 1, \tag{14.1}$$

---

[1] HARRIS (1959a). The corollary is related to a weaker result of MOYAL (1962), which, however, applies in more general spaces. It is also related to a result of SEVAST'YANOV (1958).

where $W(A)$ *is a random variable such that*

$$\mathscr{E}_x(W(A)) = \mu(x) \int_A \nu(y)\, dV(y), \tag{14.2}$$

$$\mathscr{E}_x(W(A))^2 = U(x) \Big(\int_A \nu(y)\, dV(y) \Big)^2, \tag{14.3}$$

and U *is defined by* (13.6). *If* A *and* B *are subsets of* X *such that* $V(A)$ *and* $V(B)$ *are positive, then the correlation between* $W(A)$ *and* $W(B)$ *is* 1; *i.e., the relation*

$$W(B) = \frac{\int_B \nu(y)\, dV(y)}{\int_A \nu(y)\, dV(y)} W(A)$$

holds with probability 1.

From (14.2) we see that $W(X)$ has a positive probability of being positive. Hence if $\varrho > 1$ there is a positive probability of survival. Presumably, one could prove that the conditional probability that $W(X) = 0$, given $Z_n(X) \to \infty$, is 0 (see Sec. I.8.1, Remark 1). Assuming this, we arrive at the result that $Z_n(X) \sim \varrho^n W(X)$, where $W(X)$ is random, while the proportions of the various types, when $Z_n(X)$ does not go to 0, approach nonrandom limits.

Corollary to Theorem 14.1. *If* $\varrho > 1$ *and the conditions of Theorem* 14.1 *hold, then the probability of extinction* $q(x)$ *is, for each* x, *less than* 1.

15. Determination of the extinction probability when $\rho > 1$

We next give a result insuring that the extinction probability can be determined by an iterative process, starting with any of a wide class of trial functions. We shall illustrate the use of such an iterative process by a numerical example in the next chapter.

We recall the definitions $q_n(x) = P_x\{Z_n(X) = 0\}$, $n = 0, 1, \ldots$, and $q(x) = \lim q_n(x)$.

Theorem 15.1.[1] *Suppose that the conditions of Theorem* 14.1 *hold and that there is a positive integer* N *such that* $q_N(x) > 0$ *for each* x. *Then* (a) q *is uniformly positive and uniformly less than* 1; (b) q *is the only function that is positive, uniformly less than* 1, *and satisfies the functional equation*

$$q(x) = \Phi_x^{(1)}(-\log q); \tag{15.1}$$

(c) *if* $q^{(0)}$ *is any function that is positive and uniformly less than* 1, *and if we define the sequence* $q^{(1)}, q^{(2)}, \ldots$ *by*

$$q^{(n+1)}(x) = \Phi_x^{(1)}(-\log q^{(n)}) = \Phi_x^{(n+1)}(-\log q^{(0)}), \tag{15.2}$$

then $\lim_{n\to\infty} q^{(n)}(x) = q(x)$, $x \in X$.

---

[1] HARRIS (1959a); MOYAL (1962).

Proof. Using (7.2) and the notation of Theorem 11.1, we have, for $m \geqq N$,

$$\left.\begin{aligned} q_{m+1}(x) &= \lim_{t\to\infty} \Phi_x^{(m+1)}(t s_X) \\ &= \lim_{t\to\infty} \Phi_x^{(1)}[-\log \Phi_\cdot^{(m)}(t s_X)] = \Phi_x^{(1)}(-\log q_m)\,. \end{aligned}\right\} \tag{15.3}$$

We then obtain (15.1) by letting $m \to \infty$ in (15.3).

To see that q is uniformly less than 1, we consider the random variable $W(X)$ (see Theorem 14.1). Using the Schwarz inequality conditional to the hypothesis $W(X) > 0$, we have, putting $W(X) = W$,

$$\mathscr{E}_x(W^2 \mid W > 0) \geqq [\mathscr{E}_x(W \mid W > 0)]^2,$$

or, since $\mathscr{E}_x W^n = P_x(W > 0)\,\mathscr{E}_x(W^n \mid W > 0)$, $n = 1, 2$, we have

$$P_x(W > 0) \geqq \frac{(\mathscr{E}_x W)^2}{\mathscr{E}_x W^2}\,. \tag{15.4}$$

From (14.2), (14.3), and (15.4) we see that $P_x(W > 0)$ is uniformly positive, which proves that q is uniformly < 1. The uniform positivity of q follows from Theorem 11.1, since $q_n(x)$ is a nondecreasing function of n. To prove (b) and (c), we use essentially the same argument used to demonstrate the analogous results in Theorem II.7.2, making use of Theorem 11.2. ▯

The assumptions of Theorem 15.1 could perhaps be weakened.

16. Another kind of limit theorem

The preceding limit theorems, which are natural generalizations of those for a finite number of types, do not exhaust all possibilities. We shall next give an example of a process with quite a different kind of behavior.

Let X be the real line. An object at x has probability π_n of having n children; assume that each child, independently of the others, has a probability density $f(y)$ for being at $x + y$.

Let $Z_n(x)$ be the number of objects whose positions are $\leqq x$ in the n-th generation, let Z_n be the total number in the n-th generation, and put $a_n(x) = \mathscr{E} Z_n(x)$, $b_n(x) = \mathscr{E}(Z_n(x))^2$, $m = \sum n \pi_n$, $\beta = \sum n^2 \pi_n - m$. Then, using the methods described earlier in this chapter, we obtain, if there is one initial object at 0,

$$\left.\begin{aligned} a_{n+1}(x) &= m \int_{-\infty}^{\infty} a_n(x-y) f(y)\, dy\,, \\ b_{n+1}(x) &= m \int_{-\infty}^{\infty} b_n(x-y) f(y)\, dy + \beta \left(\int_{-\infty}^{\infty} a_n(x-y) f(y)\, dy \right)^2 \\ &= m \int_{-\infty}^{\infty} b_n(x-y) f(y)\, dy + \frac{\beta}{m^2} (a_{n+1}(x))^2. \end{aligned}\right\} \tag{16.1}$$

From (16.1) we see that $a_n(x) = m^n F_n(x)$, where F_n is the cumulative distribution of the n-th convolution of f. Thus if f has mean α and standard deviation σ, then the central limit theorem implies that $m^{-n} a_n(n\alpha + x\sigma\sqrt{n}) \to H(x)$, where H is the Gaussian cumulative distribution.

We conjecture, subject to possible additional assumptions:

Proposition A. *If $m > 1$, then for each x the sequence of random variables $Z_n(n\alpha + x\sigma\sqrt{n})/m^n$ converges in probability to $H(x)W$, where W is a random variable independent of x.*

If Proposition A is true, then the ratio

$$\frac{Z_n(n\alpha + x_1\sigma\sqrt{n})}{Z_n(n\alpha + x_2\sigma\sqrt{n})} \tag{16.2}$$

should approach $H(x_1)/H(x_2)$ if $W \neq 0$. A result related to Proposition A has been given by Bartlett (1960, p. 78), and (16.2) is closely related to a result announced by Kolmogorov (1941), although only part of the proof is given in that paper. The process treated by Kolmogorov was a generalization of the nucleon cascade discussed earlier in this chapter. However, the application that Kolmogorov had in mind was the grinding of particles; in that case it is the mass, rather than the energy, of a particle that is shared by its children. In such applications, if one considers the *logarithms* of the masses (or energies), one obtains a process that appears to be similar in nature to the process we have discussed above, although not identical with it. Kolmogorov's work has recently been extended by Filippov (1961), who considers, however, the random integral $\int_{-\infty}^{x} y\,dZ(y)$ rather than $Z(x)$. Peter Ney (1961) has also studied similar processes.

If the expectation density $m(x, y)$ has a positive right eigenfunction μ (whether or not Condition 10.1 holds), then we should expect the asymptotic behavior of a branching process to be closely related to that of the Markov process whose transition probability density is $p(x, y) = m(x, y)\mu(y)/\varrho\mu(x)$. If the iterated transition $p_n(x, y)$ approaches a probability density $\pi(y)$, we can anticipate limit theorems similar to those of the cases with a finite number of types. If $p_n(x, y)$ goes to 0, then other situations, such as that of this section, may arise.

17. Processes with a continuous time parameter

Moyal (1961a) has given a treatment of a general class of branching processes with a continuous time parameter. A simple example showing the type of functional equation considered is the following. Let X be the real line and let $p_t(x, y)$ be the transition density of a temporally homogeneous Markov process. Suppose each object wanders according to this process. However, an object at x at time t has a probability $\lambda\pi_n\,dt$ of being replaced, at the same location, by n objects in the interval $(t, t+dt)$. We suppose λ to be independent of x and t. We suppose $\sum_{n=0}^{\infty}\pi_n=1$ and $\pi_1=0$, and we put $\pi(s)=\sum\pi_i s^i$.

Let $\Phi_x(s,t)$ be the moment-generating functional at time t, if initially there is one object at x. The probability that no transformation occurs between times 0 and t is $e^{-\lambda t}$, in which case the MGF is $\int e^{-s(y)}p_t(x,y)\,dy$. If the initial object is transformed at time τ, at the point y, the MGF at time t will be $\pi[\Phi_y(s,t-\tau)]$. Hence we have the integral equation

$$\left.\begin{aligned}\Phi_x(s,t)&=e^{-\lambda t}\int_{-\infty}^{\infty}e^{-s(y)}p_t(x,y)\,dy\\&\quad+\lambda\int_0^t e^{-\lambda\tau}\int_{-\infty}^{\infty}p_\tau(x,y)\,\pi[\Phi_y(s,t-\tau)]\,dy\,d\tau.\end{aligned}\right\}\qquad(17.1)$$

For further treatment and more complicated cases the reader is referred to MOYAL's paper. See also the last paragraph of Sec. 1.

NEYMAN and SCOTT (1957a) have treated populations of multiplying and migrating objects, emphasizing biological applications.

Appendix 1

Proof of Theorem 3.1. We use the following terminology, besides that of Secs. 2 and 3. Let $\mathscr{A}$ be the collection of basic sets. Let Ω' be the collection of all nonnegative integer-valued set functions $\omega(A)$ defined for $A\in\mathscr{A}$. Let Ω'' and Ω''' be respectively those set functions in Ω' that, when the sets involved are all basic, satisfy (b) of Sec. 2 or both (b) and (c). Let $C(\Omega')$ and $C^*(\Omega')$ be respectively the family of cylinder sets in Ω' and their Borel extension, and similarly for Ω'' and Ω'''.

Because of the relation between point-distributions and set functions (see Sec. 2.1), it will be sufficient to show that the functions p determine uniquely a probability measure on $C^*(\Omega''')$.

The basic theorem of KOLMOGOROV (1933, p. 27) implies that the probabilities p determine a unique probability measure P_1 on $C^*(\Omega')$. Our main task is to show that Ω''' belongs to $C^*(\Omega')$ and that $P_1(\Omega''')=1$.

Lemma 1. *$\Omega''\in C^*(\Omega')$ and $P_1(\Omega'')=1$.*

Proof. Ω'' consists of those ω satisfying all relations $\omega(A_1\cup\cdots\cup A_k)=\omega(A_1)+\cdots+\omega(A_k)$ when the A's are disjoint and basic. Each such relation has P_1-measure 1, and there are only denumerably many of them. Hence $\Omega''\in C^*(\Omega')$ and $P_1(\Omega'')=1$.

Lemma 2. *Suppose $\omega\in\Omega''$ and suppose $A_1, A_2, \ldots$ are basic sets such that $A_1\supset A_2\supset\ldots$, and $\omega(A_i)\geqq 1$, $i=1,2,\ldots$. Then there is a sequence of rational intervals $I_1\supset I_2\supset\ldots$, $I_i\subset A_i$, with $\omega(I_i)\geqq 1$, $i=1,2,\ldots$.*

The proof is left to the reader.

We now consider the following case.

Case 1. *X is the real line.* Define the following subsets of Ω'':

$$\Omega_{-\infty} = \{\omega\colon \omega \in \Omega'', \omega((-\infty, i]) \geqq 1, \qquad i = -1, -2, \ldots\};$$

$$\Omega_{\infty} = \{\omega\colon \omega \in \Omega'', \omega((i, \infty)) \geqq 1, \qquad i = 1, 2, \ldots\};$$

$$\Omega_x \ (x \text{ rational}) = \{\omega\colon \omega \in \Omega'', \omega((x, x+2^{-i}]) \geqq 1, \qquad i = 1, 2, \ldots\}.$$

Let Ω_0 be the union of $\Omega_{-\infty}$, Ω_∞, and Ω_x for all rational x.

Lemma 3. *If $\omega \in \Omega''$ and does not satisfy* (c) *of Sec.* 2, *then $\omega \in \Omega_0$. (If $\omega \in \Omega_0$ it obviously does not satisfy* (c).)

Proof. From Lemma 2 there must be a sequence of rational intervals $I_1 > I_2 > \ldots$ such that $\bigcap_i I_i = 0$ and $\omega(I_i) \geqq 1$. Suppose that one of the intervals, say I_1, is finite, the argument being similar in the contrary case. Recalling that the I_i are open on the left and closed on the right, we see from elementary properties of closed sets that there must be a j such that all the I_i, for $i \geqq j$, have the same left-hand endpoint, say x_0; otherwise the I_i would have a point in common. But then $\omega \in \Omega_{x_0}$ and hence $\omega \in \Omega_0$. □

From Lemma 3 we see that $\Omega''' = \Omega'' - \Omega_0$. However, Condition 3.1 (4) implies that $P_1(\Omega_0) = 0$. Since Ω_0 obviously belongs to $C^*(\Omega')$, we see that $\Omega''' \in C^*(\Omega')$ and $P_1(\Omega''') = 1$. A cylinder set in Ω''' is the intersection of Ω''' with a cylinder set in Ω'. Hence $B \in C^*(\Omega''')$ implies $B = B_1 \cap \Omega'''$, where $B_1 \in C^*(\Omega')$ (Halmos (1950, p. 25)). Hence we may define a measure P on $C^*(\Omega''')$ by putting $P(B) = P_1(B_1)$ for $B \in C^*(\Omega''')$. Notice also that if P is any probability measure on $C^*(\Omega''')$, then we can define a probability measure P_1 on $C^*(\Omega')$ by putting $P_1(B) = P(B \cap \Omega''')$ for any $B \in C^*(\Omega')$. If there were two different measures P determined by the functions p, then there would be two such measures P_1 on $C^*(\Omega')$, contrary to Kolmogorov's theorem. Hence our result is proved for Case 1.

The case where X is a d-dimensional Euclidean space is very similar, and the general case can be deduced from it. We shall not give the details.

Appendix 2

Measurability of random integrals. The argument for the measurability of the random integral (4.1) is typical of that used to prove measurability of other functions introduced in this chapter, being based on the idea of a *monotone class* of sets (Halmos (1950, pp. 27–28)).

Let A_1 be a fixed subset of X. If A_1 is a basic set, then $\omega(A_1)$ is obviously a measurable function of ω, in view of the way we defined the measurable sets in Ω_X (Sec. 3.1). Now let $\mathscr{D}$ be the class of sets A such that $\omega(A)$ is a measurable function of ω. Then $\mathscr{D}$ includes the Borel sets on X. The reader can deduce this by the following steps: (a) $\mathscr{D}$ includes the basic sets as just indicated, (b) $\mathscr{D}$ is a monotone class, and (c) hence $\mathscr{D}$ includes the Borel sets (Halmos (1950, p. 27, Theorem B)).

From (c) we see that the integral $\int_X h \, d\omega$ is a measurable function of ω whenever h is the indicator function of a Borel set in X. From this, by a limiting process, it is easy to show that the integral is measurable whenever h is a Borel-measurable function on X.

Appendix 3

Proof of Theorem 10.1.[1] See the remark following Def. 10.1.

If f and g are functions on X, the notation $f \succ g$ will mean $f(x) \geqq g(x)$, $x \in X$, but $f - g \neq 0$.

Case 1. The integer n_0 of Condition 10.1 is 1. Define the operator T by

$$T f(x) = \int m_1(x, y) f(y)\, dV(y).$$

Let S be the set consisting of every positive number ϱ' for which there is a bounded nonnegative function f such that

$$T f \succ \varrho' f. \tag{1}$$

Note that such an f must be positive on a set of positive V-measure.

From the conditions of the theorem, it can be seen that S is a nonempty interval and that it has a finite least upper bound ϱ. Let $\{f_n\}$ be a sequence of functions with $f_n \succ 0$, such that (1) is satisfied with $f = f_n$ and $\varrho' = \varrho_n > 0$, where $\varrho_n \to \varrho$, and such that

$$\int f_n(x)\, dV(x) = 1, \qquad n = 1, 2, \ldots. \tag{2}$$

From (1), (2), and the boundedness of m_1 we see that the f_n are bounded uniformly in n and x. It follows from a known theorem of analysis (see, e.g., BANACH (1932, pp. 130—131)) that there is a subsequence $\{f_{n_i}\}$ and a bounded function μ such that

$$\lim_{i \to \infty} \int f_{n_i}(x)\, g(x)\, dV(x) = \int \mu(x)\, g(x)\, dV(x) \tag{3}$$

for every g such that $\int |g(x)|\, dV(x) < \infty$. The function μ is obviously nonnegative; putting $g \equiv 1$, we see from (2) and (3) that $\int \mu(x)\, dV(x) = 1$.

In (1) replace f by f_{n_i}, ϱ' by ϱ_{n_i}, multiply by any bounded nonnegative function g, and integrate with respect to the differential $dV(x)$. Then

$$\int g(x) \left[\int m_1(x, y) f_{n_i}(y)\, dV(y)\right] dV(x) \geqq \varrho_{n_i} \int f_{n_i}(x)\, g(x)\, dV(x). \tag{4}$$

The inner integral on the left side of (4) is bounded and approaches the limit $\int m_1(x, y)\, \mu(y)\, dV(y)$. Hence we have

$$\int g(x) \left[\int m_1(x, y)\, \mu(y)\, dV(y)\right] dV(x) \geqq \varrho \int \mu(x)\, g(x)\, dV(x). \tag{5}$$

Since g is arbitrary, (5) implies

$$\int m_1(x, y)\, \mu(y)\, dV(y) \geqq \varrho\, \mu(x) \tag{6}$$

for almost all x (V-measure). But the equality must hold in (6) for almost all x. Otherwise, multiplying both sides of (6) by $m_1(u, x)\, dV(x)$ and integrating on x, we should have, from the strict positivity of m_1,

$$\int m_1(u, x) \left[\int m_1(x, y)\, \mu(y)\, dV(y)\right] dV(x) > \varrho \int m_1(u, x)\, \mu(x)\, dV(x) + \varepsilon, \tag{7}$$

where ε is a positive constant independent of u. But (7) would imply the existence of a $\varrho' > \varrho$ for which (1) holds with $f(x) = \int m_1(x, y)\, \mu(y)\, dV(y)$, and this is a contradiction. Hence the equality holds in (6) for almost all x. We may then change

[1] The proof is based on a characterization of ϱ that seems to have occurred independently to a number of mathematicians, including F. BOHNENBLUST, who outlined a proof to the author about 1949, and H. WIELANDT (1950). BIRKHOFF (1957) uses quite a different approach, and the approach of KREIN and RUTMAN (1948) is again different.

μ, if necessary, at a set whose V-measure is 0, in order to have equality in (6) for *every* x, and μ is then a right eigenfunction. From the equalities

$$\int m_1(x, y)\mu(y)\, dV(y) = \varrho\mu(x), \qquad \int \mu(y)\, dV(y) = 1,$$

we have $\varrho\mu(x) \geqq \inf_{x,y} m_1$, whence μ is uniformly positive.

At this point it is convenient to use probabilistic considerations suggested by the "expectation process" of Sec. II.12. Define

$$p(x, y) = \frac{m_1(x, y)\mu(y)}{\varrho\mu(x)}. \tag{8}$$

Then p is bounded and uniformly positive on X and can be considered as a Markov transition probability density, since $\int p(x, y)\, dV(y) \equiv 1$. Moreover, the iterated densities have the form

$$\left.\begin{aligned} p_n(x, y) &= \int p_{n-1}(x, u)\, p(u, y)\, dV(u) \\ &= \frac{m_n(x, y)\mu(y)}{\mu(x)\varrho^n}, \qquad n = 2, 3, \ldots, \end{aligned}\right\} \tag{9}$$

as can be verified directly.

It is known from probability theory[1] that there is a function π, the "stationary density", with $\int \pi(y)\, dV(y) = 1$, such that

$$|\pi(y) - p_n(x, y)| \leqq c\Delta^n, \qquad |\Delta| < 1, \tag{10}$$

where c and Δ are constants independent of x, y, and n, and

$$\pi(y) = \int \pi(x)\, p(x, y)\, dV(x). \tag{11}$$

From (11) and the boundedness and uniform positivity of p we see that π is bounded and uniformly positive.

If we put $\nu(y) = \pi(y)/\mu(y)$, then we see from (8) and (11) that ν satisfies the equation $\int \nu(x)\, m_1(x, y)\, dV(x) = \varrho\nu(y)$, and hence ν is a left eigenfunction. From the definition of ν and the fact that π is a probability density we have

$$\int \mu(x)\nu(x)\, dV(x) = 1.$$

From (9) and (10) the relation (10.5) follows immediately.

If f is any nonnegative right eigenfunction of m_1, corresponding to an eigenvalue ϱ_1, then from (10.5) we have

$$\left.\begin{aligned} \varrho_1^n f(x) &= \int m_n(x, y) f(y)\, dV(y) \\ &= \varrho^n \mu(x) \int \nu(y) f(y)\, dV(y)\, [1 + O(\Delta^n)]. \end{aligned}\right\} \tag{12}$$

Since $\int \nu f\, dV > 0$, we have $\varrho_1 = \varrho$ and $f = \text{constant} \times \mu$. A similar argument shows that ν is the only nonnegative left eigenfunction. Moreover, it follows from (12) that no eigenvalue exceeds ϱ in absolute value, no eigenvalue except ϱ itself has absolute value ϱ, and μ and ν are the only eigenfunctions (nonnegative or not) corresponding to the eigenvalue ϱ.

Case 2. $n_0 > 1$. Define the operator T as above. Let I denote the identity operator. From the arguments of Case 1 there is a bounded uniformly positive function μ_1

[1] The relation (10) can be proved almost exactly as in the case of Markov chains with a finite number of states (see, e.g., Hostinsky (1931, pp. 15—17 and 44—45)). See also Doob (1953, Chapter V, particularly pp. 216—217).

corresponding to an eigenvalue of the operator T^{n_0}. Let $\varrho > 0$ be an n_0-th root of this eigenvalue; we have $(T^{n_0} - \varrho^{n_0} I)\mu_1 = 0$, or $(T - \varrho I)(T^{n_0-1} + \varrho T^{n_0-2} + \cdots + \varrho^{n_0-1} I)\mu_1 = 0$. Hence the bounded uniformly positive function $\mu = (T^{n_0-1} + \cdots + \varrho^{n_0-1} I)\mu_1$ is an eigenfunction of T, corresponding to ϱ. We can then argue almost as in Case 1.

Chapter IV

Neutron branching processes (one-group theory, isotropic case)

1. Introduction

Since the discovery of nuclear fission, there has been great interest in the mathematical study of the transport and multiplication of neutrons in matter. The reactions, which are described in detail in, for example, WEINBERG and WIGNER (1958), are very complicated, and the mathematical work has been based on more or less simplified models.

In this chapter we shall treat the so-called *one-group theory*, where all neutrons are supposed to have the same constant speed or energy. We consider only the *isotropic case*; that is, when a neutron is detached from a nucleus or is "scattered" (deflected), all directions are supposed to have equal probabilities. This model receives serious attention from physicists, although the cases in which it can give quantitatively good results are very limited. Simplified though it is, the model leads to difficult problems of both computing and analysis.

Nearly all the literature on neutron transport (we shall use the term to include multiplication) deals with expected values, not fluctuations. This is partly because in most cases of interest very large numbers of neutrons are involved and partly because the treatment of fluctuations is difficult. However, there are a few situations where fluctuations appear to be of some interest because small numbers of neutrons are involved. (See SEMAT (1946, p. 372); WEINBERG and WIGNER (1958, p. 221); FEYNMAN, DE HOFFMAN, and SERBER (1956).)

In this chapter, after studying the expected numbers in successive generations, we calculate numerically the probability that a single neutron generates an infinite family in a supercritical sphere. This kind of calculation lies outside the scope of expected-value theories. We also give the asymptotic form of the probabilities for the total size of a family in the case of a critical body. After a brief discussion of the case of a continuous time parameter, we discuss a quite different method (invariant imbedding), using the invariance principles developed by AMBARTSUMIAN, CHANDRASEKHAR, BELLMAN, KALABA, and WING. (See references in Secs. 8 and 9.)

A detailed discussion of various mathematical methods that have been used to treat expected values in neutron transport can be found in WEINBERG and WIGNER (1958), or DAVISON (1957). Interesting papers on the current state of mathematical theories are contained in the Symposium volume of BIRKHOFF and WIGNER (1961).

The results of Chapter III can be applied to certain models with more complicated assumptions than those used in this chapter, but numerical results are increasingly difficult to obtain as the models become more complex. To obtain numerical results when very complicated assumptions are made, Monte Carlo (artificial sampling) procedures are often used.

2. Physical description

From the point of view of transport theory a neutron is characterized at any instant by its position, direction of motion, and energy. At any time it may collide with an atomic nucleus. When this happens it may be "scattered" (deflected), generally with a loss of energy, or it may be absorbed by the nucleus; in the latter case it may or may not cause the nucleus to break up or "fission". The fission process may result in the production of various particles that do not explicitly enter the calculations plus a random number of new neutrons (the average number is $2\frac{1}{2}$ to 3), whose energies and directions are random. The new neutrons undergo similar reactions. The probability laws for these events depend in a complicated manner on the energy of the neutrons, and also depend on the nature of the enveloping matter. Contrary to the case of the electron-photon cascade studied in Chapter VII, the energy of a child (i.e., a neutron produced by fission) may be greater than that of the parent (the neutron causing the fission).

We shall idealize the treatment by neglecting the effect of energy on the reactions (one-group theory) and by assuming that the direction of motion of a neutron just produced by fission, or of a neutron just after scattering, is independent of all past history and is distributed uniformly on the unit sphere (isotropy). According to physicists with whom the author has discussed the problem, one may base rough computations on these assumptions in a few cases, e.g., nuclear reactions in pure metallic plutonium (isotopic number 239) or uranium (isotopic number 235). Such reactions, to be sure, are of only limited interest in applications. However, the isotropic one-group case is still considered to be of basic importance.

3. Mathematical formulation of the process

We shall first treat successive generations of neutrons as a branching process of the sort discussed in Chapter III. This is physically less interesting than studying the number of neutrons at a given time. However, it is adequate to answer such questions as whether a neutron born at a given point will have infinitely many descendants. Moreover, we shall see in Sec. 7 that the study of the process in time can be related to the study of the generations.

The family of neutrons develops inside a three-dimensional "body" X, which we shall assume to be convex in shape, of finite size, and homogeneous throughout. It is assumed that a neutron originating at a point x in X moves in a straight line, whose direction is uniformly distributed over the unit sphere. The movement continues until the neutron leaves the body or suffers a collision with an atomic nucleus in the body. The probability that the neutron has a collision in any interval of length Δ is $\alpha\Delta + o(\Delta)$, where under our assumptions α is independent of the location and energy of the neutron. Then the probability that a neutron travels a distance $>\ell$ in the body without a collision is $e^{-\alpha\ell}$. If the neutron gets outside the body without a collision, it disappears from our consideration.

We call α the *cross section* and $1/\alpha$ the *mean free path*. As an example, $1/\alpha$ is about $2\frac{1}{2}$ cm. in Pu^{239} for neutrons whose energy is about $\frac{1}{2}$ million electron volts. This is in the energy range of interest in some reactions.

When a neutron has a collision, say at the point y in X, it is transformed into a random number of neutrons, each of which then moves away from y with an isotropic choice of direction independently of past history. We adopt the convention that even if a collision results only in the scattering (deflection) of a neutron, the neutron after scattering is considered to be a new neutron originating at the point of scattering, the new direction being chosen isotropically.

3.1. Transformation probabilities. When a collision occurs, we assume that $p_0, p_1, \ldots$ are the probabilities that a neutron is transformed into 0, 1, ... neutrons. The numbers p_r are assumed independent of position and past history. The probability p_1 may be large, since it includes the case of scattering without fission. We put

$$f(t) = \sum_{r=0}^{\infty} p_r t^r, \qquad |t| \leq 1. \tag{3.1}$$

3.2. The collision density. In what follows x and y are three-dimensional vectors, $x-y$ is their vector difference and $|x-y|$ is the distance between them. Since y is three-dimensional, the transformation $y \to \alpha y$ entails the transformation of the differential volume dy to $\alpha^3\, dy$.

Let x be a point in the interior of X, let S be the surface of a sphere of radius r centered at x, and let dS be an element of area on S. Suppose that the element of volume whose base is dS and whose height is dr

lies inside X. The probability that a neutron born at x will move in a direction that will take it through dS, if no collision is suffered, is $dS/4\pi r^2$. The probability that the first collision occurs at a distance from x between r and $r+dr$ is $\alpha e^{-\alpha r}\,dr$. Hence the probability that the neutron has a collision in the volume element dy at y, and no collisions between x and y, is $\alpha e^{-\alpha|y-x|}\,dy/4\pi|y-x|^2 = R(\alpha(y-x))\alpha^3\,dy$, where

$$R(y) = \frac{e^{-|y|}}{4\pi|y|^2}. \tag{3.2}$$

Since R is a probability density in infinite space and is everywhere positive, and since X is bounded, we see that $\int_X R(\alpha(y-x))\alpha^3\,dy < 1$.

3.3. Definition of the branching process. We shall now define a general branching process of the sort discussed in Chapter III. The type x of a neutron is simply the point where it is born, where the "birth" of a neutron is either its detachment from a nucleus as a result of fission or its assumption of a new direction after scattering. Because of our assumption of isotropy we need not specify the direction at birth of a neutron as part of its type, provided we deal only with questions concerning the *numbers* of neutrons in various generations and not their directions.

Following the procedure and notation of Chapter III, we suppose that Z_0 represents the 0-th generation, a single neutron born at a point x in X. The first-generation neutrons are then represented by a point-distribution Z_1, which puts n units of weight 1 at y if n first-generation neutrons are born at y. Let $\mathscr{E}_x$ denote an expected value when the initial neutron is at x. If $s(y)$ is a nonnegative function of y, the moment-generating functional of Z_1 (see Sec. III.5) is then the functional of s defined by

$$\Phi_x^{(1)}(s) = \mathscr{E}_x e^{-\int_X s(y)\,dZ_1(y)}.$$

The probability that the initial neutron has a collision at y, followed by transformation into n neutrons, is $p_n R(\alpha(y-x))\alpha^3\,dy$. The probability of leaving X without a collision is $1 - \int_X R(\alpha(y-x))\alpha^3\,dy$. Hence

$$\left.\begin{aligned}\Phi_x^{(1)}(s) &= \Big[1 - \int_X R(\alpha(y-x))\,\alpha^3\,dy\Big] \times e^0 \\ &\quad + \sum_{n=0}^{\infty} p_n \int_X R(\alpha(y-x))\,e^{-ns(y)}\alpha^3\,dy \\ &= 1 - \int_X R(\alpha(y-x))\,\alpha^3\,dy + \int_X R(\alpha(y-x))\,f(e^{-s(y)})\,\alpha^3\,dy.\end{aligned}\right\} \tag{3.3}$$

The moment-generating functional in (3.3) will then serve to define our branching process, as in Sec. III.6.

In case we wish to deal with the family of a neutron launched in some particular direction, we can begin with the children of the original neutron, since each one will be launched isotropically from its point of birth.

4. The first moment

We shall henceforth assume that $f'(1)$ and $f''(1)$ are finite and $p_0 \neq 1$.

As in Chapter III, let $M(x, A) = \mathscr{E}_x Z_1(A)$ be the expected number of first-generation neutrons born in the region A of X, if the initial neutron is at x. Put $s = \lambda s_A$ on the right side of (3.3), where s_A is the indicator function of A (see Def. III.5.1). Then

$$M(x, A) = -\left.\frac{\partial \Phi_x^{(1)}(\lambda s_A)}{\partial \lambda}\right|_{\lambda=0} = f'(1) \int\limits_A R\left(\alpha(y-x)\right)\alpha^3\, dy. \tag{4.1}$$

Thus M has the density

$$m_1(x, y) = f'(1) R\left(\alpha(y-x)\right)\alpha^3 = f'(1)\,\alpha \frac{e^{-\alpha|y-x|}}{4\pi|y-x|^2}. \tag{4.2}$$

Although R is an unbounded function, it is known (see Petrovskiĭ (1957, p. 29)) that the expectation density $m_4(x, y)$ for the fourth generation is bounded and continuous. Moreover, m_1 is uniformly positive (Def. III.10.2; recall that X is a bounded region), and hence from the relation $m_{n+1}(x, y) = \int\limits_X m_n(x, z)\, m_1(z, y)\, dz$ we see by induction that m_n is, for each n, uniformly positive. Thus the conditions of Theorem III.10.1 are satisfied, and there exist a unique positive eigenfunction μ (because of the symmetry of R, it is both a left and right eigenfunction) and a positive number ϱ satisfying

$$\varrho\mu(x) = \int\limits_X m_1(x, y)\,\mu(y)\, dy. \tag{4.3}$$

We shall use the notation $\mu_{(\alpha)}$ and $\varrho_{(\alpha)}$ when it is necessary to indicate the value of α.

If $f(y)$ is an arbitrary bounded positive function, then $\int m_n(x, y) f(y)\, dy$ is approximately proportional to $\varrho^n \mu(x)$ if n is large, giving us a possible way to compute ϱ and μ.

On the role of positivity in more general neutron problems, see Birkhoff (1961).

5. Criticality

The body X is said to be *subcritical*, *critical*, or *supercritical* according as $\varrho < 1$, $\varrho = 1$, or $\varrho > 1$.

Theorem 5.1. *If X is a fixed bounded convex body, then the eigenvalue $\varrho_{(\alpha)}$ is a strictly increasing continuous function of α that approaches 0 as $\alpha \downarrow 0$ and that approaches $f'(1)$ as $\alpha \to \infty$.*

Since changing α is equivalent to changing the scale of the body, we see that if $f'(1) \leq 1$, then every bounded convex body is subcritical, while if $f'(1)>1$, then every such body is supercritical if sufficiently magnified (α remaining fixed) and is subcritical if sufficiently shrunk.

It would be difficult to attribute this familiar result, which stems from results on integral equations. MULLIKIN (1961 a) has given a proof for slabs and spheres, based on positivity arguments, and hence applicable to more general kernels than the one considered here.

We shall not give a proof of Theorem 5.1, since it would be rather lengthy. The intuitive meaning is as follows. If α is very small, the probability is high that a neutron will be lost from the body before having a collision; hence $\varrho_{(\alpha)}$, which represents the multiplicative factor per generation, is very small. On the other hand, if α is very large, then a neutron will have a very high probability of having a collision in a very short distance; hence the factor is almost $f'(1)$, the expected number created per collision.

If $\varrho>1$, then the conditions of Theorem III.14.1 are satisfied, and there is a nonzero probability that a single neutron will initiate a family that will grow indefinitely, i.e., an explosion. If many neutrons are injected simultaneously into a body that is supercritical, then an explosion is very nearly certain to occur immediately. (The neutrons may come from spontaneous fission of atoms inside the body, or additional neutrons may be supplied from outside.) If $\varrho\leqq 1$, then the family of each neutron will be finite.

If $\varrho=1$, then the theory of Chapter III tells us that although the family of each neutron will eventually die out, the expected total number of progeny in all generations of the family is infinite. The distribution of the number in the family is considered in Sec. 6.3.

6. Fluctuations; probability of extinction; total number in the critical case

It is in principle possible to calculate the exact probability for any number of neutrons in the n-th generation, using the recurrence relations of Sec. III.7. However, this would not be a feasible computational procedure. One could also calculate the second moments using the formulas of Sec. III.13. Physicists have generally used differential equations to calculate fluctuations; see, e.g., COURANT and WALLACE (1947).

As an example of a calculation where probabilities rather than just expected values are involved, we observe that even if a body is supercritical, a single neutron may not generate an infinite family. The probability that it does not is simply the extinction probability $q(x)$ of Chapter III. Thus from equations (III.15.1) and (3.3) we see that q

satisfies the functional equation

$$q(x) = 1 - \int_X R(\alpha(y-x))\alpha^3 dy + \int_X R(\alpha(y-x)) f(q(y))\alpha^3 dy. \qquad (6.1)$$

An equation similar to (6.1) was given by SCHROEDINGER (1945), except that SCHROEDINGER begins with several neutrons at x.

The function q can be calculated numerically by successive approximations, convergence being assured by the results of Sec. III.15. We shall next present the results of such a calculation for the case in which X is a sphere.

6.1. Numerical example. In case X is a sphere of radius D, it is known (DAVISON (1957, p. 96)) that the isotropic one-group theory of neutron multiplication can be reduced to a one-dimensional problem. A neutron born at a distance $\xi < D$ from the center has the probability density $R_1(\xi, \eta)$ for a collision at a distance η from the center, where

$$R_1(\xi, \eta) = \frac{\eta}{2\xi}\left[E_1(|\xi-\eta|) - E_1(|\xi+\eta|)\right], \qquad \eta < D,$$

$$E_1(\xi) = \int_\xi^\infty \frac{e^{-t}}{t} dt, \qquad \xi > 0,$$

provided we take the constant α equal to 1 (i.e., we take the "mean free path" as the unit of length). Let $Q_D(\xi)$ be the extinction probability for a sphere of radius D when the initial neutron is a distance ξ from the center. Then equation (6.1) becomes

$$Q_D(\xi) = 1 - \int_0^D R_1(\xi, \eta)\, d\eta + \int_0^D R_1(\xi, \eta) f[Q_D(\eta)]\, d\eta. \qquad (6.2)$$

For the generating function f of Sec. 3.1 take

$$\left.\begin{aligned} f(t) &= 0.025 + 0.830t + 0.070t^2 + 0.050t^3 + 0.025t^4, \\ f'(1) &= 1.22 = \text{expected number per collision.} \end{aligned}\right\} \qquad (6.3)$$

The numbers in (6.3) are discussed in Sec. 6.2 below.

For each of several values of D ranging from 3 to 10, equation (6.2) was solved numerically by successive approximations[1]. Thus for each value of D the function $Q_D(\xi)$ was available for numerous values of ξ. We shall not tabulate all these values here, but instead tabulate, as a function of D, the values of $1 - Q_D(0)$, the probability that an initial neutron at the center of a sphere generates an infinite family, and $1 - Q_D(D)$, the probability that an initial neutron at the surface generates an infinite family. If we double the latter values, we get the probability

[1] I am indebted to ALFRED NELSON for the numerical analysis and programming of the computations, which were carried out on The RAND Corporation's JOHNNIAC digital computer.

that a neutron impinging on the sphere (i.e., isotropic inward) generates an infinite family. The author thinks the numbers in Table 1 are correct to within 1 or 2 in the *second* decimal place (this statement refers to the numerical accuracy of the solution of the equation and has no bearing on the applicability of the results to real materials).

Table 1. *Probability of infinite family as function of radius*

D (mean free paths)	$1-Q_D(0)$ (initial neutron at center)	$1-Q_D(D)$ (initial neutron at surface)
3	0.019	0.003
4	0.369	0.057
5	0.566	0.089
6	0.682	0.109
7	0.750	0.122
8	0.797	0.133
10	0.834	0.146
∞	0.8414	—

The value 0.8414 is $1-0.1586$, the latter number being the positive root less than 1 of the equation $t=f(t)$. From the results of Chapter I, this root is the probability of extinction in an infinite medium.

A graph of $1-Q_D(0)$ as a function of D suggests that the critical radius is about 2.9 mean free paths. The function appears to be concave in D and to meet the D-axis at a nonzero acute angle. In fact, T. W. MULLIKIN (1961 a) has shown that for a wide class of models, including the present one, the angle is not zero, and can be evaluated if we have computed the eigenfunction corresponding to the eigenvalue ϱ.

6.2. Further discussion of the example. The numbers used in the above computation do not represent the best approximation that one could make for some real material. However, they are somewhat similar to what might be realized in the metal Pu^{239}, except that the mean number 1.22 produced per collision is too small; about 1.4 would be more realistic[1].

A mass of Pu^{239} that is critical or slightly larger will have a few hundred spontaneous fissions per second. Thus Table 1 suggests that if the radius is, say, 30 per cent larger than the critical radius, an explosion should occur very quickly, since the individual neutron reactions are very rapid. However, spontaneous fission is evidently not considered

[1] When an atom of Pu^{239} undergoes fission as a result of a collision, the mean number of neutrons produced is about 3, but since most collisions result in scattering or absorption without fission, the mean number per collision is about 1.4. The author used a lower value because of a misunderstanding, but it did not seem worth while to repeat the calculation.

to be an adequate source of neutrons for detonating an atomic bomb, since additional neutrons are supplied. (See the article "Atomic Bomb", McGraw-Hill *Encyclopedia of Science and Technology* (1960).)

6.3. Total number of neutrons in a family when the body is critical. If the body X is critical, then, according to Theorem III.12.1, the family of a single neutron will die out with probability 1, although the expected value of the total number of its descendants in all generations is infinite. In the case in which the neutrons are assumed to move according to diffusion laws rather than according to the law postulated in Sec. 3 above (see also Sec. 8), SEVAST'YANOV (1958) has obtained an expression for the probability that there are altogether n members of the family in all generations. If we apply the methods of SEVAST'YANOV to our case, we obtain the following result[1]. Let $P_n(x)$ be the probability that there are altogether n neutrons in all generations, if the original neutron is at x. Then if X is critical we have

$$P_n(x) = c\mu(x)\, n^{-\frac{3}{2}} + O(n^{-\frac{5}{2}}), \qquad n \to \infty, \tag{6.4}$$

where $\mu(x)$ is the positive eigenfunction of Sec. 4 and

$$c = \sqrt{\frac{f'(1)\int \mu(x)\,dx}{2\pi f''(1)\int \mu^3(x)\,dx}}\,.$$

Formula (6.4) should be compared with equation (I.13.4), due to OTTER, in which we should put $\alpha = 1$ to correspond to the critical case.

7. Continuous time parameter

Neutron branching processes with a continuous time parameter are often studied by means of the Boltzmann transport equation for the expected density of neutrons (see WEINBERG and WIGNER (1958, Chapter IX)). We shall not give the equation here, but only remark that it is an integro-differential equation corresponding to the *forward differential equation* in the theory of Markov processes. There is likewise a backward form, although the author has not seen it used.

Another approach is through an *integral equation* for the density. Since the equation is derived by tracing the neutrons existing at time t back to their last preceding collision, it may be called a *last-collision* method. We shall not give this equation (see WEINBERG and WIGNER, p. 190), but shall give a rather similar one that might be called a *first-collision* equation. We have here an example of the *point of regeneration*

[1] This result is obtained by formally applying to the present case the series solution used by SEVAST'YANOV for his problem. It seems certain that the result can be made rigorous. The error term $O(n^{-\frac{5}{2}})$ is presumably bounded independently of x.

to be discussed in Chapter VI. Although either form of integral equation can be used to treat expectations, the first-collision method seems better suited to treat higher moments and probabilities.

7.1. Integral equation treatment. The following treatment is intended to be plausible rather than rigorous. It should be emphasized that *a similar treatment can be used for higher moments or for the moment-generating functional.* We continue to make the assumptions given in Secs. 2 and 3.

Let $M(x, A, t)$ be the expected number of neutrons in the region A in X at time t if originally there is one neutron at x, launched isotropically. We define M to be 0 if t is negative.

Since our equation will be hinged on the moment of first collision, when all subsequent directions will be equally probable, we need not treat directions explicitly. However, since the position of a neutron at an arbitrary time is an incomplete description of its state, the expectations M do *not* possess the semigroup property; that is, $\int M(y, A, t_1)\, d_y M(x, y, t_2)$ is *not* identical with $M(x, A, t_1+t_2)$.

Choose the unit of distance so that the constant α of Sec. 3 is 1 and the unit of time so that the speed of a neutron is 1. Let S_t be the spherical surface of radius t centered at x. The probability that the initial neutron has no collision between times 0 and t and passes through the surface element $d\Sigma$ on S_t is $R(t)\, d\Sigma$, where R is defined by (3.2). Hence the probability that the neutron is in A at t and has suffered no collisions is

$$R(t) \int_{A\cap S_t} d\Sigma = \int_{A\cap S_t} R(y-x)\, d\Sigma. \tag{7.1}$$

If the neutron suffers a collision at y before time t, it is replaced by an average of $f'(1)$ neutrons, each of which acts like the original one. Since the time to reach y is $|y-x|$, we have

$$\left.\begin{aligned} M(x, A, t) = &\int_{A\cap S_t} R(y-x)\, d\Sigma \\ &+ \int_X f'(1)\, R(y-x)\, M(y, A, t-|x-y|)\, dy. \end{aligned}\right\} \tag{7.2}$$

Equation (7.2) can be solved by iteration, with sufficient labor. Thus if we write (7.2) in the form $M = M_0 + TM$, where M_0 is the first term on the right side of (7.2) and T is an integral operator, then iteration gives $M = M_0 + TM_0 + T^2 M_0 + \cdots$, where the successive terms are the expected numbers of neutrons in A at t that have suffered 0, 1, 2, ... collisions. This grouping by collisions is a familiar procedure in many problems in probability theory.

We are interested in the behavior of M when t is large. To study this we shall take the Laplace transform of both sides of (7.2), putting

$\int_0^\infty M(x, A, t) e^{-\sigma t} dt = \tilde{M}(x, A, \sigma)$; the integral will be defined when the real part of σ is sufficiently large.

If y is a point of S_t, then $e^{-\sigma t} R(y-x)\, d\Sigma\, dt = e^{-\sigma|y-x|} R(y-x)\, dy$, since $d\Sigma\, dt$ is a volume element, and hence the Laplace transform of the first term on the right side of (7.2) is $\int_A k(x, y, \sigma)\, dy$, where we have put $k(x, y, \sigma) = R(y-x) e^{-\sigma|y-x|}$. As for the second term on the right side of (7.2), if we multiply by $e^{-\sigma t} = e^{-\sigma(t-|y-x|)-\sigma|y-x|}$, integrate on t, and observe that $M(y, A, t-|y-x|) = 0$ if $t < |y-x|$, we obtain

$$f'(1) \int_X k(x, y, \sigma) \tilde{M}(y, A, \sigma)\, dy .$$

Hence we obtain the equation

$$\tilde{M}(x, A, \sigma) = \int_A k(x, y, \sigma)\, dy + f'(1) \int_X k(x, y, \sigma) \tilde{M}(y, A, \sigma)\, dy . \qquad (7.3)$$

Let us deduce from (7.3) the asymptotic behavior of $M(x, A, t)$ as $t \to \infty$. If σ is real, the kernel $f'(1) k(x, y, \sigma)$ has a positive right eigenfunction[1] $\mu_\sigma(x)$ corresponding to the positive eigenvalue ϱ_σ. Presumably, ϱ_σ is a decreasing differentiable function of σ, with $\varrho_{-\infty} = \infty$ and $\varrho_\infty = 0$. Let σ_0 be the root of the equation $\varrho_\sigma = 1$. Then if $\sigma > \sigma_0$, we can, from the theory of integral equations, write the solution of (7.3) as a Neumann series,

$$\tilde{M}(x, A, \sigma) = \int_A \sum_{n=1}^{\infty} (f'(1))^{n-1} k_n(x, y, \sigma)\, dy , \qquad (7.4)$$

where the iterated kernels k_n are defined by $k_1(x, y, \sigma) = k(x, y, \sigma)$, $k_{n+1}(x, y, \sigma) = \int_X k(x, z, \sigma) k_n(z, y, \sigma)\, dz$. From the results of Sec. III.10 we have $(f'(1))^n k_n(x, y, \sigma) \sim (\varrho_\sigma)^n \mu_\sigma(x) \mu_\sigma(y)$, $n \to \infty$, σ fixed, $\sigma > \sigma_0$, where we have normalized so that $\int (\mu_\sigma(x))^2\, dx = 1$. Thus we should have, from (7.4),

$$\tilde{M}(x, A, \sigma) \sim [f'(1)(1-\varrho_\sigma)]^{-1} \mu_{\sigma_0}(x) \int_A \mu_{\sigma_0}(y)\, dy , \qquad \sigma \downarrow \sigma_0 . \qquad (7.5)$$

Near $\sigma = \sigma_0$ we have $1 - \varrho_\sigma \sim -(\sigma - \sigma_0)(d\varrho_\sigma/d\sigma)_{\sigma=\sigma_0}$. Hence (7.5) suggests the asymptotic form

$$M(x, A, t) \sim \frac{\mu_{\sigma_0}(x) \int_A \mu_{\sigma_0}(y)\, dy}{f'(1)\left(-\dfrac{d\varrho_\sigma}{d\sigma}\right)_{\sigma=\sigma_0}} e^{\sigma_0 t} , \qquad t \to \infty . \qquad (7.6)$$

As far as the author knows, no rigorous general proof of (7.6) exists, although it is related to the very general results of JÖRGENS (1958), and

[1] Notice the difference in the definitions of μ_σ and ϱ_σ here, and $\mu_{(\alpha)}$ and $\varrho_{(\alpha)}$ in Sec. 4.

the exponential nature of the solution is commonly assumed. See also LEHNER and WING (1955, 1956) and WEINBERG (1961, p. 5). A useful summary of the results of LEHNER and WING and of JÖRGENS is given in WING (1961).

8. Other methods

DAVISON (1957), WEINBERG and WIGNER (1958), and the papers in the Symposium volume of BIRKHOFF and WIGNER (1961) describe a number of other methods in neutron transport theory. Of these we mention only diffusion theory. Since the successive steps between collisions of a neutron comprise a random walk, we should expect the motion of a neutron to be similar to a diffusion process when the steps are small compared with the size of the body. In fact, the diffusion approximation is often used for the expected values, some ingenuity being required to choose suitable boundary conditions.

SEVAST'YANOV (1958, 1961) has discussed branching processes where particles undergo diffusion, and his results could be applied to treat fluctuation phenomena, if the diffusion approximation is to be used. (See also Sec. 6.3 above.)

9. Invariance principles[1]

Stochastic processes are often studied by means of some kind of invariance principle. Roughly speaking, we find the answer to a particular problem by considering a family of problems in which the particular one is imbedded. Consideration of the relationship between different members of the family leads to some sort of functional equation whose solution gives the answer to the whole family of problems. Different members of the family are indexed by a parameter such as time, distance, etc.

A familiar example is the study of the fortune of a gambler who wins or loses one dollar at a time through the operation of a chance device. If we want to calculate $p(x)$, the probability that he loses all his money before he attains a fortune of \$1,000 if his initial fortune is x, we relate $p(x)$ to $p(x-1)$ and $p(x+1)$ through a difference equation.

In the treatments of neutron multiplication discussed above, we have made use of functional equations in which position and time are variables, but we have always considered the body as unchanging. We arrive at new and different functional equations if we make use of invariance principles derived by considering the body as imbedded in a family of bodies. We then obtain equations in which some dimension of the body

[1] This terminology follows CHANDRASEKHAR (1950) and is not to be confused with another common use of the term having to do with limit theorems in the theory of probability.

is a variable. Such methods are suitable for dealing with certain symmetric bodies such as infinite plane slabs or spheres.

Methods of this sort have been employed by STOKES (1862), AMBARTSUMIAN (1943), and CHANDRASEKHAR (1950) in connection with the theory of radiative transfer. More recently such methods have been further developed (as "invariant imbedding") by BELLMAN, KALABA, and WING in a series of papers (e.g., 1957, 1958, and 1960; see also the papers by BELLMAN and KALABA and by WING in WEINBERG and WIGNER (1958), and PREISENDORFER (1958)).

As an example, we shall treat, without rigor, a case of one-dimensional neutron transport slightly different from that discussed in Chapter III. See Sec. III.8.2 for references to FERMI's original work on this problem.

It is possible to pass rigorously from the Boltzmann transport equation to various principles of invariance or equations derived from them; see, e.g., BUSBRIDGE (1960) and MULLIKIN (1961 b).

10. One-dimensional neutron multiplication

Suppose X is the one-dimensional interval $(0, T)$, $T>0$. Assume that a neutron traversing an interval inside X of length Δx has a probability $\alpha\Delta x+o(\Delta x)$ of suffering a collision in the interval. When a collision occurs, two new neutrons are produced, *one going to the right and one to the left*. Each new neutron continues to move until it leaves X or has a collision. We wish to calculate $p_r(T)$, the probability that a total of r neutrons emerge from X on the right, in all generations, if initially one neutron enters X from the right.

Suppose the small additional interval $(T, T+\Delta)$ is added to X, where $\Delta>0$. Then, neglecting powers of Δ higher than the first, a neutron entering at $T+\Delta$ can result in r neutrons emerging at $T+\Delta$ in the following ways: (a) The initial neutron enters $(0, T)$ having had no collisions in $(T, T+\Delta)$. Then either (a$_1$) r neutrons emerge at the right from $(0, T)$ and all cross $(T, T+\Delta)$ with no further collisions; or (a$_2$) $j\leqq r$ neutrons emerge from $(0, T)$, of which one has a collision in $(T, T+\Delta)$. Of the two neutrons produced in this collision, one emerges on the right while the other enters $(0, T)$ and eventually causes $r-j$ neutrons to emerge on the right. (b) The initial neutron has a collision in $(T, T+\Delta)$. One of the resulting neutrons emerges on the right while the other enters $(0, T)$ and eventually causes $r-1$ neutrons to emerge on the right.

Taking account of these eventualities we obtain the equations

$$p_r(T+\Delta)=(1-\alpha\Delta)\left\{p_r(T)(1-r\alpha\Delta)+\sum_{j=1}^{r}p_j(T)(j\alpha\Delta)p_{r-j}(T)\right\}$$
$$+\alpha\Delta p_{r-1}(T)+o(\Delta), \qquad r=0,1,2,\ldots;\; p_{-1}(T)\equiv 0,$$

leading to the set of differential equations

$$\left.\begin{aligned}\frac{dp_r(T)}{dT} = \alpha p_{r-1}(T) - \alpha(1+r)p_r(T) + \alpha\sum_{j=1}^{r} j p_j(T) p_{r-j}(T),\\ r=0,1,\ldots,\end{aligned}\right\} \quad (10.1)$$

where $p_{-1}\equiv 0$ and $\sum_{j=1}^{0}$ is interpreted as 0. The initial conditions are $p_0(0)=1$, $p_r(0)=0$, $r>0$. From (10.1) we can solve successively for $p_0(T)$, $p_1(T)$, ..., since for each value of r we have a differential equation that is linear in $p_r(T)$.

When T is less than the critical length (defined by analogy with the definition in Sec. 5), then $\sum p_r(T)=1$. Multiply both sides of (10.1) by r and sum from $r=1$ to $r=\infty$, obtaining for the expectation $U(T)=\sum r p_r(T)$ the equation, valid for T subcritical,

$$\frac{dU}{dT} = \alpha(1+U^2), \qquad U(0)=0. \quad (10.2)$$

The solution of (10.2) is $U(T)=\tan(\alpha T)$. Thus the critical length is $T_0=\pi/2\alpha$. It is interesting to notice that here the critical length appears as a singularity of the solution of a nonlinear functional equation — i.e., (10.2) — rather than as the value of T that makes the eigenvalue of a linear integral operator equal to 1.

Whereas a treatment of this problem by means of the appropriate form of the Boltzmann equation leads to a two-point boundary value problem, the treatment by invariant imbedding leads instead to an initial-value condition. The analogue of this fact is important in the numerical solution of certain problems.

Bellman, Kalaba, and Prestrud (1962) have applied invariant imbedding to the problem of reflection from plane slabs and have calculated extensive tables.

For further treatment of one-dimensional neutron models, see Harris (1960b), and, for results about the asymptotic behavior in the case of more general one-dimensional kernels, Conner (1961).

Chapter V

Markov branching processes (continuous time)

1. Introduction

Many authors have used Markov processes as mathematical models for the growth of biological or physical populations. Among the pioneer works are papers by McKendrick (1914 and 1926) on the spread of epidemics, Yule (1924) on the evolution of new species, Furry (1937) on showers of electrons, and FELLER (1939) on "the struggle for existence". We should also note the works of VOLTERRA (1931) and LOTKA (1934) on the interaction of competing species, without the assumption of randomness.

For physical and biological applications, it is desirable to study a great variety of stochastic processes, including many Markovian ones. However, the treatment in this chapter will be limited almost exclusively to those Markov processes that are extensions to continuous time of the Galton-Watson process; we shall call them *Markov branching processes*, "continuous time" being understood in this chapter. Although special, they are of interest in that they lead to analytic results, concerning many special cases of the Galton-Watson process, that are more precise than have been obtained for the general case. Moreover, there is a connection with a problem that has interested mathematicians since the time of ABEL: how to define the n-th iterate of a function when n is not an integer.

In spite of their simplicity, these processes have been instructive in applications, for example, in ARLEY's study of cosmic rays (1943), as well as in examples given in the other references cited here.

For discussion of applications of more general kinds of Markov processes, the reader is referred to the books of BARTLETT (1955, 1960) for numerous biological applications; NEYMAN (1956, 1961) for papers on the struggle for existence, epidemics, carcinogenesis, and related topics; BAILEY (1957) on epidemics; BLANC-LAPIERRE and FORTET (1953) for a variety of applications; and BHARUCHA-REID (1960) for numerous applications in physics and biology. The book of D'ANCONA (1954) discusses deterministic models of the struggle for existence. Other useful references are the Royal Statistical Society Symposium papers of J. E. MOYAL, M. S. BARTLETT, and D. G. KENDALL (1949).

In this chapter our mathematical model for the growth of a population is simply a random function representing its size. Hence we have no explicit mathematical mechanism to represent the relationships among the various objects, although these relationships are in our minds when we formulate the model. For example, if a death occurs, we cannot ask to *which* object it happens, nor can we discuss the ages of objects. The necessary mechanism will be established in Chapter VI, for the class of *age-dependent processes*, of which temporally homogeneous Markov branching processes will then appear as a special case.

The first general formulation of Markov branching processes (continuous time) appears to have been given by KOLMOGOROV and DMITRIEV (1947), who considered the case of several types of objects. Of course, special cases had been considered earlier. (See, e.g., Sec. 7.)

2. Markov branching processes

We assume the reader to be familiar with the elements of the theory of Markov processes whose states are integers (i.e., Markov chains; see, e.g., FELLER (1957, Chapter 17), for an introductory treatment).

In this chapter we study a random function $Z(t)$, taking nonnegative integer values, representing the size of a population. If the size of the population is known at any time, say $Z(t_1)=i$, then the subsequent history of the population is assumed to follow probability laws that depend on i and perhaps on absolute time, but not on the ages of the objects present at t_1, nor on any other properties of the objects that might be revealed if we knew the history of the population before time t_1. Thus we postulate that Z will be a Markov process. We also postulate that the families developing after time t_1 from the objects present at t_1 have no influence on one another. This is what singles out our processes from other Markov processes.

The assumption of age-independence seems a natural first step in introducing stochastic considerations into population theories, since age-dependence entails mathematical complications. The assumption seems unsuited to any kind of animal or plant (however, see Secs. 8 and 15.1), but appears to be acceptable in special cases for neutrons (Chapter IV) and electron-photon cascades (Chapter VII). As regards the independence of different objects, this is a valid assumption for the physical examples just cited, but would be of very limited usefulness for biological systems. On this point see Sec. I.7.3 and Sec. 8.

We shall study $Z(t)$ almost exclusively in terms of the transition probabilities $P_{ij}(\tau, t)=\text{Prob}\{Z(t)=j \mid Z(\tau)=i\}$, $0 \leqq \tau \leqq t$. The transition probabilities will be studied by means of two systems of differential equations satisfied by them, the *forward* and the *backward* equations of KOLMOGOROV. The equations, in turn, can be written, once we have prescribed the form of certain functions $b_i(t)$ and $p_{ij}(t)$, $i, j=0, 1, 2, \ldots$. We interpret these functions as follows. If we are given $Z(t)=i$, then the probability of a change of state in the interval $(t, t+\Delta)$ is $b_i(t)\Delta+o(\Delta)$. If we know that a change occurs at t when the state is i, then $p_{ij}(t)$ is the probability that the new state is j.

Let us suppose that an object existing at t has a probability $b(t)\Delta+o(\Delta)$ of dying in the interval $(t, t+\Delta)$, where b is a continuous nonnegative function. If it dies at any time τ, the probabilities are $p_0(\tau), p_2(\tau), p_3(\tau), \ldots$ that it is replaced by 0, 2, 3, ... objects. We omit the possibility of a death followed by replacement by a single object; because of the lack of dependence on age, this represents no change in the situation. By the same token, the death of an object and its replacement by n new ones, $n \geqq 2$, is equivalent to the continuation of the object after giving birth to $n-1$ new ones.

Note that if an object is born at time t_1, then the probability density for a life-length τ has the form

$$b(t_1+\tau)\, e^{-\int_{t_1}^{t_1+\tau} b(x)\,dx},$$

which is an exponential function if b is constant.

If i independent objects are present at t, then the probabilities of 0 deaths, 1 death, $\geqq 2$ deaths in $(t, t+\Delta)$ are, from the binomial theorem, $1-ib(t)\Delta+o(\Delta)$, $ib(t)\Delta+o(\Delta)$, and $o(\Delta)$. If one death occurs at τ, then the probability that the new population size is j, if it was i, is $p_{j-i+1}(\tau)$. Thus we are motivated to take $b_i(t)=ib(t)$ and $p_{ij}(t)=p_{j-i+1}(t)$. The appropriate forms of the forward and backward systems are then (3.3) and (3.4).

We now define a Markov branching process, but the definition should be considered in the light of the remarks that follow it.

Definition 2.1. A *Markov branching process* is a Markov chain whose states are nonnegative integers and whose transition probabilities are a solution to the *forward equations* (3.3). (See Def. 3.1.) The probabilities will then automatically be a solution to the *backward equations* (3.4), as we shall see. It is assumed that b is a continuous strictly positive function, that the p_i are continuous and nonnegative, that $p_1(t)\equiv 0$, and that $\sum_{j=0}^{\infty} p_j(t)\equiv 1$. If b and the p_i are independent of t, we speak of the *temporally homogeneous case*.

Remark 1. We allow the possibility that $\sum_j P_{ij}(\tau, t)<1$ if $\tau<t$, corresponding to a case in which Z may reach ∞ in a finite period of time. In such cases, as we shall see, the backward equations have solutions that do not satisfy the forward equations and that lack the characteristic independence property of a true branching process. We shall see that the forward equations are satisfied by only one set of transition probabilities.

Remark 2. We have *defined* the process only in the sense that we have defined the transition probabilities $P_{ij}(\tau, t)$. The reader may think of the random function $Z(t)$ as a step function whose value changes only when births or deaths occur. However, in order to validate this conception, some additional mathematical framework is required, such as that of Chapter VI. (See the discussion of separability in Doob (1953, Chapter 2).)

Remark 3. For convenience we have required $b(t)>0$. However, there would seem to be no difficulty if we should allow $b(t)\geqq 0$.

3. Equations for the probabilities

Let $Z(t)$ be a Markov branching process (Def. 2.1) with transition probabilities

$$P_{ik}(\tau, t) = \text{Prob}\big(Z(t) = k \mid Z(\tau) = i\big), \quad i, k = 0, 1, 2, \ldots; \; 0 \leqq \tau \leqq t. \tag{3.1}$$

Because of the Markovian nature of the process, these probabilities satisfy the Chapman-Kolmogorov equations:

$$P_{ik}(\tau, t_2) = \sum_{j=0}^{\infty} P_{ij}(\tau, t_1) P_{jk}(t_1, t_2), \quad i, k = 0, 1, \ldots; \; 0 \leqq \tau \leqq t_1 \leqq t_2. \tag{3.2}$$

We shall base our study of the probabilities P_{ik} on the usual *forward* and *backward* differential equations for Markov processes, in the special form appropriate to Def. 2.1. The forward equations are (here $\delta_{ik} = 1$ if $i = k$ and 0 otherwise)

$$\left.\begin{aligned} \frac{\partial P_{ik}(\tau, t)}{\partial t} &= -k b(t) P_{ik}(\tau, t) + b(t) \sum_{j=1}^{k+1} P_{ij}(\tau, t) j p_{k-j+1}(t), \\ & \qquad P_{ik}(\tau, \tau + 0) = \delta_{ik}, \end{aligned}\right\} \tag{3.3}$$

and the backward equations are

$$\left.\begin{aligned} \frac{\partial P_{ik}(\tau, t)}{\partial \tau} &= i b(\tau) P_{ik}(\tau, t) - i b(\tau) \sum_{j=i-1}^{\infty} p_{j-i+1}(\tau) P_{jk}(\tau, t), \qquad i > 0; \\ \frac{\partial P_{0k}}{\partial \tau} &= 0; \qquad P_{ik}(t - 0, t) = \delta_{ik}. \end{aligned}\right\} \tag{3.4}$$

Definition 3.1. We shall call a set of functions $\{P_{ik}(\tau, t)\}$ a *solution* to (3.3) or (3.4) if they are nonnegative, absolutely continuous in τ and t separately, satisfy (3.3) or (3.4), satisfy (3.2), and satisfy the inequalities

$$\sum_k P_{ik}(\tau, t) \leqq 1, \qquad i = 0, 1, 2, \ldots. \tag{3.5}$$

3.1. Existence of solutions. It follows from results of Feller (1940) that there is always a solution $\{P_{ik}\}$ common to (3.3) and (3.4). We shall see in Sec. 4 that any solution of (3.3) has the property

$$P_{ik}(\tau, t) = \sum_{r_1 + r_2 + \cdots + r_i = k} P_{1r_1}(\tau, t) P_{1r_2}(\tau, t) \ldots P_{1r_i}(\tau, t). \tag{3.6}$$

The meaning of (3.6) is that the *population at time t, starting with i objects at time τ, is distributed as the sum of i independent populations starting with one object.*

Kolmogorov and Dmitriev (1947) began with (3.6) and the general properties of Markov processes as axioms and thence deduced the basic equation (5.1) and then (4.3) for the case of several types. (See also Sec. 15 below.) This route was also followed in the paper of Sevast'yanov (1951).

3.2. The question of uniqueness. The question of uniqueness of a solution of (3.4) is of particular interest since it is connected with the truth of (3.6) and with the question whether the equality holds in (3.5).

FELLER (1940) studied chains where the *inequality* holds in (3.5) when $t > \tau$. In our case this can happen if $Z(t)$ can increase so rapidly that it reaches ∞ in a finite period of time. Let us call this the *explosive* case. (This is *not* analogous to the supercritical case in nuclear fission.)

As shown by DOOB (1945), if the inequality holds in (3.5), there is a whole family of processes satisfying the Kolmogorov *backward* equations ((3.4) in our case). We obtain such processes by letting $Z(t)$ start anew with some finite value whenever it reaches ∞. There are many ways in which the new start can be made. Thus if there are solutions for which $\sum_k P_{ik} < 1$, then (3.4) has many solutions, although (3.3) still has only one, as we shall see. In case the solution of (3.3) satisfies $\sum_k P_{ik} \equiv 1$, then (3.4) has only one solution (Sec. 4).

However, *there is always only one solution of* (3.4) *that satisfies* (3.6), *i.e., that has the characteristic property of a branching process.* Let us now explain this situation (a proof will be given in Theorem 4.2).

Suppose we have the explosive case with $Z(0) = 2$ and let us extend the process (in one of many possible ways) by agreeing that whenever $Z(t)$ reaches infinity it returns immediately to 2 and begins again.

The probabilities for this extended process satisfy the backward, but *not* the forward, equations.

Now in the extended process the relation (3.6) does not hold; i.e., different branches of the family do *not* act independently. It is true that starting with $Z(0) = 2$ is equivalent to starting with two independent families of size 1. However, as soon as *either* family reaches infinity, the rules require that the total population size returns to 2. Hence an explosion in one family affects the other one, and different objects are not independent.

3.3. A lemma. The following lemma will be useful. It can be proved by putting $i = k = 1$ in (3.3), obtaining the inequality $\partial P_{11}/\partial t \geqq -b(t) P_{11}$. We leave the details of the proof to the reader.

Lemma 3.1. *If* $\{P_{ik}\}$ *is a solution of* (3.3) *then*

$$P_{11}(\tau, t) \geqq \exp\left[-\int_\tau^t b(x)\, dx\right].$$

We next introduce generating functions, which will be the main analytic tool in the rest of this chapter. They have long been used for differential-difference equations (see BATEMAN (1943)). The first use in connection with the differential-difference equations of continuous-time Markov chains may have been by C. PALM (see ARLEY and BORCHSENIUS (1945)).

4. Generating functions

For the moment we shall suppose $\{P_{ik}(\tau, t)\}$ to be any solution of (3.3); a proof of uniqueness will come later.

Definitions 4.1. Let $h(s,t)=\sum_{k=0}^{\infty} p_k(t)\, s^k$, $|s|\leqq 1$, where the p_i are the functions of Def. 2.1. Let

$$F_i(s,\tau,t)=\sum_{k=0}^{\infty} P_{ik}(\tau,t)\, s^k, \qquad |s|\leqq 1, \tag{4.1}$$

where the functions P_{ik} are a solution of (3.3). (See the definition of *solution* in the preceding section.) At this point we must admit the possibility that $F_i(1,\tau,t)<1$ if $\tau<t$, corresponding to the "explosive case" mentioned in the preceding section.

If we multiply both sides of (3.3) by s^k and sum on k, we obtain formally the equations[1]

$$\left.\begin{aligned} \frac{\partial F_i(s,\tau,t)}{\partial t} &= b(t)\,[h(s,t)-s]\,\frac{\partial F_i(s,\tau,t)}{\partial s}, \\ i=0,1,2,\ldots;\ & F_i(s,\tau,\tau+0)=s^i. \end{aligned}\right\} \tag{4.2}$$

We have the following result.

Theorem 4.1. *There is just one solution* $\{P_{ik}\}$ *of* (3.3). *The generating functions* F_i *defined by* (4.1) *satisfy* (4.2) *for* $|s|<1$, *and* $F_i=(F_1)^i$, $i=0,1,2,\ldots$. *Accordingly, the branching-process property* (3.6) *is true.*

The proof of Theorem 4.1 is given in Appendix 1 to this chapter.

From the results of Feller (1940) we know that the solution of (3.3), now known to be unique, must also satisfy (3.4). Put $i=1$ in (3.4), multiply both sides by s^k, and sum on k. Using the relation $F_i=(F_1)^i$, we then obtain formally[2]

$$\frac{\partial F_1}{\partial \tau}=-b(\tau)\,[h(F_1,\tau)-F_1], \qquad F_1(s,t-0,t)=s, \qquad t>0. \tag{4.3}$$

We have the following result concerning equations (3.4).

Theorem 4.2. *Equations* (3.4) *have just one solution satisfying the branching-process property* (3.6), *and this is the unique solution of* (3.3). *The generating function* F_1 *defined by* (4.1) *with* $i=1$ *satisfies* (4.3) *for* $|s|\leqq 1$.

The proof is in Appendix 1.

The following property of F_1 will be useful in the sequel. We assume $|s|\leqq 1$.

---

[1] Apparently first given in this degree of generality by Bartlett (1949). Sevast'yanov (1951) gave the corresponding equations for the case of several types.

[2] First given by Kolmogorov and Dmitriev (1947) for the more general case of several types.

Lemma 4.1. *The generating function F_1 satisfies the inequality* $|F_1(s, \tau, t)| \leqq (1 - P_{11}(\tau, t))(1 - |s|) + |s|$. *If τ and t are confined to a finite interval, then there is a constant c, $0 < c < 1$, such that*

$$|F_1(s, \tau, t)| \leqq 1 - c(1 - |s|).$$

We leave the proof of this lemma (using Lemma 3.1) to the reader.

The differential equations for the generating function have been studied in the temporally homogeneous case from the point of view of semigroups by Bharucha-Reid and Rubin (1958) and Bharucha-Reid (1961).

4.1. Condition that the probabilities add to 1. The following simple condition insures that the unique solution of (3.3) satisfies $\sum_k P_{ik}(\tau, t) \equiv 1$. In this case the solution is the one and only solution of the system (3.4).

Condition 4.1. The series $\sum r p_r(t)$ converges uniformly on each finite t interval. (The sum is then necessarily a continuous function of t.)

To prove that the probabilities add to 1, let t be fixed and let $F_1(1, \tau, t) = G(\tau)$. Then G satisfies the ordinary differential equation $dG/d\tau = -b(\tau)\big(h(G, \tau) - G\big)$ with $G(t-0) = 1$, and $0 \leqq G \leqq 1$. The right side of the differential equation satisfies a Lipschitz condition in the region $-1 \leqq G \leqq 1$, $0 \leqq \tau \leqq t$, by virtue of Condition 4.1 and the definition of h. Hence the solution is unique, and since 1 is a solution we must have $G(\tau) \equiv 1$.

According to Feller (1940), the solution $\{P_{ik}\}$ of (3.3) is a *minimal* solution of (3.4), meaning that any other solution $\{P^*_{ik}\}$ satisfies $P^*_{ik}(\tau, t) \geqq P_{ik}(\tau, t)$, $i, k = 0, 1, \ldots$. But since

$$1 \geqq \sum_k P^*_{ik}(\tau, t) \geqq \sum_k P_{ik}(\tau, t) = 1,$$

we see that $P^*_{ik} = P_{ik}$.

In Sec. 9 we shall give necessary and sufficient conditions for the probabilities to add to 1 in the temporally homogeneous case.

The imposition of Condition 4.1 is simply the most obvious way of preventing objects from multiplying so rapidly that the population size reaches infinity in a finite time interval.

5. Iterative property of F_1; the imbedded Galton-Watson process

The generating function F_1 has the following property.

Theorem 5.1. *Let $F_1(s, \tau, t)$ be the generating function defined by* (4.1) *with $i = 1$. Then F_1 satisfies the functional relationship*

$$F_1(s, t_1, t_3) = F_1[F_1(s, t_2, t_3), t_1, t_2], \qquad 0 \leqq t_1 \leqq t_2 \leqq t_3;\ |s| \leqq 1. \tag{5.1}$$

We sketch the proof of this result, which is familiar from the theory of differential equations. See also the remarks following formula (3.6).

Let t_3 be fixed and suppose $|s|<1$. Then $F_1(s, \tau, t_3)$ is a solution of (4.3) for $0 \leqq \tau \leqq t_3$. Suppose $F_1(s, t_2, t_3)=\sigma$; then $|\sigma|<1$ (Lemma 4.1). Now $F_1(\sigma, \tau, t_2)$ is a solution of the differential equation in (4.3), defined for $0 \leqq \tau \leqq t_2$, with the initial condition $F_1(\sigma, t_2-0, t_2)=\sigma=F_1(s, t_2, t_3)$. Since both solutions are bounded by a constant <1 (Lemma 4.1), the uniqueness argument near the end of Appendix 1 shows that they are equal for $0 \leqq \tau \leqq t_2$. This proves (5.1) when $|s|<1$. The result for $|s|=1$ then follows from continuity.

5.1. Imbedded Galton-Watson processes. Let us now assume that Condition 4.1 holds, so that $F_1(s, \tau, t)$ is a probability generating function in the usual sense. For simplicity of exposition let us consider the temporally homogeneous case in which $F_1(s, \tau, t+\tau)=F_1(s, t)$. Let Δ be a fixed positive number, and let $Z_n=Z(n\Delta)$, $n=0, 1, \ldots,$ where $Z(t)$ is a Markov branching process corresponding to F_1, and $Z(0)$ is taken to be an arbitrary nonrandom integer. Using the temporal homogeneity of the process and the relation $F_i(s, t)=(F_1(s, t))^i$ (see Theorem 4.1) we have

$$\left.\begin{aligned} \mathscr{E}(s^{Z_{n+1}} \mid Z_n=i) &= \mathscr{E}(s^{Z(\Delta)} \mid Z(0)=i) \\ &= F_i(s, \Delta)=(F_1(s, \Delta))^i=[\mathscr{E}(s^{Z(\Delta)} \mid Z(0)=1)]^i. \end{aligned}\right\} \tag{5.2}$$

But (5.2) is the defining property of a Galton-Watson process, and we see that the sequence $\{Z(n\Delta)\}$ is a Galton-Watson process. Thus the generating function $F_1(s, n\Delta)$ is the n-th iterate of $F_1(s, \Delta)$, a fact that we could also deduce from (5.1).

If the process is not temporally homogeneous, or if we consider a sequence $\{Z(t_i)\}$, where the t_i are not equally spaced, analogous results hold except that we must consider Galton-Watson processes with variable generating functions (see Sec. I.13.5).

5.2. Fractional iteration. Consider the temporally homogeneous case in which $b(t)=b=$ constant, $h(s, t)=h(s)$, and $F_i(s, \tau, t)=F_i(s, t-\tau)$. In this case in place of (5.1) we have

$$F_1[F_1(s, t_1), t_2]=F_1(s, t_1+t_2), \qquad t_1, t_2 \geqq 0, \tag{5.3}$$

and we have pointed out that $F_1(s, n)$ is the n-th iterate of $F_1(s, 1)$; let us denote the latter function by $f(s)$. Since (5.3) expresses the characteristic property of iteration when t_1 and t_2 are integers (compare (I.2.5)), we may consider $F_1(s, t)$ as a *fractional iterate* of f even when t is not an integer.

A classical problem of analysis is, given a function $f(s)$, to find a function $F(s, t)$, with $F(s, 1)=f(s)$, satisfying (5.3). If f is analytic in the neighborhood of a fixed point s_0, $f(s_0)=s_0$, and if $0<|f'(s_0)|<1$, KOENIGS (1884) showed how the problem may be solved. The limit

$$\lim_{n\to\infty}\frac{f_n(s_0)-s_0}{(f'(s_0))^n}$$

can be shown to exist for s near s_0; call it $A(s)$. Then A satisfies the functional equation $A(f(s))=f'(s_0)A(s)$, whence $f(s)=A^{-1}[f'(s_0)A(s)]$ and $f_n(s)=A^{-1}[(f'(s_0))^n A(s)]$. The latter expression then serves as a definition of f_n when n is not an integer. Another approach is via ABEL's functional equation, described in ABEL (1881, posthumous). This is illustrated by using (I.11.4); if f is given and if π can be found satisfying (I.11.4), then we put $F(s,t)=\pi^{-1}[t+\pi(s)]$. A résumé of some of the classical work on the problem may be found in HADAMARD (1944).

If f is a probability generating function, it is natural to ask *whether we can find* $F(s,t)$ *satisfying* (5.3), *such that* F *is a probability generating function for each* t. We also impose a regularity condition; we shall require that the limit

$$\lim_{t\to 0}\frac{F(s,t)-s}{t}=\xi(s) \tag{5.4}$$

exists uniformly for $|s|\leqq 1$. If such a function F can be found, we shall say that f *belongs to the Class* C.

If f belongs to the Class C, and if $f(s)\not\equiv s$, it can be shown that $\xi(s)$ must be a power series of the form $b(-s+p_0+p_2s^2+p_3s^3+\cdots)$, $b>0$, $p_0+p_2+p_3+\cdots=1$, $p_i\geqq 0$. Hence using (5.3) and (5.4) and letting $t_1\to 0$ or $t_2\to 0$ we obtain, respectively,

$$\frac{\partial F(s,t)}{\partial t}=\xi(s)\frac{\partial F(s,t)}{\partial s} \tag{5.5}$$

and

$$\frac{\partial F(s,t)}{\partial t}=\xi[F(s,t)]. \tag{5.6}$$

These are the equations for the generating function of a Markov branching process in the temporally homogeneous case, and thus if f belongs to the Class C, the Galton-Watson process corresponding to f can be imbedded in a Markov branching process.

If we put $F(s,1)=f(s)$ and equate the right sides of (5.5) and (5.6) we obtain the functional equation

$$f'(s)\xi(s)=\xi[f(s)]. \tag{5.7}$$

If f belongs to the Class C (in which case it can be seen that $f'(0)>0$) and if $f(0)=0$, then (5.7) shows that $p_0=0$, and the coefficients $p_2, p_3, \ldots$ can be determined by differentiating (5.7) repeatedly at $s=0$.

It is obvious that many generating functions f do not belong to the Class C; for example, polynomials of degree $\geqq 2$ do not.

It can be shown that if f is a generating function with $f(0)=0$ and $0<f'(0)<1$, then a necessary and sufficient condition that f belongs to C is that the numbers $p_2, p_3, \ldots$, determined by differentiating (5.7), are nonnegative; and furthermore that if f is entire and $f(0)=0$, then f does not belong to C (HARRIS (1951)). However, the general problem of determining when a given f belongs to C is unsolved.

6. Moments[1]

We define the k-th factorial moment of $Z(t)$, $k=1, 2, \ldots$ by

$$m_k(\tau, t)=\mathscr{E}[Z(t)(Z(t)-1) \ldots (Z(t)-k+1)\,|\,Z(\tau)=1], \qquad 0\leqq\tau\leqq t.$$

If the series $\sum_{n=0}^{\infty} n^k p_n(t)$ converges uniformly on each finite t interval, then $m_k(\tau, t)$ is finite and can be determined by differentiating (4.2) or (4.3) k times with respect to s at $s=1$. (See Appendix 2.) For example, we have the following equation for m_1:

$$\frac{\partial m_1(\tau, t)}{\partial \tau}=b(\tau)\left[1-\frac{\partial h(1, \tau)}{\partial s}\right] m_1(\tau, t), \qquad 0\leqq\tau\leqq t;\ m_1(t-0, t)=1. \quad (6.1)$$

If $k>1$, then $m_k(t-0, t)=0$.

The solution of (6.1) is

$$m_1(\tau, t)=\exp\left\{\int_\tau^t b(u)\left(\sum_{j=1}^{\infty} j p_j(u)-1\right) du\right\}, \qquad 0\leqq\tau\leqq t. \qquad (6.2)$$

For convenience we give the first two moments in the temporally homogeneous case.

Definition 6.1. In the temporally homogeneous case, let $a=b(h'(1)-1)$.

Theorem 6.1. *In the temporally homogeneous case, taking $Z(0)=1$, we have*

$$\mathscr{E}Z(t)=e^{at}, \qquad (6.3)$$

$$\text{Variance } Z(t)=\left(\frac{h''(1)-h'(1)+1}{h'(1)-1}\right) e^{at}(e^{at}-1), \qquad a\neq 0, \qquad (6.4)$$

$$\text{Variance } Z(t)=h''(1)\, bt, \qquad a=0. \qquad (6.5)$$

7. Example: the birth-and-death process

The most familiar Markov branching process is the birth-and-death process, where it is supposed that any object existing at time t has a probability $\mu(t)\,dt$ of dying in the time interval $(t, t+dt)$ and a probability $\lambda(t)\,dt$ of disappearing and being replaced by two new objects in the

[1] The results of this section were given by SEVAST'YANOV (1951) and in part in earlier papers by a number of authors.

same time interval (the latter event is equivalent from our point of view to remaining and giving birth to one additional object). In the notation of preceding sections, a birth-and-death process has $p_i(t)=0$ for $i=1$ or $i>2$, $p_0(t)=\mu(t)/(\lambda(t)+\mu(t))$, $p_2(t)=\lambda(t)/(\lambda(t)+\mu(t))$, and $b(t)=\lambda(t)+\mu(t)$.

We assume that b is always strictly positive, although the contrary case can be treated. Equation (4.3) now becomes the Riccati-type equation

$$\frac{\partial F_1(s,\tau,t)}{\partial\tau}=-(1-F_1)\left[\mu(\tau)-\lambda(\tau)F_1\right]; \qquad F_1(s,t-0,t)=s. \tag{7.1}$$

The solution of (7.1) was given by D. G. KENDALL (1948a). Putting

$$\varrho(\tau,t)=\int_\tau^t[\mu(x)-\lambda(x)]\,dx,$$

$$W(\tau,t)=e^{-\varrho}\left[1+\int_\tau^t e^{\varrho(\tau,x)}\mu(x)\,dx\right],$$

$$\xi=1-\frac{e^{-\varrho}}{W}, \qquad \eta=1-\frac{1}{W},$$

KENDALL shows that we can write the solution of (7.1) as

$$F_1(s,\tau,t)=\frac{\xi+(1-\xi-\eta)s}{1-\eta s}, \tag{7.2}$$

leading to the following expressions for the probabilities:

$$P_{10}(\tau,t)=\xi, \qquad P_{1n}=(1-P_{10})(1-\eta)\eta^{n-1}, \qquad n=1,2,\ldots. \tag{7.3}$$

If $Z(\tau)=1$, then the first moment is given by $m_1(\tau,t)=e^{-\varrho}$, and the variance is $e^{-\varrho}(2W-1-e^{-\varrho})$.

It follows from (7.3) and the definition of ξ that the probability of extinction is 1 (i.e., $\lim_{t\to\infty}P_{10}(\tau,t)=1$ for each $\tau\geqq 0$) if and only if

$$\int_0^\infty\mu(t)\,e^{\varrho(0,t)}\,dt=\infty.$$

The first treatment of the birth-and-death process seems to have been that by YULE (1924) in the special case $\mu=0$, $\lambda=$constant. YULE's application will be described below. FURRY (1937) independently devised this model as a simplified description of the multiplication of particles in cosmic-ray showers. FELLER (1939) treated the case in which λ and μ are nonzero constants, obtaining the first and second moments. FELLER proposed the model as an extension of deterministic models for the growth of populations. According to BARTLETT (1949), PALM first gave the probabilities explicitly in the case λ constant, μ constant$\neq 0$ (1943, unpublished). ARLEY (1943, pp. 109ff.) treated the case $\lambda=$constant,

μ = constant $\times t$, as a preliminary step in his mathematical model for multiplication in cosmic-ray showers. The increase of μ with t represents the declining energy of a particle as it suffers collisions.

The birth-and-death process does not seem to be a good model for the spread of an *epidemic* in a finite population, since when a large proportion of the population has been infected, we cannot suppose that the rate of new infections is independent of past history. However, KENDALL (1956) has shown how one can combine a birth-and-death process, used to represent the *early* stages of an epidemic, with a deterministic model used to represent the later stages.

8. YULE'S problem

In the application of YULE (1924), $Z(t)$ represents the number of species in some genus. It is assumed that if the number of species existing at time t is i, then $\lambda i\, dt$ is the probability that a new species originates by a mutation in the time interval $(t, t+dt)$. It is assumed that species do not die except as a result of cataclysms, which are not included in the model.

The creation of new species at a rate proportional to the number of existing species and independent of their ages may appear somewhat plausible if we suppose that (a) a species rather quickly reaches a maximum size determined by its environment and thereafter does not grow in size, and (b) different species interfere only slightly with one another. The second supposition appears in some cases to be supported by certain remarks of DARWIN[1] (*Origin of Species*, Sixth Edition, Chapter IV, section entitled "Divergence of Character"). YULE was properly cautious about his assumptions. Still, he found that the theory fit the observed data very well, "in fact better than one has any right to expect" (p. 27).

YULE supposes that the number of genera in a family of genera is the same kind of process as $Z(t)$, except that λ is replaced by another constant g. Thus the expected number of genera at time t is, say, $N_0 e^{gt}$.

YULE used his model to estimate certain parameters that are impossible to determine by direct observation. As a simple example, let us estimate the ratio λ/g. Since geological records are not adequate for measuring λ and g directly, we must make the estimate by observing *presently existing* genera.

From (7.3), the probability that there are n species in a genus at time t, if there is 1 species at time 0, is

$$P_{1n}(t) = e^{-\lambda t}(1-e^{-\lambda t})^{n-1}, \qquad n=1,2,\dots. \tag{8.1}$$

[1] YULE does not adduce these remarks.

Since $gN_0 e^{gx}\,dx$ is the expected number of new genera created in $(x, x+dx)$, the expected number of genera at time t having n species is

$$N_n^* = \int_{-\infty}^{t} g N_0 e^{gx} P_{1n}(t-x)\,dx, \tag{8.2}$$

if we assume the process to have been active since $t=-\infty$. One can calculate the quantities N_n^* explicitly. Here we point out only that *the proportion of monotypes* (i.e., of genera with one species) *is given by*

$$P^* = \frac{N_1^*}{\sum N_n^*} = \frac{g}{g+\lambda}. \tag{8.3}$$

Suppose we observe a large family of genera and find that the observed proportion of monotypes is the number $\hat{P}$. Then presumably $\hat{P}$ is close to P^*, and hence we can estimate

$$\frac{\lambda}{g} \sim \frac{1}{\hat{P}} - 1.$$

One of YULE's examples is a family of beetles with 627 genera comprising 9,997 species, 34.29% of the genera having one species. Hence we estimate $\lambda/g \sim 1.9$. Actually, YULE takes into account the fact that the time since the process began is not infinite and shows how to estimate this time; this refinement changes the estimate of λ/g only slightly, however, in the example just cited.

9. The temporally homogeneous case

We treat only the temporally homogeneous case below (except Sec. 11.4). Hence b and h are independent of t; the probabilities P_{ij} and generating function F_i are functions of $t-\tau$; and $\partial F_i/\partial\tau = -\partial F_i/\partial t$. We write $F_1(s,t)$, $h(s)$, etc., in place of $F_1(s,\tau,t+\tau)$, $h(s,t)$, etc.

Equation (4.3) now becomes

$$\frac{\partial F_1(s,t)}{\partial t} = b\left[h(F_1(s,t)) - F_1(s,t)\right], \qquad F_1(s,0)=s;\ t\geqq 0;\ |s|\leqq 1. \tag{9.1}$$

If $-1<s<1$, then (9.1) yields the relation

$$\int_{s}^{F_1(s,t)} \frac{dx}{h(x)-x} = bt, \tag{9.2}$$

provided $h(x)-x$ does not vanish for any x between s and $F(s,t)$ inclusive.

Let us now state a necessary and sufficient condition that the probabilities should add to 1. (See Sec. 4.1.)

Theorem 9.1.[1] *In the temporally homogeneous case, in order that the generating function F_1 should satisfy $F_1(1,t)\equiv 1$, it is necessary and*

[1] This result appears without proof in ZOLOTAREV (1957) and is attributed to E. B. DYNKIN.

sufficient that for each $\varepsilon > 0$ *the integral*

$$\int_{1-\varepsilon}^{1} \frac{du}{h(u) - u} \tag{9.3}$$

should diverge.

Corollary. *If* $h'(1) < \infty$, *then* $F_1(1, t) \equiv 1$.

Proof. Choose s_0, $0 < s_0 < 1$, such that $h(x) - x$ does not vanish for any x with $s_0 \leqq x < 1$. Choose s_1 and t_1 so that $s_0 < s_1 < 1$, $t_1 > 0$, and $s_1 - bt_1 > s_0$. From (9.1) we see that $|\partial F_1/\partial t| \leqq b$ and hence $F_1(s, t) \geqq s - bt > s_0$ for $s_1 \leqq s < 1$, $0 \leqq t \leqq t_1$. From Lemma 4.1 we see that $|F_1(s, t)| < 1$ for $|s| < 1$. Hence we can apply (9.2), obtaining

$$\int_{s}^{F_1(s, t_1)} \frac{dx}{h(x) - x} = bt_1, \qquad s_1 \leqq s < 1. \tag{9.4}$$

In order for the integral in (9.4) to retain the value bt_1 as $s \to 1$, we must have $F_1(1, t_1) < 1$ if the integral in (9.3) converges, since t_1 is positive; and we must have $F_1(1, t_1) = 1$ if the integral diverges, since t_1 is finite.

It is a consequence of a known result of the theory of Markov chains (and can also be shown from (5.1)) that either $F_1(1, t) = 1$ for all $t \geqq 0$ or $F_1(1, t) < 1$ for all $t > 0$. Hence the result is proved. □

The corollary follows from the theorem, since

$$h(u) - u = \big(h'(1) - 1\big)(u - 1) + o(u - 1), \qquad u \to 1.$$

Assumption for the sequel. In the remainder of this chapter we shall assume, unless the contrary is stated, that $h'(1) < \infty$.

10. Extinction probability

From (3.4) we see that $P_{00}(t) \equiv 1$. Using (3.2) we have

$$P_{k0}(t_1 + t_2) \geqq P_{k0}(t_1) P_{00}(t_2) = P_{k0}(t_1).$$

Hence $P_{k0}(t)$ is a nondecreasing function of t, $k = 0, 1, 2, \ldots$. From the relation $F_k(0, t) = \big(F_1(0, t)\big)^k$ (see Theorem 4.1) we have $P_{k0}(t) = \big(P_{10}(t)\big)^k$.

Definition 10.1. The *extinction probability* q is defined by $q = \lim_{t \to \infty} P_{10}(t)$.

We interpret q as the probability that the family of a single object will eventually die out. We can deduce the value of q from the underlying conception of our Markov process as a branching process, as described at the beginning of Sec. 2. When a single object is transformed, the expected number of objects by which it is replaced is $h'(1)$, and we should expect the family always to die out or to have a chance not to die out according as $h'(1) \leqq 1$ or $h'(1) > 1$. We are thus led to the following result, by analogy with Theorem I.6.1.

Theorem 10.1.[1] *The probability of extinction q is the smallest nonnegative root of the equation $s=h(s)$. Hence $q=1$ if and only if $h'(1)\leqq 1$.*

We defer the proof of Theorem 10.1 to the following chapter, where we shall prove a corresponding proposition of which this theorem is a special case.

It is natural to equate "extinction" to the event $\{Z(t)=0$ for all sufficiently large $t\}$. A rigorous basis for this equivalence is provided in Chapter VI.

If $t_0>0$ and if $Z(0)=1$, the sequence $Z(t_0)$, $Z(2t_0)$, $Z(3t_0)$, ... is a Galton-Watson process whose generating function is $F_1(s, t_0)$ (Sec. 5.1). The extinction probability for this process is, from Theorem I.6.1, the smallest nonnegative root of $s=F_1(s, t_0)$. On the other hand, we have $\lim P(Z(nt_0)=0|Z(0)=1)=\lim P_{10}(nt_0)=q$, from Def. 10.1. Hence we have the following result.

Theorem 10.2. *The extinction probability q is the smallest nonnegative root of the equation $s=F_1(s, t)$, where t may have any positive value.*

11. Asymptotic results

We can obtain asymptotic results for Markov branching processes that are more precise than those available for general Galton-Watson processes. These results are, of course, applicable to those Galton-Watson processes that happen to be imbeddable in a Markov branching process (see Sec. 5.1).

As before, put $h(s)=\sum_{r=0}^{\infty} p_r s^r$ and $a=b(h'(1)-1)$, where the quantities p_r and b, defined in Sec. 2, are now independent of time. We continue to assume $h'(1)<\infty$. Since e^{at} is the expected number of objects at time t, beginning with one object at time 0, it is natural to expect differences in behavior according as $a<0$, $a=0$, or $a>0$ (i.e., $h'(1)<1$, $h'(1)=1$, or $h'(1)>1$).

We shall assume $Z(0)=1$ unless the contrary is stated.

11.1. Asymptotic results when $h'(1)<1$. If $h'(1)<1$ and $h''(1)<\infty$ then

$$\frac{1}{h(s)-s}=\frac{1}{(1-h'(1))(1-s)}+B(s), \qquad 0\leqq s<1,$$

where B is a bounded function. Let us define a function $K(s)$, $0\leqq s<1$, by

$$K(s)=\int_1^s\left[\frac{1}{h(x)-x}-\frac{1}{(1-h'(1))(1-x)}\right]dx-\frac{\log(1-s)}{1-h'(1)}. \tag{11.1}$$

[1] Given by SEVAST'YANOV (1951).

Then $K'(s)=1/(h(s)-s)$ and we may write the solution of (9.1) as

$$F_1(s,t)=K^{-1}[K(s)+bt], \qquad 0\leqq s<1, t\geqq 0. \tag{11.2}$$

If $h''(1)<\infty$ then the integrand in (11.1) is bounded as $x\to 1-0$, and hence $K(s)=-\log(1-s)/(1-h'(1))+O(1-s)$. From this it follows that $1-K^{-1}(u)=e^{-(1-h'(1))u}[1+O(e^{-(1-h'(1))u})]$ as $u\to\infty$. Hence we can determine the form of F_1 as $t\to\infty$, leading to the following result.

Theorem 11.1.[1] *Suppose $h'(1)<1$ (i.e., $a<0$) and $h''(1)<\infty$. Then*

$$1-F_1(0,t)=1-P_{10}(t)=e^{-(1-h'(1))K(0)}e^{at}+O(e^{2at}), \qquad t\to\infty, \tag{11.3}$$

where K is defined by (11.1). Moreover, the conditional probability generating function of $Z(t)$, given $Z(t)\neq 0$, has the limiting form

$$g(s)=1-\exp\left\{-(1-h'(1))\int_0^s\frac{du}{h(u)-u}\right\}.$$

11.2. Asymptotic results when $h'(1)=1$. By methods similar to those of the preceding section, although more involved, we obtain the following result.

Theorem 11.2.[1] *Suppose $h'(1)=1$, $h'''(1)<\infty$. Then*

$$P(Z(t)>0)\sim\frac{2}{bh''(1)t} \tag{11.4}$$

and

$$\lim_{t\to\infty}P\left\{\frac{2Z(t)}{bh''(1)t}>u\,|\,Z(t)>0\right\}=e^{-u}, \qquad u\geqq 0. \tag{11.5}$$

11.3. Asymptotic results when $h'(1)>1$. Using (6.3) and the relation $F_i=(F_1)^i$, we have

$$\mathscr{E}(Z(t+\Delta)\,|\,Z(t)=i)=\mathscr{E}(Z(\Delta)\,|\,Z(0)=i)=\partial(F_1)^i/\partial t|_{s=1,\,t=\Delta}=ie^{a\Delta}, \ \Delta\geqq 0.$$

Hence if $\mathscr{E}[(Z(t))^2\,|\,Z(0)=1]$ is finite, we have

$$\mathscr{E}[Z(t+\Delta)Z(t)\,|\,Z(0)=1]=e^{a\Delta}\mathscr{E}[(Z(t))^2\,|\,Z(0)=1]. \tag{11.6}$$

If $h''(1)<\infty$ and $h'(1)>1$, we can then use (6.4), (11.6), and arguments similar to those of Sec. I.8, to show that $Z(t)e^{-at}$ converges in probability and in mean square to a random variable W. It will follow from more general results in Chapter VI that there is also convergence with probability 1. Compare also Remark 3 in Sec. I.8.1.

If $\varphi(s)$ is the moment-generating function of W,

$$\varphi(s)=\mathscr{E}[e^{-sW}\,|\,Z(0)=1],$$

[1] Sevast'yanov (1951).

we have the following expression for its functional inverse[1]:

$$\varphi^{-1}(u)=(1-u)\exp\left\{\int_1^u\left[\frac{h'(1)-1}{h(x)-x}+\frac{1}{1-x}\right]dx\right\}, \qquad q<u\leqq 1, \quad (11.7)$$

where q is the extinction probability of Theorem 10.1. We omit the proof.

Remark. The distribution function corresponding to φ has a density except for a possible discontinuity at 0. This follows from Theorem I.8.3, since the limiting distribution of $Z(t)e^{-at}$ must be the same as the limiting distribution of $Z(n)e^{-an}$ as $n\to\infty$ through integer values.

We can show from (11.7) that φ has the asymptotic behavior

$$\varphi(s)-q\sim A s^{-B}, \qquad s\to\infty,$$

where $B=(1-h'(q))/(h'(1)-1)$, and A can be evaluated from (11.7) as a definite integral. Hence we can use a well-known Tauberian theorem (see, e.g., WIDDER (1941, p. 192)) to find the behavior of $P\{W\leqq u\}$ as $u\downarrow 0$. Thus for the case of continuous time we can get a sharper result than for discrete time. (See the next-to-last paragraph of Sec. I.8.2.)

11.4. Extensions. ZOLOTAREV (1957) has examined the situation when $h''(1)$ or $h'''(1)$ is infinite. It is of particular interest that when $h'(1)=1$ and $h(s)$ satisfies a certain regularity condition near $s=1$ that allows $h''(1)=\infty$, there arises a new class of limiting distributions for $Z(t)$, different from the exponential distribution.

SEVAST'YANOV (1957a, 1959) extended the results of Sec. 11.2 to show that (11.5) may be replaced by a stronger statement. Roughly speaking, the conditional distribution of $Z(t)/t$ is almost exponential if t is large, if $h'(1)$ is close to 1, and if $bh''(1)$ and $bh'''(1)$ are suitably restricted. More general results of this nature have been given by ČISTYAKOV (1961).

ČISTYAKOV and MARKOVA (1962) have obtained asymptotic results for certain cases without temporal homogeneity.

12. Stationary measures

In Sec. I.11 we proved the existence of a stationary measure for the Galton-Watson branching process (omitting the state 0). In the case of continuous time we can make the results more precise.

We use the following lemma about power series.

Lemma 12.1. *Let $f(s)$ be a probability generating function with $f(0)>0$. Then the function $1/[f(s)-s]$ has, near $s=0$, an expansion in powers of s with nonnegative coefficients.*

---

[1] SEVAST'YANOV (1951); HARRIS (1951).

Proof.[1] We use the identity

$$\frac{1}{f(s)-s}=\frac{1}{(1-s)\,[1-\Psi(s)]}\,, \tag{12.1}$$

where $\Psi(s)=\big(1-f(s)\big)/(1-s)$. The function in (12.1) is analytic at $s=0$ if $p_0>0$. The function Ψ is analytic at $s=0$, and its power series expansion has nonnegative coefficients,

$$\Psi(s)=1-p_0+(1-p_0-p_1)\,s+(1-p_0-p_1-p_2)\,s^2+\cdots.$$

Hence $[f(s)-s]^{-1}=(1+s+s^2+\cdots)(1+\Psi+\Psi^2+\cdots)$ is likewise a power series with nonnegative coefficients. ▯

Theorem 12.1. *Suppose $h(0)>0$. Then the function*

$$\pi(s)=\int_0^s \frac{dx}{h(x)-x}$$

is a power series in s with nonnegative coefficients, $\pi(s)=\sum_{r=1}^{\infty}\pi_r s^r$, and

$$\pi[F_1(s,t)]=\pi(s)+bt \tag{12.2}$$

whenever $|s|<q$, where q is the extinction probability.

Corollary. *The coefficients π_r are a stationary measure for $Z(t)$; i.e.,*

$$\pi_i=\sum_{r=1}^{\infty}\pi_r\,P_{ri}(t),\qquad t\geqq 0;\ i=1,2,\ldots. \tag{12.3}$$

Proof. From Lemma 12.1 and the definition of π we see that π is a power series with nonnegative coefficients, and obviously $\pi(0)=0$. Since $h(x)-x>0$ for $0\leqq x<q$, the singularity of π of smallest modulus on the real axis is at $s=q>0$, and hence π is analytic for $|s|<q$ (see Titchmarsh (1939, p. 214)). From Theorem 10.1 it follows that $|F_1(s,t)|\leqq F_1(|s|,t)<q$ when $|s|<q$. Thus we can divide both sides of (9.1) by $h(F_1)-F_1$ when $|s|<q$, and integration then leads to (12.2).

The corollary follows if we equate coefficients of powers of s in (12.2), recalling (Theorem 4.1) that $F_i=(F_1)^i$, $i=1,2,\ldots$. ▯

The relation (12.2) is well known from theories of continuous iteration (Hadamard (1944)).

Except for a constant multiplicative factor, there is only one set of nonnegative numbers π_i, not all 0, satisfying (12.3) for all $t\geqq 0$. This is of some interest, since there is not always uniqueness in the case of discrete time. However, we omit the proof.

[1] I am indebted to Professors Darling and Karlin for pointing out this proof, which is simpler than the original one.

13. Examples

13.1. The birth-and-death process. In the case of the birth-and-death process of Sec. 7 we have $h(s)=\frac{\mu}{\lambda+\mu}+\left(\frac{\lambda}{\lambda+\mu}\right)s^2$. From (7.2) we see that F_1 is, for each t, a fractional linear function of the sort considered in Sec. I.7.1, where we have the correspondence $c=\eta$, $(1-b-c)/(1-c)=\xi$, $s_0=\xi/\eta$. Here b is not to be confused with the parameter mentioned at the beginning of Sec. 9. If $\mu \neq 0$, the function π of Theorem 12.1 is given by

$$\left.\begin{aligned} \pi(s) &= \frac{1+\frac{\lambda}{\mu}}{1-\frac{\lambda}{\mu}} \log\left(\frac{1-\frac{\lambda}{\mu}s}{1-s}\right), \qquad \mu \neq 0, \mu \neq \lambda, \\ \pi(s) &= \frac{2s}{1-s}, \qquad \mu=\lambda>0. \end{aligned}\right\} \tag{13.1}$$

We see from (13.1) that (a) if $\lambda=\mu$, the coefficients π_i of the stationary measure are all equal; (b) if $\lambda<\mu$, then π_i is proportional to $[1-(\lambda/\mu)^i]/i$; (c) if $\lambda>\mu$, then π_i is proportional to $[(\lambda/\mu)^i-1]/i$. In order to verify that these values correspond to those of Sec. I.11.6, note that $1/s_0=\eta/\xi$, which by direct calculation from the definitions in Sec. 7 is equal to λ/μ.

13.2. Another example. Suppose $h(s)=s^{k+1}$, where k is a positive integer. Take the parameter b of Sec. 9 equal to 1. In this case it can be verified from (9.2) that

$$F_1(s,t)=s\left[e^{kt}-(e^{kt}-1)s^k\right]^{-1/k}.$$

Thus we obtain the generating functions of Sec. I.7.2 if we identify m with e^k. The limiting distribution of $Z(t)/\mathscr{E}Z(t)$ has the density (I.8.11).

13.3. A case in which $F_1(1,t)<1$. Suppose $h(s)=1-\sqrt{1-s}$, and take $b=1$. From Theorem 9.1 we see that $F_1(1,t)<1$, $t>0$. Applying (9.2), we find

$$F_1(s,t)=1-\left[1-e^{-t/2}+e^{-t/2}\sqrt{1-s}\right]^2.$$

Then $F_1(1,t)=P\{Z(t)<\infty\}=1-(1-e^{-t/2})^2$. A class of similar examples was given by ZOLOTAREV (1957).

14. Individual probabilities

Refer to Secs. 2, 3, 4, and 6 for notation.

In Secs. 11.2 and 11.3 we gave limit theorems about the *cumulative* distribution of $Z(t)$ when $h'(1)=1$ and $h'(1)>1$, respectively. We shall now give two results of ČISTYAKOV (1957) about the *individual* probabilities. The result of Sec. 11.1 already provides a theorem about individual probabilities in case $h'(1)<1$ (see Theorem I.9.1).

We use the following terminology. Let

$$P_j^*(t) = P\{Z(t) = j \mid Z(t) > 0, Z(0) = 1\}.$$

Let Δ be 1 if $p_0 > 0$. Otherwise let Δ be the greatest common divisor of all numbers $n \geqq 1$ such that $p_{n+1} > 0$. In case $h'(1) > 1$ we let $\varkappa(x)$, $x > 0$, be the probability density of the limiting distribution of $Z(t)/e^{at}$. This density exists, according to Theorem I.8.3, and $\int_0^\infty \varkappa(x)\,dx = 1 - q$, where q is the extinction probability. Let $0 < c_1 < c_2$ be arbitrary positive constants.

Theorem 14.1. *Suppose* $h'(1) = 1$, $h^{\text{iv}}(1) < \infty$. *Then*

$$P_j^*(t) = \frac{2}{b t h''(1)} \left[e^{-2j/bth''(1)} + O(t^{-\frac{1}{2}} \log t)\right], \qquad t \to \infty,\ c_1 < \frac{j}{t} < c_2.$$

Theorem 14.2. *Suppose* $h'(1) > 1$, $h''(1) < \infty$. *Then for* $t \to \infty$ *with* $c_1 < n e^{-at} < c_2$ *we have*

$$P_n^*(t) = \frac{\Delta e^{-at}}{1-q} \left[\varkappa(n e^{-at}) + o(1)\right]$$

if $n - 1$ *is divisible by* Δ, *and* $P_n^*(t) = 0$ *otherwise.* (The proofs of Theorems 14.1 and 14.2 are given by Čistyakov (1957).)

Remarks. It seems likely that $\varkappa(x) > 0$ for each $x > 0$. If we could have $\varkappa(x_0) = 0$, Theorem 14.2 would not give a sharp result for $n e^{-at} = x_0$. As far as the author knows, this question has not been settled.

It is not clear to what extent Čistyakov's results apply in the case of Galton-Watson processes with discrete time. (See Sec. I.8.4.) There seems to be no doubt that they apply, with the proper change of notation, if $p_1, p_2, p_3, \ldots$ (as defined in Chapter I) are all strictly positive. The definition of Δ is given by Čistyakov in a different but equivalent form.

Zolotarev (1954) has given limit theorems for the form of $P_j^*(t)$ as $t \to \infty$ with j fixed.

15. Processes with several types

Processes with k different types of objects can be treated by methods similar to those used above. Suppose that an object of type i existing at time t has a probability $\beta_i(t)\Delta + o(\Delta)$ of being transformed during the interval $(t, t+\Delta)$, $\Delta > 0$, and let $h^{(i)}(s_1, \ldots, s_k, t)$ be the generating function of the numbers of the different types born at a transformation. If $F^{(i)}(s_1, \ldots, s_k, t)$ is the generating function for the numbers of the different types at time t, if there is one object of type i at time 0, then we can deduce the pair of systems of equations[1], for $i=1, \ldots, k$,

$$\frac{\partial F^{(i)}}{\partial t} = \sum_{j=1}^{k} \beta_j(t)\,[h^{(j)}(s_1, \ldots, s_k, t) - s_j]\,\frac{\partial F^{(i)}}{\partial s_j}, \tag{15.1}$$

$$\frac{\partial F^{(i)}}{\partial \tau} = -\beta_i(\tau)\,[h^{(i)}(F^{(1)}, \ldots, F^{(k)}, \tau) - F^{(i)}], \tag{15.2}$$

with $F^{(i)}(s_1, \ldots, s_k, \tau, \tau+0) = F^{(i)}(s_1, \ldots, s_k, t-0, t) = s_i$. A result of ARLEY (1943, Theorem 1, p. 81) insures that if the β_j are continuous functions and if the $h^{(i)}$ are polynomials in $s_1, s_2, \ldots$ whose coefficients are continuous in t, then $F^{(i)}(1, \ldots, 1, \tau, t) \equiv 1$.

For each value of i we can (in principle) solve (15.1) as a single partial differential equation for $F^{(i)}$. On the other hand, the equations in (15.2) must in general be treated as a *system* of ordinary differential equations.

The treatment of moments can, of course, be carried out using matrix methods as in Chapter II.

JIŘINA (1957), ČISTYAKOV (1959a, 1959b, 1960, 1961), and SAVIN and ČISTYAKOV (1962) have proved analogues, for processes with several types, of the limit theorems of Sec. 11.

15.1. Example: the multiphase birth process. We have pointed out in Sec. 2 that, in the temporally homogeneous case, the Markov processes discussed in this chapter correspond to family trees where each object has the exponential life-length density be^{-bt}. This density function does not describe well the life span of biological objects, such as bacteria. For example, the life-length (generation time) of many kinds of bacteria has a density function with a well-marked mode (KELLY and RAHN (1932), POWELL (1955), TOBIAS (1961)).

In Chapter VI we shall formulate a model that deals with branching processes for objects with nonexponential life-lengths. Here, however, we consider a method due to D. G. KENDALL (1948b) that allows the treatment of certain processes of this sort as multitype Markov branching processes.

Let us suppose that (a) an object goes through k stages during its life, where k is a nonrandom positive integer; (b) the lengths of successive stages are independent random variables; (c) the length of each stage has the probability density be^{-bt}; (d) at the end of the k-th stage the object is transformed into two new objects, each of which then begins stage 1. Then the total life-length of an object has the probability density $b^k t^{k-1} e^{-bt}/\Gamma(k)$. The mean and variance of the life-length are adjusted by varying k and b. The stages do not necessarily correspond to real events in the life of the object, but are introduced as a mathe-

[1] See the footnotes to Sec. 4.

matical device so that we can continue to deal with exponential distributions.

KENDALL has treated this process as a multitype branching process, where the number of the i-th type at time t is the number of objects in stage i. An object that is in stage i at time t has a probability $b\,dt$ of being transformed during $(t, t+dt)$; this is the characteristic property of the exponential distribution. An object of type $i<k$ is transformed into one object of type $i+1$, while an object of type k is transformed into two objects of type 1.

Although equations (15.1) or (15.2) apparently cannot be solved explicitly in this case, KENDALL has studied the moments. In particular, the expected number of objects of all types, starting with one object of type 1, is $e^{-bt}\sum_{n=0}^{\infty} 2^n \sum_{j=0}^{k-1} \frac{(bt)^{nk+j}}{(nk+j)!}$. The variance can also be evaluated, and the asymptotic form of the variance of the total population size was determined by KENDALL. However, we shall consider this in the context of Chapter VI.

15.2. Chemical chain reactions. The common gaseous elements usually exist in a state of two atoms to the molecule; thus hydrogen and chlorine exist usually as H_2 and Cl_2. However, when hydrogen and chlorine are mixed in the presence of light, resulting in the production of hydrogen chloride gas (HCl), an important role is played by hydrogen and chlorine in their monatomic forms, H and Cl. The reaction is presumed to proceed as follows: Cl_2 + light = 2 Cl; Cl + H_2 = H + HCl; H + Cl_2 = Cl + HCl; Cl + H_2 = H + HCl; .... Eventually the chain is broken when the reaction reaches the wall of the container, or sometimes before. However, there may also be *branching*. This happens when an HCl molecule produced by the reaction H + Cl_2 is unusually active and creates, by collision with a Cl_2 molecule, two new Cl atoms that start their own chains.

SEMENOFF (1935, especially Chapter III) discussed such problems, although without formulating them as Markov branching processes in our sense. SEVAST'YANOV (1951) and SINGER (1953) have formulated models for chemical chain reactions as Markov branching processes.

A more recent discussion of chemical chain reactions (not in a Markov chain formulation) has been given by FROST and PEARSON (1961, Chapter 10, especially pp. 258—259). The stochastic aspects of chemical reactions (not necessarily chain reactions) have been treated by BARTHOLOMAY (e.g., 1958), who suggested a formulation in terms of Markov processes. According to KOLMOGOROV and DMITRIEV (1947), M. A. LEONTOVICH, in 1935, deduced the differential equations for the probabilities in a Markov chain formulation of certain chemical reactions.

16. Additional topics

16.1. Birth-and-death processes (generalized). The name *birth-and-death process* has come to be applied to any Markov process, with nonnegative integer states, whose only transitions are changes of $+1$ or -1. Such a process is not in general a branching process.

In the temporally homogeneous case we have the interesting representation

$$P_{ij}(t)=\pi_j\int_0^\infty e^{-xt}Q_i(x)\,Q_j(x)\,d\Psi(x), \qquad i,j=0,1,\dots,$$

where the functions Q_i are polynomials that are orthogonal with respect to the probability distribution Ψ, and the numbers π_j are sometimes a stationary measure for the process.

We shall not enter into a discussion of this representation, but refer the reader to papers by LEDERMAN and REUTER (1954), KARLIN and McGREGOR (1957a, 1957b, 1959, 1960), and KENDALL (1958). A simple direct treatment for the case of a finite number of states has been given by KRAMER (1959).

16.2. Diffusion model. Suppose that a population is large enough so that we can represent its size as a continuous random function $X(t)$. Let us suppose that X is a Markovian diffusion process (BARTLETT (1955, pp. 85ff.)) and that if $X(t)=x$, then the mean and variance of the increment $X(t+\Delta)-X(t)$ are each proportional to $x\Delta$ when Δ is small and positive, i.e.,

$$\mathcal{E}[X(t+\Delta)-X(t)\,|\,X(t)=x]=a\,x\Delta+o(\Delta), \qquad x>0$$

$$\mathcal{E}[(X(t+\Delta)-X(t))^2\,|\,X(t)=x]=2b\,x\Delta+o(\Delta), \qquad x>0,$$

where a and b are constants, $-\infty<a<\infty$, $b>0$. If $X(0)=\xi>0$, the probability density $u(x,t)$ of $X(t)$ satisfies

$$\left.\begin{aligned}\frac{\partial u(x,t)}{\partial t}&=b\frac{\partial^2}{\partial x^2}[x u(x,t)]-a\frac{\partial}{\partial x}[x u(x,t)],\\ &0<x<\infty;\; u(x,0)=\delta(x-\xi),\end{aligned}\right\} \tag{16.1}$$

where δ is the function of DIRAC.

The mean and variance of $X(t)$ are given by

$$\mathcal{E}X(t)=\xi e^{at}, \qquad \operatorname{Var}X(t)=\frac{2b\xi}{a}e^{at}(e^{at}-1),$$

(or $\operatorname{Var}X=2\xi bt$ if $a=0$). This is the same form as in the case of a branching process. The solution u of (16.1) has been determined explicitly by FELLER (1939), who also showed (1951) that the diffusion equation (16.1) can be deduced directly from the functional equations $f_{n+1}(s)=f[f_n(s)]$ for the Galton-Watson process, under the assumption that the population is large enough so that it may be considered a continuous variable. See also SEVAST'YANOV (1959, Theorem 3).

16.3. Estimation of parameters. In the temporally homogeneous case, the random variables $Z(t), Z(2t), Z(3t), \ldots$, t fixed, are a Galton-Watson process; hence the remarks of Sec. I.13.4 are applicable for such a sequence.

In the temporally homogeneous case of the birth-and-death process of Sec. 7, the parameters λ and μ can be estimated in several ways, depending on the circumstances. The estimates are complicated if they are based on the observation of Z *during a fixed time interval*. However, they assume a simpler form if they are based on the history of the process *until a fixed number N of jumps occur* (or until extinction occurs if this happens before the N-th jump). Estimates in this case may be obtained as follows. (In the rest of this section we do *not* require $Z(0)=1$.)

Let N be a fixed positive integer and let T be the time at which Z undergoes its N-th jump or becomes 0, whichever is sooner. Let N' be the number of jumps undergone by Z up to and including the time T. Then, according to MORAN (1951),

$$\mathscr{E}\left\{\left(\frac{1}{N'}\right)\int_0^T Z(t)\,dt\right\} = \frac{1}{\lambda+\mu}. \tag{16.2}$$

The random variable $2(\lambda+\mu)\int_0^T Z(t)\,dt$ has a chi-square distribution with $2k$ degrees of freedom under the condition that $N'=k\leqq N$.

The maximum likelihood estimate of $\lambda/(\lambda+\mu)$ is given by

$$\hat{p} = \frac{k-Z(0)}{2k}$$

if the population becomes extinct on the k-th jump, $k\leqq N$; and

$$\hat{p} = \frac{N-Z(0)+Z(T)}{2N}$$

if $Z(T)>0$ (MORAN (1953)).

For further discussion of estimation problems and references to other work, the reader is referred to MORAN (1953), ANSCOMBE (1953), BARTLETT (1955, Chapter 8), and DARWIN (1956). ARMITAGE (1952) has discussed estimation in models for mutations. BHARUCHA-REID (1958) has discussed problems of testing hypotheses about birth-and-death processes.

16.4. Immigration. Let $F_1(s,t)$ be the generating function for a temporally homogeneous branching process, starting with one object at time 0. If independent families originate, or immigrate, at times

$t_1 < t_2 < \cdots < t_n < t$, then the generating function for the number of objects at time t is the product $F_1(s, t-t_1) F_1(s, t-t_2) \ldots F_1(s, t-t_n)$. We suppose that $t_1, t_2, \ldots, t_n$ are the times at which events occur in a Poisson process with rate ν per unit of time. By a known property of the Poisson process, if we are given the condition that exactly n events occur in $(0, t)$, then the times $t_1, t_2, \ldots, t_n$ are distributed as the smallest, next smallest, ..., largest of n independent random variables uniformly distributed on $(0, t)$. Hence the generating function for the number of objects at time t is

$$e^{-\nu t} \sum_{n=0}^{\infty} \frac{(\nu t)^n}{n!} \left(\frac{1}{t} \int_0^t F_1(s, t_1)\, dt_1 \right)^n = \exp\left\{ -\nu t + \nu \int_0^t F_1(s, t_1)\, dt_1 \right\}. \quad (16.3)$$

(See BARTLETT (1955, p. 77).) For additional results on birth-and-death processes with immigration, see KARLIN and MCGREGOR (1958).

SEVAST'YANOV (1957b) has given limit theorems for certain processes with immigration.

16.5. Continuous state space. JIŘINA (1958) has considered vector-valued Markov processes that are analogous to branching processes, but whose components are continuous rather than integral.

16.6. The maximum of $Z(t)$. The random function[1] Z can undergo jumps of $-1, 1, 2, 3, \ldots$; the probabilities for some of these jumps may be 0. If we consider only the successive states of the process, without regard to the times at which they occur, we obtain the "imbedded Markov chain" $\zeta_0, \zeta_1, \zeta_2, \ldots$, where $P(\zeta_{n+1}=j \mid \zeta_n = i) = p_{j-i+1}, i=1, 2, \ldots$; $j=i-1, i+1, i+2, \ldots$, and $P(\zeta_{n+1}=0 \mid \zeta_n=0)=1$. Notice that if $\zeta_n = i \neq 0$, then the probability distribution of $\zeta_{n+1}-\zeta_n$ is independent of i.

The maximum of $Z(t)$ for $0 \leqq t < \infty$ is the same as the maximum of ζ_n for $n=0, 1, \ldots$. Concerning this maximum we have the following result, due to ZOLOTAREV (1954). We assume $p_0 > 0$.

Theorem 16.1. *Let* $\eta = \max_{0 \leqq n < \infty} \zeta_n$. *Then* $P(\eta \leqq n \mid \zeta_0 = k, \eta < \infty) = u_{n-k}/q^k u_n$, $1 \leqq k \leqq n$, *where* q *is the extinction probability and* $u_0 + u_1 s + u_2 s^2 + \cdots$ *is the power series expansion of the function* $p_0/(h(s)-s)$.

The proof is given by ZOLOTAREV (1954). Note that this is a result about random walks and can also be derived from results such as those given by KEMPERMAN (1950, 1961, Chapter 5).

Further results about the maximum have been given by URBANIK (1956).

[1] See Remark 2 following Def. 2.1.

Appendix 1

Proof of Theorems 4.1 and 4.2. Let us suppose that τ is fixed and that $\tau \leqq t \leqq \tau + T$, where T is an arbitrary fixed positive number. We shall also consider i fixed.

From the definition of F_i and the properties of a solution of (3.3) (see Def. 3.1 and Sec. 3.1), we see that the series $\sum_k P_{ik} s^k$ converges absolutely and uniformly in t for fixed s, $|s| < 1$, and is hence continuous in t. Likewise, using (3.3), we see that the series $\sum_k (\partial P_{ik}/\partial t) s^k$ converges absolutely and uniformly to a continuous function of t, and is hence equal to $\partial F_i/\partial t$. The justification of (4.2) is then an exercise in manipulating power series and is left to the reader. We must now consider the uniqueness of the solution of (4.2).

Let s_0 be fixed, $0 < s_0 < 1$. Let

$$A = \sup_{\substack{\tau \leqq t \leqq \tau + T \\ -s_0 \leqq s \leqq s_0}} |b(t)(h(s,t) - s)|, \tag{1}$$

and let n be an integer large enough so that $s_0 - A\,T/n > 0$. Note that

$$\sup |b(t)(\partial h/\partial s - 1)| < \infty,$$

where s and t have the same range as in (1). Then from a known result in the theory of partial differential equations (KAMKE (1944, p. 34)), (4.2) has exactly

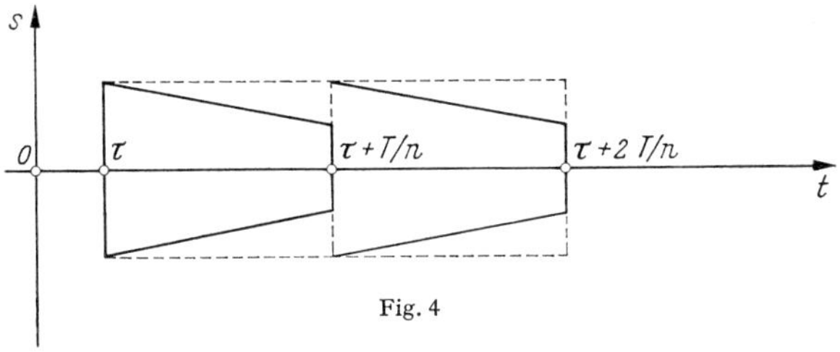

Fig. 4

one solution in the trapezoid $\tau \leqq t \leqq \tau + T/n$, $-(s_0 - A(t - \tau)) \leqq s \leqq s_0 - A(t - \tau)$, taking the initial value $F_i(s, \tau, \tau + 0) = s^i$. (See Fig. 4.)

If F_i^* is a solution that is a power series in s, convergent for $|s| \leqq s_0$, then we must have $F_i^* = F_i$ in the rectangle $\tau \leqq t \leqq \tau + T/n$, $-s_0 \leqq s \leqq s_0$, since $F_i = F_i^*$ in the trapezoid. We can then use the same argument to show that the uniqueness property holds throughout the next rectangle of width T/n, and so on until we reach $\tau + T$. Since T is arbitrary we see that (4.2) can have only one solution in the region $-s_0 < s < s_0$, $t \geqq 0$, that is for each t a power series in s convergent for $|s| \leqq s_0$.

If (3.3) had a second solution P_{ik}^*, the corresponding generating function F_i^* would be a power series solution of (4.2) that would have to coincide with F_i, and hence $P_{ik} = P_{ik}^*$. Thus we have proved the uniqueness of the solution of (3.3).

If F_1 is a solution of (4.2) with the initial value s, then it is readily verified that $(F_1)^i$ is a solution with the initial value s^i. From the uniqueness property, we thus have $F_i = (F_1)^i$, and this proves (3.6).

Now suppose that t is fixed and $|s| \leqq 1$. From (3.4) we see that

$$\sum_k |P'_{ik}(\tau, t)| \leqq i\,b(\tau),$$

where the prime denotes differentiation with respect to τ. Hence

$$\sum_k \int_0^t |P'_{ik}(x, t)|\, dx = \int_0^t \sum_k |P'_{ik}(x, t)|\, dx < \infty,$$

and hence $\lim_{N\to\infty} \sum_{k=N}^{\infty} \int_0^t |P'_{ik}(x, t)|\, dx = 0$. Then if $i < N_1 < N_2$ we have

$$\left|\sum_{k=N_1}^{N_2} P_{ik}(\tau, t)\right| = \left|\int_t^\tau \left(\sum_{N_1}^{N_2} P'_{ik}(x, t)\right) dx\right| \leqq \int_0^t \left(\sum_{N_1}^{N_2} |P'_{ik}(x, t)|\right) dx$$

$$\leqq \int_0^t \left(\sum_{N_1}^{\infty}\right) dx = \sum_{N_1}^{\infty} \int_0^t |P'_{ik}(x, t)|\, dx.$$

The last sum goes to 0 as $N_1 \to \infty$. Hence the series $\sum_k P_{ik}(\tau, t)$ is, for fixed t, convergent uniformly in τ for $0 \leqq \tau \leqq t$, and the sum is a continuous function of τ. Hence if we sum the right side of (3.4) on k, the first series converges uniformly. To sum the second series, observe that the equality $F_i(s, \tau, t) = (F_1(s, \tau, t))^i$ holds, by continuity, when $|s| = 1$, since it holds for $|s| < 1$. Hence

$$\left.\begin{aligned} \sum_{k=0}^{\infty} f_k(\tau) &= \sum_{k=0}^{\infty} \sum_{j=i-1}^{\infty} p_{j-i+1}(\tau) P_{jk}(\tau, t) = \sum_{j=i-1}^{\infty} p_{j-i+1}(\tau) F_j(1, \tau, t) \\ &= F_1^{i-1}(1, \tau, t)\, h[F_1(1, \tau, t), \tau], \qquad i = 1, 2, \ldots, \end{aligned}\right\} \tag{2}$$

where f_k is defined by the context in (2). Since the last term in (2) is continuous in τ, the series $\sum f_k(\tau)$ converges uniformly in τ by Dini's theorem. Hence $\sum_k |P'_{ik}(\tau, t)|$ converges uniformly in τ, and hence $\partial F_1(s, \tau, t)/\partial\tau = \sum_k (P'_{ik}(\tau, t))\, s^k$, $|s| \leqq 1$. The reader can now complete the justification of (4.3).

If $|s| < 1$ then, from Lemma 4.1, $|F_1(s, \tau, t)|$ is bounded by a constant $\alpha < 1$, provided τ and t are confined to a finite interval; α may depend on s.

If s is fixed, $|s| < 1$, and t is fixed, then the right side of (4.3) satisfies a Lipschitz condition in the region $-\alpha \leqq F_1 \leqq \alpha$, $0 \leqq \tau \leqq t$, where α is any constant less than 1, since $|h(y_2, \tau) - h(y_1, \tau)| \leqq |y_2 - y_1| \sum r p_r(\tau)\, \alpha^r$ if $|y_1| \leqq \alpha$, $|y_2| \leqq \alpha$. Hence (4.3) can have for each s with $|s| < 1$ only one solution that is bounded by some constant less than 1. But if (3.4) had a second solution P^*_{ik} satisfying (3.6), the corresponding generating function F^*_1 would be a solution of (4.3) which, by the same reasoning as above, would be bounded by a constant less than 1, and hence $F^*_1 = F_1$ and $P^*_{1k} = P_{1k}$, $k = 0, 1, 2, \ldots$. Since $P_{i0}, P_{i1}, P_{i2}, \ldots$ are all determined by $P_{10}, P_{11}, P_{12}, \ldots$ through (3.6), the uniqueness property is proved.

Appendix 2

Proof of (6.1). The argument is sufficiently well illustrated by the case $k = 1$. Define, for $|s| < 1$,

$$m(s, \tau, t) = \sum_{k=1}^{\infty} k P_{1k}(\tau, t)\, s^{k-1} = \frac{\partial F_1}{\partial s}.$$

Then from (4.3) we have, putting $h'(s, \tau) = \partial h(s, \tau)/\partial s$,

$$\frac{\partial}{\partial s} \frac{\partial F_1}{\partial \tau} = \frac{\partial m}{\partial \tau} = -b(\tau)\,[h'(F_1, \tau) - 1]\, m, \qquad |s| < 1, \tag{1}$$

the interchange of derivatives in (1) being valid for $|s|<1$ because of the uniform absolute convergence of the series. The condition $P_{ik}(t-0, t)=\delta_{ik}$ implies $m(s, t-0, t)=1$, and the solution of (1) is thus

$$m(s, \tau, t)=\exp\left\{\int_{\tau}^{t} b(x)\,[h'(F_1(s, x, t), x)-1]\,dx\right\}, \qquad |s|<1. \tag{2}$$

If we let $s \uparrow 1$, the condition that $\sum j p_j(x)$ converges uniformly implies that the integrand in (2) is bounded and approaches a bounded limit, and hence

$$m_1(\tau, t)=\exp\left\{\int_{\tau}^{t} b(x)\left[\sum_{j=1}^{\infty} j p_j(x)-1\right] dx\right\}. \tag{3}$$

This is just the solution of (6.1).

Chapter VI

Age-dependent branching processes

1. Introduction

An object born at time 0 has a random life-length (or generation time) ℓ with the probability distribution $G(t)=\operatorname{Prob}(\ell \leqq t)$. At the end of its life it is replaced by a random number of similar objects of age 0, the probability being p_r that the new number of objects is r. The probabilities p_r are assumed to be independent of absolute time, of the age of an object at the time it is replaced, and of the number of other objects present. The process continues as long as objects are present.

Let $Z(t)$ be the number of objects present at time t. Then $Z(t)$ is a random function that we shall call an *age-dependent branching process*, and we reserve this terminology for the type of process described above, although certain more general ones will be discussed briefly. The reason for the term "age-dependent" is that the probability $dG(\tau)/[1-G(\tau)]$ that an object, living at age τ, dies in the age interval $(\tau, \tau+d\tau)$ is in general a *nonconstant* function of the age τ.

In general $Z(t)$ is not Markovian, although we shall see in Sec. 11 that when G is exponential, Z becomes a Markov process of the type studied in Chapter V. It does seem intuitively plausible that we obtain a Markov process, in an extended sense, if we describe the state of the population at time t not simply by the number of objects present but by a list of the ages of all objects. The set of ages will then constitute a point-distribution of the sort studied in Chapter III. This Markovian point of view will be adopted only briefly and heuristically in Sec. 27, although we shall consider in some detail the age structure of a population.

Age-dependent processes can be used as mathematical models for certain phases of the multiplication of colonies of bacteria, or other creatures that reproduce by fission, and can be used to study the distribution of ages or quantities related to age, such as the mitotic index. As we shall see (Sec. 23), the basic assumptions of the simplest models are questionable, but improved models can be constructed along similar lines (Sec. 28). Age-dependent processes might also be used to study the multiplication of neutrons in an infinite medium, where the cross section of a neutron depends on its energy. (See Chapter IV.) Age-dependent birth-and-death processes (Sec. 29) might be used in certain circumstances as models for the reproduction of higher forms of animals.

The processes studied in most of this chapter were first considered in joint papers by RICHARD BELLMAN and the author (1948, 1952). The age-dependent birth-and-death processes treated briefly in Sec. 29 seem first to have been studied as stochastic processes by D. G. KENDALL (1949).

We might consider the work of this chapter as a stochastic extension of renewal theory, which has been used for many years in deterministic studies of population growth. In fact, all the moments for age-dependent branching processes satisfy renewal-type integral equations. Stochastic fluctuations of the sort incorporated into age-dependent processes might be important in small populations, or in dealing with the descendants of a single object in a large population. (See Sec. I.7.3.) One must then consider whether any of our asymptotic theorems can be valid before the population becomes so large that our assumptions cease to apply. The author cannot answer this question, although in the case of bacteria, there is sometimes a "logarithmic" phase of growth (see Sec. 23), in which the behavior is consistent with our asymptotic theory of exponential growth (Sec. 21).

The actual analytic treatment of moments and probabilities begins in Sec. 15 (based on equation (7.3)), and readers interested primarily in this aspect need not read the first fourteen sections in detail.

2. Family histories

We shall now deal in an explicit mathematical fashion with family histories. This is more satisfying than the sort of treatment used for Markov processes in Chapter V, because we can give explicit expression to our intuitive notions about the process. Moreover, it enables us to deal rigorously with such questions as the distribution of ages in the population. The procedure frees us from the need to prove existence theorems for functional equations (see Sec. 9) and can sometimes be helpful in obtaining analytic results (Sec. 15.1).

Treatments of family trees have been given by EVERETT and ULAM (1948c, d) for Galton-Watson processes with several types, independently by OTTER (1949) for Galton-Watson processes with one type, and by

URBANIK (1955a, 1958) for Markov branching processes. In each case the authors defined a probability measure on a collection of family trees.

Our procedure is to define a probability measure on "family histories", each one of which defines uniquely a family tree but contains additional information also. This redundancy does not hurt and makes the problem simpler from the measure-theoretic point of view, since we deal simply with a denumerable collection of independent random variables. However, there are results of intrinsic interest in the above treatments. For example, in the papers of OTTER and of EVERETT and ULAM, family trees (with constant generations times) are treated as a topological space.

2.1. Identification of objects in a family.

Definition 2.1. Let $\mathscr{I}$ be the collection of all elements ι, where ι is either 0 or a finite sequence of positive integers $i_1 i_2 \ldots i_k$. The collection $\mathscr{I}$ is denumerably infinite. When appropriate we can assume the elements ι to be enumerated in a sequence $\iota_1, \iota_2, \ldots$, beginning, say, with 0, 1, 2, 11, 3, 21, 12, 111, ... The exact ordering is immaterial but is supposed to be fixed.

Beginning with an ancestor, denoted by $\langle 0 \rangle$, we denote his first, second, ..., i-th, ... child by $\langle 1 \rangle, \langle 2 \rangle, \ldots, \langle i \rangle, \ldots$. The j-th child of the i-th child is denoted by $\langle ij \rangle$, and in general $\langle i_1 i_2 \ldots i_k \rangle$ denotes the i_k-th child of the ... of the i_2-th child of the i_1-th child of the ancestor.

Since we deal chiefly with the case in which all children appear simultaneously, there is no obvious physical interpretation for the meaning of "first", "second", ... child. We may, e.g., suppose that the ordering is performed by a chance device independently of the properties of the objects in the process.

2.2. Description of a family. If we prescribe the length of life and number of children of each object $\langle \iota \rangle$, then we can trace out the history of a family, assuming that the time of birth of a child coincides with the time of death of the parent.

Definitions 2.2. By a *family history* ω, we mean a sequence $\omega = (\ell, \nu; \ell_1, \nu_1; \ell_2, \nu_2; \ell_{11}, \nu_{11}; \ldots)$, where the pairs ℓ_ι, ν_ι appear in a definite order of enumeration, as suggested in Def. 2.1; we have written ℓ, ν rather than ℓ_0, ν_0. The number ℓ_ι may have any nonnegative value and represents the length of life of $\langle \iota \rangle$, while ν_ι has any nonnegative integer value and represents the number of children of $\langle \iota \rangle$. The collection of all family histories is denoted by Ω.

Each family history defines a family tree in the following way. The ancestor is born at time $t=0$ and lives until $t=\ell$, at which time it is replaced by ν children. The i-th child lives until $t=\ell+\ell_i$ and is then replaced by ν_i children, etc. For example, in Fig. 5, we have $\nu=2$, $\nu_1=0, \nu_2=3, \ldots$.

Remark. If, say, $\nu=3$, then ℓ_4 has no physical meaning, since it is the length of life of an object that is not born. Thus each ω contains redundant information concerning objects that, like "dream children", have no real existence, and two different ω's may correspond to the same family tree.

We denote the components of a given ω by $\ell(\omega)$, $\nu(\omega)$, $\ell_1(\omega)$, etc.

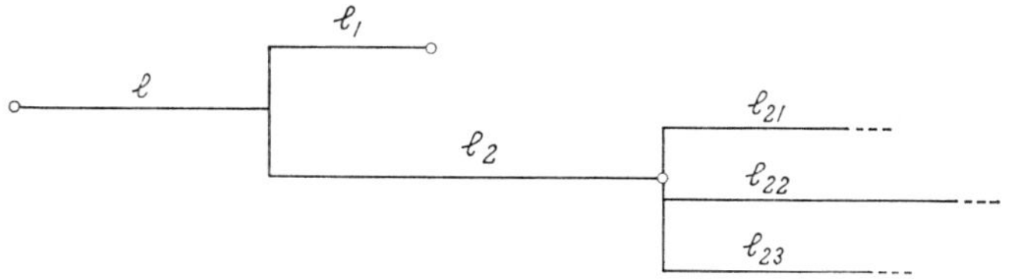

Fig. 5. Initial stages of a family tree

2.3. The generations. If ω is prescribed, then certain objects are and certain ones are not included in the family tree. The ancestor $\langle 0\rangle$ is always included. Since the ancestor has ν children, the objects $\langle 1\rangle$, $\langle 2\rangle, \ldots, \langle \nu\rangle$ are included. Likewise $\langle ij\rangle$ is included if $i\leqq\nu$ and $j\leqq\nu_i$, and so on.

Definitions 2.3. For each ω in Ω, we define a sequence $I_0(\omega), I_1(\omega), \ldots$, where I_k is a collection of objects $\langle\iota\rangle$ called the k-th generation. The 0-th generation $I_0(\omega)$ is the ancestor $\langle 0\rangle$, and $I_1(\omega)$ is the set of all objects $\langle i\rangle$ with $1\leqq i\leqq\nu(\omega)$. The succeeding generations are defined inductively: $I_k(\omega)$ is the set of all objects $\langle i_1 i_2 \ldots i_k\rangle$ such that $\langle i_1 i_2 \ldots i_{k-1}\rangle$ belongs to $I_{k-1}(\omega)$, and such that $\nu_{i_1 i_2 \ldots i_{k-1}}(\omega)\geqq i_k$. The set of objects $\bigcup\limits_{k=0}^{\infty} I_k(\omega)$ is called the *family* $I(\omega)$.

Remark. If $\langle i\rangle$ belongs to $I_1(\omega)$, then $i\leqq\nu(\omega)$; hence $I_1(\omega)$ is finite for each ω. Similarly we see that each of the collections $I_2(\omega), I_3(\omega), \ldots$ is finite, although $I(\omega)$ may be infinite. From the definition it follows that if $I_k(\omega)$ is empty, then so are $I_{k+1}(\omega), I_{k+2}(\omega), \ldots$.

Definition 2.4. We shall say that the family $I(\omega)$ contains an *infinite line of descent* if there is an infinite sequence of positive integers $i_1, i_2, \ldots$ such that $\langle i_1 i_2 \ldots i_k\rangle$ belongs to $I_k(\omega)$ for each $k=1, 2, \ldots$.

If $I(\omega)$ contains an infinite line of descent, it is obvious that none of the generations $I_k(\omega)$ is empty. We leave it to the reader to construct an example showing that the converse proposition would not be true if an object could have infinitely many children. However, the converse is true under our present hypotheses, as the following result shows.

Theorem 2.1.[1] *Suppose that for some ω each of the generations $I_1(\omega), I_2(\omega), \ldots$ is nonempty. Then the family $I(\omega)$ contains an infinite line of descent.*

Proof. Let ω be fixed. Call the object $\langle\iota\rangle=\langle i_1 \ldots i_k\rangle$ an *ancestor* of the object $\langle\iota'\rangle=\langle i_1 \ldots i_k \ldots i_{k+j}\rangle, j\geqq 1$, if both belong to the family $I(\omega)$, and call $\langle\iota'\rangle$ a *descendant* of $\langle\iota\rangle$. It is understood that $\langle 0\rangle$ is an ancestor of every other object in the family. Suppose the object $\langle\iota^0\rangle=\langle i_1^0 \ldots i_k^0\rangle$ is an ancestor of infinitely many objects. (The notation is appropriately modified if $\iota^0=0$.) For any descendant $\langle i_1^0 \ldots i_k^0 \ldots i_{k+j}\rangle$ we must have $i_{k+1}\leqq \nu_{\iota^0}(\omega)<\infty$. Hence there is an integer $i_{k+1}^0\leqq \nu_{\iota^0}(\omega)$ such that $\langle\iota^0\rangle$ is an ancestor of infinitely many objects of the form $\langle i_1^0 \ldots i_k^0\, i_{k+1}^0 \ldots\rangle$. Then $\langle i_1^0 \ldots i_k^0\, i_{k+1}^0\rangle$ is an ancestor of infinitely many objects. Since $\langle 0\rangle$ is an ancestor of infinitely many objects, we see by induction that there is an infinite sequence $i_1^0, i_2^0, \ldots$ such that $\langle i_1^0 \ldots i_k^0\rangle$ is an ancestor of $\langle i_1^0 \ldots i_k^0\, i_{k+1}^0\rangle$, $k=1, 2, \ldots$. This sequence provides an infinite line of descent. ∎

3. The number of objects at a given time

If the object $\langle\iota\rangle=\langle i_1 \ldots i_k\rangle$ belongs to the family $I(\omega)$, it is born[2] at the time $t'=\ell+\ell_{i_1}+\cdots+\ell_{i_1 i_2 \ldots i_{k-1}}$ and dies at the time $t''=t'+\ell_{i_1 \ldots i_k}$. If $t'\leqq t<t''$, then the age of the object at t is $t-t'$. Thus, if at time t we count the objects that are alive and have ages $\leqq y$, then $\langle\iota\rangle=\langle i_1 \ldots i_k\rangle$ is counted if and only if the following conditions hold (there are obvious modifications if $\iota=0$):

$$\left.\begin{aligned} i_1\leqq \nu,\ i_2\leqq \nu_{i_1},\ \ldots,\ i_k\leqq \nu_{i_1 i_2 \ldots i_{k-1}}, \\ t-y\leqq \ell+\ell_{i_1}+\ell_{i_1 i_2}+\cdots+\ell_{i_1 \ldots i_{k-1}}\leqq t, \\ \ell+\ell_{i_1}+\ell_{i_1 i_2}+\cdots+\ell_{i_1 \ldots i_{k-1}}+\ell_{i_1 \ldots i_k}>t. \end{aligned}\right\} \tag{3.1}$$

The first line in (3.1) says that $\langle\iota\rangle$ belongs to the k-th generation; the second line says that $\langle\iota\rangle$ was born between $t-y$ and t; the third line says that $\langle\iota\rangle$ dies after time t. Notice that an object is not counted at time t if it dies at t (even if it was just born at t), but it is counted if it is born at t and has a nonzero life-length.

This discussion motivates the following definitions.

Definitions 3.1. For each ι define $Z_\iota(y, t, \omega)$ to be 1 if (3.1) holds and to be 0 otherwise, $0\leqq y\leqq\infty$, $0\leqq t<\infty$. Put $Z_\iota(t, \omega)=Z_\iota(\infty, t, \omega)$. Define $Z(y, t, \omega)=\sum_\iota Z_\iota(y, t, \omega)$ and $Z(t, \omega)=Z(\infty, t, \omega)=\sum Z_\iota(t, \omega)$.

[1] URBANIK (1958, p. 240).

[2] Note that we do not speak of the time of birth of an object unless it actually belongs to the family $I(\omega)$.

Thus $Z(t, \omega)$ is the total number of objects alive at t; $Z(y, t, \omega)$ is the total number of age $\leqq y$ at t; and $Z_\iota(y, t, \omega)$ is 1 if $\langle\iota\rangle$ is alive and of age $\leqq y$ at t and 0 otherwise.

We admit the possibility that $Z(y, t, \omega) = \infty$ for some values of y, t, and ω.

Remark 1. Since $\ell = 0$ is allowed, $Z(0, \omega)$ is not always 1; it may be 0, finite and greater than 1, or infinite.

Remark 2. We may have $Z(t, \omega) = 0$ for some $t \geqq 0$ and some ω, even if the family $I(\omega)$ is infinite, as in the example at the beginning of Sec. 5.1. Once we introduce a probability measure, we shall see that such events have zero probability.

We leave it to the reader to prove the following result, which is an easy consequence of the definition of $Z(t, \omega)$.

Theorem 3.1. *If $Z(t_0, \omega_0) = 0$ for some t_0 and some ω_0, then $Z(t, \omega_0) = 0$ for all $t > t_0$.*

We shall sometimes write $Z(t)$, $Z(y, t)$, etc., rather than $Z(t, \omega)$, $Z(y, t, \omega)$, etc.

4. The probability measure P

We now introduce a probability measure on the space Ω of family histories ω. Our probability measure corresponds to the assumption that all objects have the same distribution of life-length and the same distribution of the number of children and all are independent. For convenience we shall on occasion use the same notation for random variables and particular values; e.g., we shall speak of the life-length ℓ_ι of $\langle\iota\rangle$, rather than introducing a random variable L_ι such that $L_\iota(\omega) = \ell_\iota$.

Definitions 4.1. (See Def. 2.2.) The probability measure P on Ω is defined by two assumptions. (1) The random variables ℓ_ι are independent, each having the same probability distribution:

$$P(\ell_\iota \leqq t) = G(t); \tag{4.1}$$

it is assumed that G is any probability distribution for which $G(0-) = 0$, $G(0+) < 1$. We take $G(t)$ to be continuous on the right. (2) The ν's are independent of each other and of the ℓ's, each having the generating function

$$h(s) = \sum_{r=0}^{\infty} p_r s^r = \sum P(\nu_\iota = r) s^r, \tag{4.2}$$

where $p_i \geqq 0$, $\sum p_i = 1$, and *the trivial cases $p_1 = 1$ and $p_0 = 1$ are always excluded.* We put $m = h'(1) \leqq \infty$.

We denote the k-th convolution of G with itself by G_k, and $G_1 = G$. Thus $G_k(t) = \int_{0-}^{t+} G_{k-1}(t-y)\, dG(y)$.

Remark on measurability. Since ω corresponds to a denumerable family of independent real-valued random variables, the basic theorem of KOLMOGOROV (1933, p. 27) insures that the above assumptions determine uniquely a countably additive probability measure P on the Borel extension of the cylinder sets in Ω (i.e., on the "measurable sets" in Ω). From the definition of $Z(t, \omega)$, it can be seen that Z is a measurable function of (t, ω), where the measurable (t, ω) sets are those generated by "rectangles" of the form $A \times B$, A being a Borel t-set and B a measurable set in Ω.

5. Sizes of the generations

In Sec. 2.3 we defined the successive generations $I_0, I_1, I_2, \ldots$. We now show that the sizes of the successive generations form a Galton-Watson branching process.

Definition 5.1. Let $\zeta_k = \zeta_k(\omega)$ be the number of objects in the k-th generation I_k, $k = 0, 1, \ldots$.

Theorem 5.1. *The random variables ζ_k are a Galton-Watson branching process with generating function $h(s)$. (See (4.2).)*

Proof. For simplicity of notation let us consider only $\zeta_0, \zeta_1, \zeta_2$, and ζ_3; the argument for the general case is similar. We have $\zeta_0(\omega) = 1$ and

$$\left.\begin{aligned} \zeta_1(\omega) &= \sum_{i=1}^{\nu(\omega)} 1 = \nu(\omega), \quad \zeta_2(\omega) = \sum_{i=1}^{\nu(\omega)} \sum_{j=1}^{\nu_i(\omega)} 1, \\ \zeta_3(\omega) &= \sum_{i=1}^{\nu(\omega)} \sum_{j=1}^{\nu_i(\omega)} \sum_{k=1}^{\nu_{ij}(\omega)} 1 = \sum_{i=1}^{\nu} \sum_{j=1}^{\nu_i} \nu_{ij}. \end{aligned}\right\} \tag{5.1}$$

We use the fact that if an event A is a union of disjoint events $A_1, A_2, \ldots$, if $P(A) > 0$, and if f is a random variable whose expectation exists, then $\mathscr{E}(f|A) = \sum_i \mathscr{E}(f|A_i) P(A_i|A)$, where we omit any terms for which $P(A_i) = 0$. Let r and ϱ be fixed nonnegative integers and let A be the event $\{\zeta_1 = r, \zeta_2 = \varrho\}$; we suppose $P(A) > 0$. Suppose $|s| < 1$ and let us calculate $\mathscr{E}\{s^{\zeta_3}|A\}$. We observe that A is a union of events $\{\nu = r, \nu_1 = \varrho_1, \ldots, \nu_r = \varrho_r\}$, where $\varrho_1, \ldots, \varrho_r$ are nonnegative integers whose sum is ϱ. Hence

$$\left.\begin{aligned} &\mathscr{E}\{s^{\zeta_3}|A\} = \sum_{\varrho_1 + \cdots + \varrho_r = \varrho} \mathscr{E}\{s^{\zeta_3}|\nu = r, \nu_1 = \varrho_1, \ldots, \nu_r = \varrho_r\} \\ &\times P\{\nu = r, \nu_1 = \varrho_1, \ldots, \nu_r = \varrho_r|A\} = \sum \mathscr{E}\left\{s^{\sum_{i=1}^{r} \sum_{j=1}^{\varrho_i} \nu_{ij}} \Big| \text{---}\right\} P\{\text{---}|A\}. \end{aligned}\right\} \tag{5.2}$$

The random variables ν_{ij} in (5.2) are independent; there are ϱ of them, as we see from (5.1) and the conditions on $\nu, \nu_1, \ldots, \nu_r$. Hence the last term in (5.2) is $(h(s))^{\varrho}\sum P\{\text{---}|A\}=h(s)^{\varrho}$. In other words the relation $\mathscr{E}\{s^{\zeta_3}|\zeta_1,\zeta_2\}=(h(s))^{\zeta_2}$ holds with probability 1, and this is the characteristic property of the Galton-Watson process. ☐

In case $G(t)$ has a single step of size 1 at $t_0>0$, then with probability 1 we have $Z(nt_0)=\zeta_n$, $n=0, 1, 2, \ldots$, and hence the sequence $Z(0)$, $Z(t_0)$, $Z(2t_0)$, ... is a Galton-Watson process.

5.1. Equivalence of $\{\zeta_n>0$, all $n\}$ and $\{Z(t)>0$, all $t\}$; probability of extinction. If a family has only a finite number of nonempty generations, then the number of objects is a bounded function of time that eventually becomes and remains 0; i.e., if for any ω we have $\zeta_n(\omega)\to 0$, then $Z(t,\omega)$ is a bounded function of t that is 0 for sufficiently large t. This is obvious from the definitions.

If there are infinitely many nonempty generations, we should expect the family to last infinitely long in time. Actually, there are exceptions. For example, if each object in the n-th generation has a life-length of 2^{-n}, then $Z(t)$ is 0 for $t>2$. However, the following result shows that the probability of such occurrences is 0.

Theorem 5.2.[1] *Let A be the event $\{\zeta_n>0$ for each $n\}$ and let B be the event $\{0<Z(t)\leqq\infty$ for each $t\geqq 0\}$. If $P(A)>0$, then $P(B|A)=1$.*

Corollary. *The probability of extinction, i.e., of the event $\{Z(t)=0$ for all sufficiently large $t\}$, is the smallest nonnegative root q of the equation $s=h(s)$.*

Proof. If $X_1, X_2, \ldots$ are independent random variables, each having the distribution G (see Def. 4.1), then the law of large numbers implies that for each $t\geqq 0$ we have

$$\lim_{k\to\infty}\operatorname{Prob}(X_1+\cdots+X_k>t)=\lim\left(1-G_k(t)\right)=1. \tag{5.3}$$

Let k be a fixed positive integer and let $\iota_1, \iota_2, \ldots$ be an enumeration of all k-th-order indices $\iota=i_1 i_2\ldots i_k$. (This is, of course, not the enumeration $\iota_1, \iota_2, \ldots$ of Def. 2.1.) Let D_j be the event that $\langle\iota_1\rangle$, $\langle\iota_2\rangle, \ldots, \langle\iota_{j-1}\rangle$ do not belong to $I(\omega)$ but $\langle\iota_j\rangle$ does belong. (See Def. 2.3.) Since $\sum_{j=1}^{\infty}P(D_j|A)=1$, we have, for each $t\geqq 0$,

$$\left.\begin{aligned} P(Z(t)>0|A)&=\sum_{j=1}^{\infty}P(Z(t)>0, D_j|A)\\ &\geqq\sum_{j=1}^{\infty}P\{\langle\iota_j\rangle \text{ dies after } t, D_j|A\}. \end{aligned}\right\} \tag{5.4}$$

Now the statement that D_j occurs and $\langle\iota_j\rangle$ dies after t is equivalent to the statement that D_j occurs and (putting $\iota_j=i_1 i_2\ldots i_k$) $\ell+\ell_{i_1}+\ell_{i_1 i_2}+\cdots+\ell_{i_1 i_2\ldots i_k}>t$. Since neither D_j nor A depends on the ℓ's, the

[1] This result was stated, and a proof indicated, by BARTLETT (1949, p. 219).

last line in (5.4) becomes

$$\left(1-G_{k+1}(t)\right)\sum_{j=1}^{\infty} P(D_j|A)=1-G_{k+1}(t).$$

Since k is arbitrary we see that for each fixed t we have $P\left(Z(t)>0\,|\,A\right)=1$. The desired result now follows from Theorem 3.1. □

The proof of the corollary is left to the reader.

6. Expression of $Z(t, \omega)$ as a sum of objects in subfamilies

If the initial object dies at or before time t, then the objects present at t are its children or their descendants. Theorem 6.1 below expresses this in terms of our mathematical formalism.

Definition 6.1. For each $\omega=(\ell, \nu; \ell_1, \nu_1; \ell_2, \nu_2; \ell_{11}, \nu_{11}; \ldots)$ and each $i=1, 2, \ldots,$ define $\omega_i=(\ell_i, \nu_i; \ell_{i1}, \nu_{i1}; \ell_{i2}, \nu_{i2}; \ell_{i11}, \nu_{i11}; \ldots)$. We can interpret ω_i as the family history of $\langle i\rangle$ and its descendants, although if $\nu<i$ then this family is not actually realized.

In what follows, if $\iota=i_1 i_2 \ldots i_k$ then $i\iota$ denotes $i i_1 i_2 \ldots i_k$; if $\iota=0$ then $i\iota$ denotes simply i.

Remark. The coordinates in ω_i comprise a subset of the coordinates of ω, namely, $\ell_{i\iota_1}, \nu_{i\iota_1}, \ell_{i\iota_2}, \nu_{i\iota_2}, \ldots,$ where $\iota_1, \iota_2, \ldots$ runs through $\mathscr{I}$ in the order prescribed in Def. 2.1. Altogether, the coordinates in $\omega_1, \omega_2, \omega_3, \ldots$ comprise, with no duplications, all coordinates in ω except ℓ and ν. Hence each ω is equivalent to a sequence $(\ell, \nu; \omega_1, \omega_2, \ldots)$.

Theorem 6.1.[1] *If* $0\leqq\ell(\omega)\leqq t$ *and* $\nu(\omega)>0$ *then*

$$Z(t, \omega)=\sum_{i=1}^{\nu} Z(t-\ell, \omega_i). \tag{6.1}$$

Proof. We show first that (see Def. 3.1)

$$Z_{i\iota}(t, \omega)=Z_\iota(t-\ell, \omega_i), \qquad 0\leqq\ell\leqq t,\ 1\leqq i\leqq\nu. \tag{6.2}$$

For convenience suppose $\iota=i_1 i_2$; the cases in which ι is zero or a sequence of any finite length are very similar. From (3.1), we see that $Z_{i\iota}(t, \omega)=1$ is equivalent to A: $i\leqq\nu, i_1\leqq\nu_i, i_2\leqq\nu_{ii_1}, \ell+\ell_i+\ell_{ii_1}\leqq t,$ $\ell+\ell_i+\ell_{ii_1}+\ell_{ii_1i_2}>t$. Now put $\omega_i=\omega'=(\ell', \nu'; \ell'_1, \nu'_1; \ldots)$, where $\ell'_\iota=\ell_{i\iota}$ and $\nu'_\iota=\nu_{i\iota}$. Then, also from (3.1), we see that $Z_\iota(t-\ell, \omega')=1$ is equivalent to B: $i_1\leqq\nu', i_2\leqq\nu'_{i_1}, \ell'+\ell'_{i_1}\leqq t-\ell, \ell'+\ell'_{i_1}+\ell'_{i_1i_2}>t-\ell$. From the definitions of the primed coordinates we see that B is equivalent to A except that A has the extra condition $i\leqq\nu$, which, however, is assumed already in (6.2). Hence the equation in (6.2) holds. If we sum both

[1] This is analogous to a relation that is very familiar for the Galton-Watson process.

sides of (6.2) over i, $1 \leqq i \leqq \nu$, and over all ι, then $i\iota$ ranges, with no duplications, over all nonzero indices in the set $\mathscr{I}$ for which the first coordinate is $\leqq \nu$. Since $Z_0(t, \omega) = 0$ if $\ell \leqq t$, the sum on the left side of (6.2) is thus $Z(t, \omega)$. If we sum the right side of (6.2) first on ι and then on i, we obtain the right side of (6.1). Hence the theorem is proved. □

7. Integral equation for the generating function

The generating function plays an important role in the treatment of age-dependent processes.

Definition 7.1. Let

$$F(s,t) = \sum_{r=0}^{\infty} P\{Z(t) = r\} s^r, \qquad t \geqq 0, |s| \leqq 1. \tag{7.1}$$

Since we have not yet eliminated cases in which $Z(t)$ may be infinite, we may have $F(1, t) < 1$. However, the generating function of a finite sum of independent random variables is still the product of the individual generating functions, even if positively infinite values are allowed. Note also that we need *not* have $F(s, 0) = s$, since $G(0+)$ may be positive.

Let us note the alternative expression

$$F(s,t) = \mathscr{E} s^{Z(t)} = \int_{\Omega} s^{Z(t,\omega)} \, dP(\omega), \tag{7.2}$$

where we interpret $0^0 = 1$, and $s^\infty = 0$ even if $|s| = 1$.

We now derive the integral equation that is basic for studying F.

Theorem 7.1.[1] *The generating function F satisfies the integral equation*

$$F(s,t) = s[1 - G(t)] + \int_{0-}^{t+} h[F(s, t-u)] \, dG(u), \qquad |s| \leqq 1, t \geqq 0, \tag{7.3}$$

where G and h are the functions defined by (4.1) *and* (4.2).

Proof. Referring to (7.2), write

$$F(s,t) = \int_{\Omega} s^{Z(t,\omega)} \, dP = \int_{\ell > t} s^Z \, dP + \sum_{k=0}^{\infty} \int_{\ell \leqq t, \nu = k} s^Z \, dP. \tag{7.4}$$

Since $Z(t, \omega) = 1$ if $\ell > t$, we have $\int_{\ell > t} s^Z dP = sP\{\ell > t\} = s[1 - G(t)]$.

Referring to the remark in Sec. 6, let us consider Ω as a product space $\Omega' \times \Omega_1 \times \Omega_2 \times \ldots$ of points $(\ell, \nu; \omega_1, \omega_2, \omega_3, \ldots)$. Let P' be the probability measure on the pair (ℓ, ν) and let P_i be the probability measure on Ω_i. Now refer to (6.1). If ℓ is fixed, then $Z(t - \ell, \omega_i)$ is a function on Ω_i, and hence if k is any positive integer we have

$$\int_{\ell \leqq t, \nu = k} s^Z \, dP = \int_{\ell \leqq t, \nu = k} dP'(\ell, \nu) \int_{\Omega_1} s^{Z(t-\ell, \omega_1)} \, dP_1 \ldots \int_{\Omega_k} s^{Z(t-\ell, \omega_k)} \, dP_k. \tag{7.5}$$

[1] Bellman and Harris (1948, 1952). The proof was given only for the case of binary splitting.

Now each of the integrals $\int_{\Omega_i} s^{Z(t-\ell,\,\omega_i)}\,dP_i$ is equal to $F(s, t-\ell)$, since the probability measure $dP_i(\omega_i)$ is exactly the same as $dP(\omega)$. Hence (7.5) is $p_k \int_{0-}^{t+0} [F(s, t-u)]^k\, dG(u)$, and this can be seen directly to be true if $k=0$. Putting this value in the right side of (7.4), we obtain the desired result. □

Remark. The above proof is essentially the following argument, spelled out in detail. If the initial object dies after t, the probability of this being $1-G(t)$, then the generating function at t is s. If the initial object dies at $u \leqq t$ and is replaced by r objects, the probability of this being $p_r\, dG(u)$, then the generating function at time t is $[F(s, t-u)]^r$. Summing the various terms, we obtain (7.3).

7.1. A special case. Suppose $h(s)=s^2$ and $G'(t)=[b^k/(k-1)!]\,t^{k-1}e^{-bt}$, where $b>0$ and k is a positive integer. Then $F(s,t)$ is identical with the generating function of the number of objects of all types in a multiphase birth process (see Sec. V.15.1). This was shown by KENDALL (1952). Presumably, not only the generating functions but the processes are the same.

8. The point of regeneration

PALM (1943), in connection with a study of telephone traffic, introduced the idea of a *Gleichgewichtspunkt* (point of equilibrium) in a stochastic process. This is closely related to the more special "recurrent events" discussed by FELLER (1949), to the "point of regeneration" discussed by BARTLETT and KENDALL (1951, p. 184), and to the "regenerative processes" discussed by SMITH (1955). Roughly speaking, a point of regeneration is an event in the history of a process with the following property: If we know that the event has just occurred and know the state of the process just after the occurrence, then knowledge of the history of the process prior to the event is of no further help in predicting the future.

In an age-dependent branching process, the death of the ancestor seems to be a point of regeneration for the process and to have something to do with our ability to write down a functional equation such as (7.3). In fact, PALM derived integral equations of the renewal type (i.e., similar to (15.1)) for the processes he discussed. Actually, it seems to the author that the existence of a point of regeneration for a process is by no means sufficient to insure the existence of a useful functional equation, but the question is rather vague, and we shall not attempt to discuss it further. In any case, there are many processes not of the Markov type, where integral equations more or less similar to (7.3) can be employed. We also note that JÁNOSSY (1950b) employed such equations in connection with studies of cosmic rays; see also Sec. VII.9.1.

9. Construction and properties of $F(s, t)$

An existence proof for a solution of (7.3) is unnecessary, since we defined $F(s,t)$ by probabilistic methods and showed that it satisfies (7.3). However, we want to prove that the solution, with suitable qualifications, is unique, and in the process of doing this we shall provide two different sequences of approximants converging to F. Each sequence has a probabilistic interpretation.

In what follows the integral $\int_0^t [\,]\,dG$ *will always be interpreted as* $\int_{0-}^{t+} [\,]\,dG$.

Theorem 9.1. *Define the functions* $F_k(s,t)$, $k=0,1,2,\ldots$, *by*

$$F_0(s,t) \equiv 1, \tag{9.1}$$

$$\left.\begin{aligned} F_{k+1}(s,t) &= s\big(1-G(t)\big) + \int_0^t h[F_k(s,t-u)]\,dG(u), \\ &k=0,1,2,\ldots;\ |s|\leqq 1;\ t\geqq 0. \end{aligned}\right\} \tag{9.2}$$

Then $F_{k+1}(s,t) \leqq F_k(s,t)$ *if* $0\leqq s\leqq 1$, *and* $\lim_{k\to\infty} F_k(s,t) = F(s,t)$, $|s|<1, t\geqq 0$, *where F is the generating function defined by* (7.1).

Proof. Define a sequence of random functions as follows. Let $Z^{(0)}(t)\equiv 0$ and for each $k=1,2,\ldots$ let $Z^{(k)}(t)$ be the number of objects in the generations $0,1,\ldots,k-1$ that are alive at time t. Then the generating function of $Z^{(0)}(t)$ is obviously the function F_0 above. Starting with (6.2) the reader can show that (6.1) holds if we replace Z by $Z^{(k)}$ on the left side and by $Z^{(k-1)}$ on the right side, $k=1,2,\ldots$. Then, by slightly modifying the derivation of (7.3), it is seen that the generating function of $Z^{(k+1)}$ is related to that of $Z^{(k)}$ by the recurrence relation (9.2); that is, $F_k(s,t)$ is the generating function of $Z^{(k)}(t)$. Hence

$$F_k(s,t) = \mathscr{E} s^{Z^{(k)}(t)}, \qquad |s|<1, \tag{9.3}$$

provided we define $0^0=1$.

Now for each ω and t we have $Z^{(0)}(t,\omega)\leqq Z^{(1)}(t,\omega)\leqq\cdots$ and $Z^{(k)}(t,\omega)\to Z(t,\omega)$, which may be infinite. Hence

$$s^{Z_k(t,\omega)} \downarrow s^{Z(t,\omega)}, \qquad 0<s<1,$$

provided we interpret s^∞ as 0. Hence $F_k(s,t)$, from (9.3), is a nonincreasing function of k and

$$F(s,t) = \mathscr{E} s^{Z(t)} = \lim_{k\to\infty} \mathscr{E} s^{Z^{(k)}(t)} = \lim_{k\to\infty} F_k(s,t), \qquad 0<s<1.$$

Since F_k is a power series in s for $|s|\leqq 1$, with nonnegative coefficients, and the functions F_k are bounded by 1 when $|s|\leqq 1$, we see that $F_k(s,t)\to F(s,t)$ for $|s|<1$ (TITCHMARSH (1939, p. 168)). ◻

Remark. In general we cannot assert $F_k \to F$ when $|s|=1$. For example, if $P\{Z(t)=\infty\}>0$, then $F(1,t)<1$ but $F_k(1,t)=1, k=0,1,2,\ldots$.

9.1. Another sequence converging to a solution of (7.3). Define the functions $F_k^*(s,t)$ by

$$\left.\begin{aligned} &F_0^*(s,t)\equiv 0,\\ &F_{k+1}^*(s,t)=s[1-G(t)]+\int_0^t h[F_k^*(s,t-u)]\,dG(u),\\ &\qquad\qquad k=0,1,2,\ldots;\ |s|\leqq 1;\ t\geqq 0. \end{aligned}\right\} \tag{9.4}$$

This was the sequence used by LEVINSON (1960) in proving an existence theorem for a solution of (7.3) under certain conditions. We see by induction that the F_k^* are power series in s for $|s|\leqq 1$, with nonnegative coefficients, that $|F_k^*|\leqq 1$, and that if $0\leqq s\leqq 1$ then $F_0^*(s,t)\leqq F_1^*(s,t)\leqq\cdots$. Hence $F_k^*(s,t)\uparrow F^*(s,t)$, where F^* is a solution of (7.3), and F^* is a power series in s, convergent for $|s|\leqq 1$.

We shall see later that $F^*(s,t)=F(s,t)$ for any $|s|\leqq 1$, $t\geqq 0$.

The following result shows that the functions F_k^* have a probabilistic interpretation.

Theorem 9.2. *For each $k=0,1,2,\ldots$, define $\eta_k(t,\omega)=\infty$ if at least one object in the k-th generation is born at or before t; otherwise let $\eta_k(t)=Z(t)<\infty$. Then F_k^* is the generating function of η_k; that is, $F_k^*(s,t)=\sum_{r=0}^{\infty} P\{\eta_k(t)=r\}s^r$, $|s|\leqq 1$.*

Thus $F_k^*(1,t)$ is the probability that $\eta_k(t)$ is finite, i.e., the probability that no object in the k-th generation is born at or before t.

Sketch of proof of Theorem 9.2. Let $\varepsilon_k(t)=\varepsilon_k(t,\omega)$ be defined as 0 or 1 according as $\eta_k(t)$ is infinite or finite. Then if $|s|<1$ we have $s^{\eta_k(t)}=\varepsilon_k(t)\,s^{Z(t)}$. Now we have the relation, derived similarly to (6.1),

$$\varepsilon_k(t,\omega)=\prod_{i=1}^{\nu}\varepsilon_{k-1}(t-\ell,\omega_i),$$

if $\ell\leqq t$ and $\nu\geqq 1$. Hence, using (6.1), we obtain, for $\ell\leqq t$ and $\nu\geqq 1$,

$$s^{\eta_k(t)}=\prod_{i=1}^{\nu}\varepsilon_{k-1}(t-\ell,\omega_i)\,s^{Z(t-\ell,\omega_i)}. \tag{9.5}$$

From (9.5), by an argument similar to the one leading to (9.2), we deduce that the generating function of η_{k+1} is related to that of η_k by the recurrence relation (9.4). Since the generating function of η_0 is $F_0^*\equiv 0$, the theorem follows. □

9.2. Behavior of $F(0,t)$. Refer to the corollary to Theorem 5.2 for the definition of the extinction probability q. We can summarize as follows the most important points about $F(0,t)=P\{Z(t)=0\}$.

Theorem 9.3. *If the extinction probability q is 0, then $F(0,t)\equiv 0$. If $q>0$, then $F(0,t)<q$ for each $t\geqq 0$, $F(0,t)$ is a nondecreasing function of t, and* $\lim_{t\to\infty} F(0,t)=q$.

Proof. The only assertion requiring comment is the inequality $F(0,t)<q$. Let t_0 be the least upper bound of the values t such that $F(0,t)<q$, and suppose $t_0<\infty$. Put $s=0$ in (7.3), obtaining $\left(\text{recall that } \int_0^t \text{ means } \int_{0-}^{t+}\right)$

$$F(0,t)=q=\int_0^t h[F(0,t-u)]\,dG(u), \qquad t>t_0. \tag{9.6}$$

Let $t\downarrow t_0$ in (9.6). Since $F(0,t-u)$ does not increase as $t\downarrow t_0$, we have

$$0<q=\lim_{t\downarrow t_0}\int_0^t h[F(0,t-u)]\,dG(u)=\int_0^{t_0} h[F(0,t_0-u+0)]\,dG(u). \tag{9.7}$$

Since the last integral in (9.7) is $\leqq G(t_0+)q$, we see that $G(t_0+)=1$, which in turn implies $t_0>0$, from Def. 4.1. From (9.7) we obtain

$$\int_0^{t_0}\{q-h[F(0,t_0-u+0)]\}\,dG(u)=0. \tag{9.8}$$

Now the integrand in (9.8) is strictly positive for $u>0$, and also we must have $G(t_0+)-G(0+)>0$. But this is impossible, since the integral is zero. Hence we must have $t_0=\infty$. □

9.3. Uniqueness[1]. Let s be a complex number, $|s|\leqq 1$, and let $f_s(t)$ be a function of t defined for $t\geqq 0$. We say that $f_s(t)$ is an *s-solution* of (7.3) if it satisfies (7.3) and if $|f_s(t)|\leqq 1$ for $t\geqq 0$.

Theorem 9.4. *For each s, $0\leqq s<1$, there is only one nonnegative s-solution of (7.3), namely, $F(s,t)$.*

Proof. Suppose s_1 is fixed, $0\leqq s_1<1$. Let $f(t)$ be an s_1-solution. From (9.1) and (9.2) we see by induction that $f(t)\leqq F_0(s_1,t)$, $f(t)\leqq F_1(s_1,t)$, $f(t)\leqq F_2(s_1,t)$, ..., and hence $f(t)\leqq F(s_1,t)$. Similarly from (9.4) we obtain $F^*(s_1,t)\leqq f(t)$. Hence if $F^*(s_1,t)=F(s_1,t)$, there is only one s_1-solution. Since F^* and F are power series converging for $|s|\leqq 1$, the theorem will follow if we show that $F^*(s,t)=F(s,t)$ for each s, $0\leqq s\leqq s_0$, where s_0 is a positive number to be specified below.

Since F is convex in s for each t, we have

$$\left.\begin{aligned} F(s,t)&\leqq F(0,t)+s\big(F(1,t)-F(0,t)\big)\\ &\leqq F(0,t)+s\big(1-F(0,t)\big).\end{aligned}\right\} \tag{9.9}$$

[1] Proved for $G(0+)=0$, for $h(s)=s^2$ by Bellman and Harris (1952) and for general h by Levinson (1960). The present proof is more complicated because we allow $G(0+)>0$.

First suppose that the extinction probability q is less than 1. In this case $h'(q)<1$. Pick $\varepsilon>0$ such that $h'(q+\varepsilon)<1$, whence $h'(s)\leqq h'(q+\varepsilon)<1$ for $0\leqq s\leqq q+\varepsilon$. Pick $s_0>0$ sufficiently small so that $F(s,t)\leqq q+\varepsilon$, $0\leqq s\leqq s_0$, $t\geqq 0$. This is possible by (9.9). Then

$$\left.\begin{aligned}0\leqq F(s,t)-F^*(s,t)&=\int_0^t\left[h\big(F(s,t-u)\big)-h\big(F^*(s,t-u)\big)\right]dG(u)\\&\leqq h'(q+\varepsilon)\int_0^t\left[F(s,t-u)-F^*(s,t-u)\right]dG(u),\qquad 0\leqq s\leqq s_0.\end{aligned}\right\}\tag{9.10}$$

Putting $M(s,T)=\sup\limits_{0\leqq t\leqq T}[F(s,t)-F^*(s,t)]$, we obtain from (9.10), for $0\leqq s\leqq s_0$,

$$M(s,T)\leqq G(T)\,h'(q+\varepsilon)\,M(s,T),\qquad T>0.\tag{9.11}$$

Since $G(T)\leqq 1$ and $h'(q+\varepsilon)<1$, equation (9.11) is impossible unless $M(s,T)$ is 0 for each $T>0$.

If $q=1$, in which case $h'(1)\leqq 1$, the argument needs a slight modification. From Theorem 9.3 and (9.9) we see that for any fixed T we can find $s_0>0$ and $\delta>0$ such that $F(s,t)\leqq 1-\delta<1$ for $0\leqq s\leqq s_0$, $0\leqq t\leqq T$. Then (9.10) and (9.11) hold for $0\leqq t\leqq T$ if we replace $h'(q+\varepsilon)$ by $h'(1-\delta)<1$, and hence $M(s,T)=0$, $0\leqq s\leqq s_0$. Since T is arbitrary, the proof is complete. □

There are cases where $P\big(Z(t)=\infty\big)>0$ for $t\geqq 0$, in which case $F(1,t)<1$. In such cases equation (7.3) has, for $s=1$, the two distinct solutions 1 and $F(1,t)$. Hence without further conditions the uniqueness property for (7.3) does not hold when $s=1$. Section V.13.3 furnishes an example of such a case, since Markov branching processes are a special case of age-dependent processes. (See Sec. 11.)

9.4. Another property of F. The following results are stated without proof. The reader may wish to prove them as research exercises, using Lemma 4 of the Appendix.

Theorem 9.5. *For each $k=1, 2, \ldots$ we have*

$$\lim_{t\to\infty}P\{Z(t)=k\}=0.$$

Corollary. *If $|s|<1$, then* $\lim\limits_{t\to\infty}F(s,t)=q$.

9.5. Calculation of the probabilities. The probabilities $P_n(t)=P\{Z(t)=n\}$ are the coefficients of the power series expansion of $F(s,t)$. In case $p_0=0$, we can also calculate the P_n recursively. For example, if $h(s)=s^2$, we have $P_1(t)=1-G(t)$ and

$$P_n(t)=\sum_{j=1}^{n-1}\int_0^t P_j(t-\tau)\,P_{n-j}(t-\tau)\,dG(\tau),\qquad n=2,3,\ldots.$$

10. Joint distribution of $Z(t_1), Z(t_2), \ldots, Z(t_k)$

Definition 10.1. Let $F_k(s_1, \ldots, s_k; t; \tau_1, \ldots, \tau_{k-1})$ be the joint generating function of $Z(t), Z(t+\tau_1), Z(t+\tau_1+\tau_2), \ldots, Z(t+\tau_1+\tau_2+\cdots+\tau_{k-1})$; $t, \tau_1, \ldots, \tau_{k-1} \geqq 0$, $k=2, 3, \ldots$. (F_k is not the same as in Theorem 9.1.)

For each of the generating functions F_k there is an integral equation similar to (7.3), derived in a similar manner. The following equation for F_2 is typical:

$$\left.\begin{aligned} F_2(s_1, s_2; t; \tau) &= \int_{0-}^{t+0} h[F_2(s_1, s_2; t-u; \tau)]\, dG(u) \\ &\quad + s_1 \int_{t+0}^{t+\tau+0} h[F(s_2, t+\tau-u)]\, dG(u) \\ &\quad + s_1 s_2 [1-G(t+\tau)], \qquad |s_1|, |s_2| \leqq 1;\ 0 \leqq t, \tau < \infty. \end{aligned}\right\} \tag{10.1}$$

We shall use (10.1) later to calculate the covariance of $Z(t)$ and $Z(t+\tau)$. However, the first use of (10.1) and its higher-order counterparts will be to prove that Z is a Markov process when $G(t)$ has the form $1-e^{-bt}$.

11. Markovian character of Z in the exponential case[1]

In the exponential case, when $G'(t) = b e^{-bt}$, we shall see that the integral equation (7.3) reduces to the differential equation for the generating function in the Markovian case studied in Chapter V. This is not sufficient to show that Z is Markovian, since it does not tell us the joint probabilities at several different times. However, we can obtain the desired result from equation (10.1) and its higher-order counterparts.

Theorem 11.1. *If $G(t) = 1-e^{-bt}$, then the generating function F satisfies*

$$\frac{\partial F(s,t)}{\partial t} = b[h(F(s,t)) - F(s,t)], \qquad t \geqq 0;\ |s| \leqq 1;\ F(s,0) = s. \tag{11.1}$$

If $F(1,t)=1$, all $t \geqq 0$, then Z is a Markov branching process, and any sequence $Z(0), Z(\Delta), Z(2\Delta), \ldots$ is a Galton-Watson process.

Two remarks are in order before the proof:

(1) Equation (11.1) is identical with equation (V.9.1) with one minor exception. In Chapter V we required $p_1 = h'(0) = 0$, while in the present chapter $p_1 > 0$ is allowed. However, this need cause no trouble. It is easy to verify that if F is the solution of (11.1), it is likewise the solution of the same equation with b replaced by $b(1-p_1)$ and $h(s)$

[1] The result that $Z(t)$ is a Markovian random function in the exponential case is essentially Theorem 1 of Urbanik (1958), although the results are not strictly comparable. The present method is different from that of Urbanik.

replaced by $(h(s)-p_1 s)/(1-p_1)$. In other words, we simply ignore the replacement of an object by a single object. The same remark applies to higher-order generating functions, such as F_2 in (11.4) below.

(2) Presumably, Theorem 11.1 is still true even if $F(1, t)<1$, provided we extend the notion of a Galton-Watson process to include random variables with infinite values. In Sec. 13 we shall give conditions assuring that $F(1, t)=1$.

Proof of Theorem 11.1. In the exponential case (7.3) has the form

$$F(s, t)=s e^{-bt}+b e^{-bt}\int_0^t h[F(s, u)]\, e^{bu}\, du. \tag{11.2}$$

From (11.2) we see that $F(s, 0)=s$. If we differentiate both sides of (11.2) with respect to t and compare the result with the undifferentiated equation, we obtain (11.1) immediately.

We next treat (10.1) in a manner that is typical for the higher-order generating functions. Equation (10.1) becomes

$$\left.\begin{aligned} F_2(s_1, s_2; t; \tau) &= b e^{-bt}\int_0^t h[F_2(s_1, s_2; u; \tau)]\, e^{bu}\, du \\ &\quad + b s_1 e^{-bt-b\tau}\int_0^\tau h[F(s_2, u)]\, e^{bu}\, du \\ &\quad + s_1 s_2 e^{-bt-b\tau}. \end{aligned}\right\} \tag{11.3}$$

Differentiation of (11.3) with respect to t now yields

$$\frac{\partial F_2}{\partial t}=-b F_2+b h(F_2); \tag{11.4}$$

that is, F_2 satisfies the same differential equation as F. The initial condition for F_2 is $F_2(s_1, s_2; 0; \tau)=\mathscr{E}[s_1^{Z(0)} s_2^{Z(\tau)}]=s_1 F(s_2, \tau)=\sigma$, say.

If $0\leqq s_1<1$ then $|\sigma|<1$. Now $F(\sigma, t)$ is a solution of (11.4) and, from (9.9) and Theorem 9.3, $|F(\sigma, t)|<1$. Hence because of uniqueness (Chapter V, Appendix 1), we must have

$$F_2(s_1, s_2; t; \tau)=F[s_1 F(s_2, \tau), t]; \tag{11.5}$$

since both sides of (11.5) are analytic for $|s_1|<1$, $|s_2|<1$, the equality must hold in this whole region.

In order to interpret (11.5), let us recall the result of Sec. I.13.1: In a Galton-Watson branching process with generating function $f(s)$, the joint generating function of Z_1 and Z_2 is $f[s_1 f(s_2)]$. In the non-temporally-homogeneous case, with generating function $f^{(i)}$ for the progeny of an object in the i-th generation, the joint generating function is, analogously, $f^{(0)}[s_1 f^{(1)}(s_2)]$. Hence the meaning of (11.5) is that $Z(t)$ and $Z(t+\tau)$ are jointly distributed as the first and second generations in a Galton-Watson process, where $F(s, t)$ is the generating function of one generation and $F(s, \tau)$ is that of the next. Thus (see Sec. V.5.1) $Z(t)$ and $Z(t+\tau)$ are distributed jointly as in a Markov branching process. A similar argument applies to higher-order distributions. Hence the result is proved. □

12. A property of the random functions; nonincreasing character of $F(1, t)$

In the next section we shall state conditions insuring that $Z(t)$ is finite with probability 1. However, we first establish a property of the process in those cases in which the contrary can be true.

Definition 12.1. Let $\overline{Z}(t)=\overline{Z}(t, \omega)$ be the total number of objects in $I(\omega)$ that are born up to and including time t, including the initial object $\langle 0\rangle$, and including objects whose life-length is 0. (See Def. 2.3.)

There exist family histories ω such that for three particular times $t_1<t_2<t_3$ we have $Z(t_1)=\infty$, $Z(t_2)<\infty$, $Z(t_3)=\infty$. Thus it is not completely obvious that $F(1, t)$, which is the probability that $Z(t)$ is finite, is a nonincreasing function of t. However, this property will be a consequence of the following result.

Theorem 12.1. *For each t we have* $P\{\overline{Z}(t)=\infty, Z(t)<\infty\}=0$.

Corollary. $P\{Z(t)=\infty\}=P\{\overline{Z}(t)=\infty\}$.

Proof. Let t be fixed. Define η_k as in Theorem 9.2, and let $\eta(t, \omega)=\lim_{k\to\infty} \eta_k(t, \omega)\leqq\infty$, the limit existing because η_k is a nonincreasing function of k. Let Ω_1, Ω_2, and Ω_3 be the sets of family histories ω satisfying, respectively, $\{\overline{Z}(t, \omega)<\infty\}$, $\{\overline{Z}(t, \omega)=\infty, Z(t, \omega)<\infty\}$, and $\{\overline{Z}(t, \omega)=\infty, Z(t, \omega)=\infty\}$. Then $\Omega=\Omega_1\cup\Omega_2\cup\Omega_3$. Moreover, $\eta(t, \omega)=Z(t, \omega)=\infty$ if $\omega\in\Omega_3$, and $\eta(t, \omega)=Z(t, \omega)<\infty$ if $\omega\in\Omega_1$; while $Z(t, \omega)<\infty$ and $\eta(t, \omega)=\infty$ if $\omega\in\Omega_2$. Using Sec. 9.1 and Theorems 9.2 and 9.4 we see that if $0<s<1$ then

$$\int_\Omega s^Z\, dP=F(s, t)=\lim F_k^*(s, t)=\lim\int_\Omega s^{\eta_k}\, dP=\int_\Omega s^\eta\, dP. \qquad (12.1)$$

Since $s^\eta<s^Z$ when $\omega\in\Omega_2$, we must have $P(\Omega_2)=0$, which proves the theorem. □

It is obvious that $P\{\overline{Z}(t)<\infty\}$ is a nonincreasing function of t, and since $P\{\overline{Z}(t)<\infty\}=P\{Z(t)<\infty\}=F(1, t)$, we see that $F(1, t)$ is nonincreasing.

13. Conditions for the sequel; finiteness of $Z(t)$ and $\mathscr{E}Z(t)$

Throughout the rest of this chapter we shall assume the following conditions to hold. As we shall see, they imply that $\mathscr{E}Z(t)<\infty$, and as a consequence $P\{Z(t)<\infty\}=1$ for each t.

Conditions 13.1. The distribution G and the generating function h satisfy the conditions of Def. 4.1; in addition $G(0+)=0$ and $m=h'(1)$ is finite.

Definition 13.1. Let $M(t)=\mathscr{E}Z(t)=\int_\Omega Z(t,\omega)\,dP(\omega)$.

Note. After we have shown $Z(t)$ to be finite with probability 1, we can make the equivalent definition $M(t)=\sum_r rP\{Z(t)=r\}=\partial F(1,t)/\partial s$.

Theorem 13.1. *For each t, we have $\mathscr{E}\overline{Z}(t)<\infty$.*

Corollary. *Since $Z(t)\leqq\overline{Z}(t)$, we have $M(t)<\infty$ and $P\{Z(t)<\infty\}=1$.*

Proof. (See Defs. 2.2 and 2.3.) For each $\iota\neq 0$, say $\iota=i_1 i_2 \ldots i_k$, let u_ι be a random variable that is 1 if $\langle\iota\rangle$ is in the family $I(\omega)$, i.e., if $\langle\iota\rangle$ is ever born, and 0 otherwise, and let v_ι be a random variable that is 1 if $\ell+\ell_{i_1}+\ell_{i_1 i_2}+\cdots+\ell_{i_1 i_2\ldots i_{k-1}}\leqq t$, and 0 otherwise. (If $k=1$, the sum is simply ℓ.) Then $\langle\iota\rangle$ is born at or before t if and only if $u_\iota v_\iota=1$, and

$$\overline{Z}(t)=1+\sum_{k=1}^{\infty}\sum_{i_1,\ldots,i_k=1}^{\infty} u_{i_1\ldots i_k} v_{i_1\ldots i_k}.$$

The u's and v's are independent. Moreover, $\mathscr{E}v_\iota=G_k(t)$, where G_k is the k-th convolution of G, and

$$\mathscr{E}u_\iota=P(\nu\geqq i_1)\,P(\nu_{i_1}\geqq i_2)\,\ldots\,P(\nu_{i_1\ldots i_{k-1}}\geqq i_k).$$

Then

$$\left.\begin{aligned}\mathscr{E}\overline{Z}(t)&=1+\sum_{k=1}^{\infty}G_k(t)\sum_{i_1}P(\nu\geqq i_1)\sum_{i_2}P(\nu_{i_1}\geqq i_2)\ldots\sum_{i_k}P(\nu_{i_1\ldots i_{k-1}}\geqq i_k)\\&=1+\sum_{k=1}^{\infty}G_k(t)\,(h'(1))^k.\end{aligned}\right\}\qquad(13.1)$$

Applying Lemma 1 of the Appendix to this chapter, we see that $\mathscr{E}\overline{Z}(t)$ is finite. □

14. Properties of the sample functions

Still assuming Conditions 13.1, we consider the properties of Z as a function of t for fixed ω. We shall show that with probability 1 the random functions are step functions with only a finite of jumps on any finite interval.

Definition 14.1. Let Ω_0 be the collection of family histories determined as follows: ω is in Ω_0 if either (a) for some ι the life-length $\ell_\iota(\omega)$ is 0 or (b) for some integer n the total number of births $\overline{Z}(n,\omega)$ is infinite. *From the condition $G(0+)=0$ and Theorem 13.1 we have $P(\Omega_0)=0$.*

Theorem 14.1. *For each $\omega\notin\Omega_0$, $Z(t,\omega)$ is a step function in t, taking integer values, with only a finite number of jumps on any finite interval.*

Proof. For every ω in Ω, without exception, $\overline{Z}(t,\omega)$ is nondecreasing in t and takes integer or infinite values. Hence $\overline{Z}(t,\omega)$ is a finite nondecreasing integer-valued function of t whenever $\omega \notin \Omega_0$. Let $\iota = i_1 i_2 \ldots i_k$ and let $Z_D(t, \omega)$ be the number of deaths up to and including time t; i.e., the object $\langle\iota\rangle$ contributes 1 to Z_D if $\langle\iota\rangle \in I(\omega)$ and if $\ell + \ell_{i_1} + \cdots + \ell_{i_1 i_2 \ldots i_k} \leqq t$. Then $Z_D(t, \omega) \leqq \overline{Z}(t, \omega)$ and $Z_D(t, \omega)$ is a nondecreasing finite integer-valued step function of t whenever $\omega \notin \Omega_0$. Moreover, we have[1] $Z(t, \omega) = \overline{Z}(t, \omega) - Z_D(t, \omega)$. Hence Z satisfies the assertion of the theorem. □

Remark. We leave it to the reader to show that if $p_0 = 0$, then with probability 1 the random function $Z(t, \omega)$ is a nondecreasing function of t.

15. Integral equation for $M(t) = \mathscr{E} Z(t)$; monotone character of M

The so-called renewal equation that we shall derive next from our stochastic theory was obtained in earlier studies by means of deterministic arguments. Some historical discussion is given by Lotka (1939b).

See again the beginning of Sec. 13.

Theorem 15.1.[2] *The expectation* $M(t)$ *satisfies the renewal equation*

$$M(t) = 1 - G(t) + m \int_0^t M(t-u)\, dG(u), \tag{15.1}$$

where $m = h'(1)$. M *is bounded on each finite t-interval and is the only solution of* (15.1) *having this property.* (*The upper limit of the integral is interpreted as* $t + 0$.)

Proof. We could obtain the proof most quickly by using the fact, already proved probabilistically, that $M(t) = \mathscr{E} Z(t) < \infty$. However, in order to have a procedure suitable for higher moments, it is best to proceed from equation (7.3). If $|s| < 1$, we can differentiate both sides of (7.3) with respect to s, since F is a convergent power series for $|s| < 1$. Hence

$$\left.\begin{aligned} \frac{\partial F(s,t)}{\partial s} &= 1 - G(t) + \int_0^t h'[F(s, t-u)] \frac{\partial F(s, t-u)}{\partial s}\, dG(u) \\ &\leqq 1 - G(t) + m \int_0^t \frac{\partial F(s, t-u)}{\partial s}\, dG(u), \qquad 0 \leqq s < 1. \end{aligned}\right\} \tag{15.2}$$

Iteration of the inequality (15.2) leads to the inequality

$$\left.\begin{aligned} \frac{\partial F(s,t)}{\partial s} \leqq 1 - G(t) &+ m[1 - G(t)] * dG(t) + m^2[1 - G(t)] * dG_2(t) \\ &+ \cdots + m^k[1 - G(t)] * dG_k(t) \\ &+ m^{k+1} \frac{\partial F(s,t)}{\partial s} * dG_{k+1}(t), \qquad k = 1, 2, \ldots, \end{aligned}\right\} \tag{15.3}$$

[1] This decomposition has been used in unpublished work of R. N. Snow.

[2] Stated without proof in Bellman and Harris (1948); proved by Levinson (1960).

where $f(t) * dG_j(t)$ means $\int_0^t f(t-u)\, dG_j(u)$. According to formula (2) and Lemma 1 of the Appendix, the function $H(t) = \sum_{k=1}^{\infty} m^{k-1} G_k(t)$ is bounded on every finite t-interval. Hence, since $\partial F/\partial s$ is a bounded function of t for fixed s, $|s|<1$, the last term in (15.3) $\to 0$ as $k\to\infty$. Letting $k\to\infty$ in (15.3), we obtain

$$\frac{\partial F(s,t)}{\partial s} \leqq 1 - G(t) + m\int_0^t [1 - G(t-u)]\, dH(u), \qquad 0 \leqq s < 1. \quad (15.4)$$

As $s \uparrow 1$, $\partial F/\partial s$ increases, the limit being $M(t)$. However, from (15.4) and Lemma 1 of the Appendix, the limit is bounded on each finite t-interval. Hence we can let $s\uparrow 1$ in the equality in (15.2), obtaining formula (15.1).

The uniqueness of the solution can be deduced from general results on integral equations, or from results of FELLER (1941). □

15.1. Monotone character of M. If $m=1$, then the solution of (15.1) is obviously $M(t)\equiv 1$.

If $m\neq 1$ we employ a probabilistic argument to show that M is nondecreasing if $m>1$ and nonincreasing if $m<1$. We base the argument on the observation that the function h enters into (15.1) only through the constant $m=h'(1)$. Hence if we consider, instead of Z, an age-dependent branching process Z_1 with the same G but with h replaced by any h_1 such that $h_1'(1)=m$, then $Z_1(t)$ will have the same expected value as $Z(t)$.

In case $m>1$ choose any h_1 such that $h_1(0)=0$ and $h_1'(1)=m$. The corresponding random function Z_1 is nondecreasing in t, with probability 1, by the remark in Sec. 14. Hence its expectation is nondecreasing in t.

In case $m<1$, take $h_1(s)=1-m+ms$. Then Z_1 is obviously nonincreasing, with probability 1, and hence its expectation is nonincreasing.

The result for $m>1$ was proved by LEVINSON (1960), using a purely analytic method.

16. Calculation of M

In general the expectation M can be calculated only by numerical or series methods. (See, e.g., FELLER (1941), LOTKA (1948).) However, we can obtain an expression for the Laplace transform of M simply by taking the Laplace transform of both sides of (15.1) and using the relation

$$\int_0^\infty e^{-st}\left[\int_0^t f(t-y)\, dG(y)\right] dt = \int_0^\infty e^{-st} f(t)\, dt \int_0^\infty e^{-st}\, dG(t).$$

We obtain the equation

$$\int_0^\infty e^{-st} M(t)\,dt = \frac{1-\int_0^\infty e^{-st}\,dG(t)}{s\left[1-m\int_0^\infty e^{-st}\,dG(t)\right]}, \tag{16.1}$$

which, from the theory of the renewal equation, is valid when the real part of s is sufficiently large.

M can be found explicitly in the important case of the *multiphase birth process* (Secs. 7.1 and V.15.1). In this case (16.1) leads to the expression

$$\int_0^\infty e^{-st} M(t)\,dt = \frac{[(b+s)^k - b^k]/s}{(b+s)^k - 2b^k}. \tag{16.2}$$

The zeros of the denominator in (16.2) are given by $\alpha_r = b(2^{1/k} e^{2\pi i r/k} - 1)$, $r=0, 1, \ldots, k-1$. Accordingly, the right side of (16.2) has the form $\sum_{r=0}^{k-1} A_r/(s-\alpha_r)$, where the A_r can be determined by the usual rules for partial fraction decompositions. Hence $M(t) = \sum A_r e^{\alpha_r t}$. Since α_0 has a greater real part than any of the other α's, we have the asymptotic expression (obtained by Kendall (1948b) in another manner)

$$M(t) \sim A_0 e^{\alpha_0 t} = \frac{2^{1/k}}{2k(2^{1/k}-1)}\, e^{b(2^{1/k}-1)t}, \qquad t \to \infty. \tag{16.3}$$

This asymptotic exponential expression for the mean population size suggests the Malthusian law of population growth. We shall see in the next section that an asymptotic exponential form holds for many age-dependent processes.

17. Asymptotic behavior of M; the Malthusian parameter

The behavior of $M(t)$ as $t \to \infty$ can be determined in most cases of interest from known results on the integral equation of renewal theory. We first show that if $h'(1) = m > 1$, then $M(t)$ is asymptotically exponential, with the sole exception of the case where G is a lattice distribution (see Def. 17.1 below); in this case there is an analogous result for values of t on the lattice.

Definition 17.1. We say that G is a *Δ-lattice* distribution if it is constant except for jumps that are located at positive integer multiples of some positive number Δ, and if Δ is the largest such number. Any Δ-lattice distribution may be called simply a lattice distribution.

Theorem 17.1. *Suppose $m > 1$. Define the constant α as the positive root of the equation*

$$m\int_0^\infty e^{-\alpha t}\,dG(t) = 1. \tag{17.1}$$

If G is not a lattice distribution then

$$M(t) \sim n_1 e^{\alpha t}, \qquad t \to \infty, \tag{17.2}$$

where

$$n_1 = \frac{m-1}{\alpha m^2 \int_0^\infty t e^{-\alpha t} dG(t)}. \tag{17.3}$$

The theorem is a special case of Lemma 2 of the Appendix, which is based on a result of SMITH (1954), which in turn can be deduced from WIENER's general Tauberian theorem. The result can also be deduced from a Tauberian theorem of IKEHARA as in the case $m=2$ treated by BELLMAN and HARRIS (1952).

In case G is a Δ-lattice distribution then $M(t)$ is constant on each interval $(0, \Delta)$, $(\Delta, 2\Delta), \ldots$, and its value on the interval $(r\Delta, (r+1)\Delta)$ is given asymptotically by

$$\left.\begin{aligned} M(r\Delta+0) &\sim \frac{\Delta(m-1)e^{\alpha r\Delta}}{m^2(1-e^{-\alpha\Delta})\int_0^\infty t e^{-\alpha t} dG(t)}, \\ r &\to \infty \quad \text{through integer values.} \end{aligned}\right\} \tag{17.4}$$

Formula (17.4) can be deduced from Theorem 2 of TÄCKLIND (1945).

Theorem 17.2. *If G has a density $G'(t) = g(t)$, and if $\int_0^\infty (g(t))^p \, dt < \infty$ for some $p > 1$, then Theorem 17.1 can be strengthened to yield*

$$M(t) = n_1 e^{\alpha t}[1 + O(e^{-\varepsilon t})], \qquad t \to \infty, \tag{17.5}$$

for some $\varepsilon > 0$.

The proof follows from Lemma 3 of the Appendix[1].

We have already seen that if $m=1$, then $M(t) \equiv 1$. If $m<1$, then there may be no real number α satisfying (17.1); such a number would have to be negative, but the Laplace transform $\int_0^\infty e^{-st} \, dG(t)$ may not even exist when the real part of s is negative. However, if α does exist and if certain additional conditions hold, then (17.2) is still true.

Theorem 17.3. *Suppose $m < 1$, G is not a lattice distribution, and there exists a real α (necessarily negative) satisfying (17.1). Suppose in addition that $\int_0^\infty t^2 e^{-\alpha t} \, dG(t) < \infty$. Then (17.2) holds, with n_1 again defined by (17.3).*

Proof. Since $\alpha < 0$, we have $e^{-\alpha t}[1 - G(t)] \leq \int_t^\infty e^{-\alpha x} dG(x)$, and hence $e^{-\alpha t}[1 - G(t)] \to 0$ as $t \to \infty$. If $0 < t < T$, we obtain by integration

[1] BELLMAN and HARRIS (1952).

by parts

$$\left.\begin{aligned}&\int_t^T e^{-\alpha x}\,dG(x)+\big(1-G(T)\big)e^{-\alpha T}\\&=-\alpha\int_t^T e^{-\alpha x}\big(1-G(x)\big)\,dx+e^{-\alpha t}\big(1-G(t)\big).\end{aligned}\right\} \tag{17.6}$$

Letting $T\to\infty$ in (17.6) we see that $e^{-\alpha x}\big(1-G(x)\big)$ is integrable on $(0,\infty)$. Hence the condition (b) in Lemma 2 of the Appendix is satisfied, with $f(t)=1-G(t)$, and the theorem follows from this lemma. □

We follow R. A. FISHER (1930a, Chapter 2) in referring to the constant α as the *Malthusian parameter* of the population, since it represents the geometric rate of increase postulated by MALTHUS. In Sec. 24 we shall show how the defining equation (17.1) for α may be deduced by elementary arguments.

18. Second moments

Definition 18.1. Let $M_2(t,\tau)=\mathscr{E}[Z(t)Z(t+\tau)]$, $t\geqq 0$, $\tau\geqq 0$.

If $h''(1)<\infty$, we can differentiate equation (10.1) once with respect to s_1 and once with respect to s_2. Putting $s_1=s_2=1$, we then obtain

$$\left.\begin{aligned}M_2(t,\tau)&=m\int_0^t M_2(t-u,\tau)\,dG(u)\\&\quad+h''(1)\int_0^t M(t-u)M(t+\tau-u)\,dG(u)\\&\quad+m\int_t^{t+\tau} M(t+\tau-u)\,dG(u)+1-G(t+\tau).\end{aligned}\right\} \tag{18.1}$$

(Interpret t and $t+\tau$ as $t+0$ and $t+\tau+0$ in the limits of integration.)

The justification of (18.1) is similar to that of (15.1).

We have the following asymptotic result in case $m>1$.

Theorem 18.1.[1] *If $m=h'(1)>1$, if $h''(1)<\infty$, and if G is not a lattice distribution, then*

$$M_2(t,\tau)=\frac{h''(1)\,n_1^2\int_0^\infty e^{-2\alpha u}\,dG(u)}{1-m\int_0^\infty e^{-2\alpha u}\,dG(u)}\,e^{\alpha\tau+2\alpha t}[1+o(1)],\qquad t\to\infty, \tag{18.2}$$

where $\lim_{t\to\infty} o(1)=0$ uniformly in τ. (Note. The denominator in (18.2) is positive because $\int_0^\infty e^{-2\alpha t}\,dG(t)<\int_0^\infty e^{-\alpha t}\,dG(t)=1/m$.)

[1] BELLMAN and HARRIS (1952).

Proof. If τ is fixed, then (18.1) is a renewal equation for the unknown function $M_2(t, \tau)$. Multiply both sides of (18.1) by $e^{-2\alpha t-\alpha\tau}$ and put

$$\left.\begin{aligned} &\overline{K}(t,\tau)=M_2(t,\tau)\,e^{-2\alpha t-\alpha\tau}, \qquad \overline{m}=m\int_0^\infty e^{-2\alpha t}\,dG(t)<1,\\ &d\overline{G}(t)=\frac{e^{-2\alpha t}\,dG(t)}{\int_0^\infty e^{-2\alpha t}\,dG(t)},\\ &\overline{H}(\overline{m},t)=\overline{G}(t)+\overline{m}\,\overline{G}_2(t)+\overline{m}^2\overline{G}_3(t)+\cdots, \end{aligned}\right\} \tag{18.3}$$

where $\overline{G}_n$ is the n-th convolution of $\overline{G}$. Then (18.1) takes the form

$$\overline{K}(t,\tau)=\overline{f}(t,\tau)+\overline{m}\int_0^t \overline{K}(t-u,\tau)\,d\overline{G}(u), \tag{18.4}$$

where, using (17.2), we see that $\lim_{t\to\infty}\overline{f}(t,\tau)=n_1^2 h''(1)\int_0^\infty e^{-2\alpha t}dG(t)$, uniformly for $\tau\geqq 0$. Using Lemma 4 of the Appendix, with $\overline{K}$, $\overline{m}$, $\overline{G}$, $\overline{H}$, and $\overline{f}$ in place of K, m, G, H, and f, we obtain the desired result. □

Remark. There is a corresponding result when G is a lattice distribution.

Lemma 18.1. *Under the conditions of Theorem 17.2, the remainder $o(1)$ in (18.2) is bounded by a term of the form $ae^{-\varepsilon_1 t}$, where a and ε_1 are positive constants independent of t and τ.*

The proof is left to the reader. It follows from the application to equation (18.4) of Theorem 17.2 and Lemmas 4 and 5 of the Appendix.

19. Mean convergence of $Z(t)/n_1 e^{\alpha t}$

It would be desirable to know whether, for $m\leqq 1$, there are conditional limit theorems for $Z(t)$ analogous to those for the Galton-Watson process. No such results are known. On the other hand, when $m>1$ the asymptotic behavior of the second moment enables us to prove the convergence in mean square of $Z(t)/\mathscr{E}Z(t)$ or equivalently of $Z(t)/n_1 e^{\alpha t}$. This will be done in Theorem 19.1, which forms the first step in the demonstration (completed in Secs. 20 and 21) that $Z(t)\sim n_1 e^{\alpha t} W$ as $t\to\infty$, where W is a random variable independent of t that (excepting events of probability 0) is 0 if and only if $Z(t)\to 0$.

Definition 19.1. Let $W_t=Z(t)/n_1 e^{\alpha t}$.

Lemma 19.1. *If $m>1$, $h''(1)<\infty$, and if G is not a lattice distribution, then*

$$\lim_{t\to\infty}\mathscr{E}[W_{t+\tau}-W_t]^2=0 \quad \textit{uniformly in } \tau, \qquad \tau\geqq 0. \tag{19.1}$$

The proof follows directly from Theorem 18.1 and is omitted.

Theorem 19.1.[1] *If $m>1$, $h''(1)<\infty$, and G is not a lattice distribution, then W_t converges in mean square to a random variable W as $t\to\infty$, and*

$$\left.\begin{aligned} &\mathscr{E}W=1,\\ &\text{Variance } W=\frac{(m+h''(1))\int_0^\infty e^{-2\alpha t}\,dG(t)-1}{1-m\int_0^\infty e^{-2\alpha t}\,dG(t)}.\end{aligned}\right\} \tag{19.2}$$

The variance of W is positive.

Proof. The mean convergence of W_t follows from (19.1). The values of the mean and variance are consequences of the asymptotic forms for $\mathscr{E}Z(t)$ and $\mathscr{E}[Z(t)Z(t+\tau)]$ found above, since

$$\mathscr{E}W=\lim_{t\to\infty}\mathscr{E}\left[\frac{Z(t)}{n_1e^{\alpha t}}\right],$$

$$\mathscr{E}W^2=\lim_{t\to\infty}\mathscr{E}(W_t)^2=\lim_{t\to\infty}\mathscr{E}\left[\frac{Z^2(t)}{n_1^2e^{2\alpha t}}\right].$$

In order to see that Variance $W>0$, we observe, using the Schwarz inequality, that

$$\left.\begin{aligned} &(m+h''(1))\int_0^\infty e^{-2\alpha t}\,dG(t)=\sum r^2p_r\int_0^\infty e^{-2\alpha t}\,dG(t)\\ &\geqq(\sum rp_r)^2\left(\int_0^\infty e^{-\alpha t}\,dG(t)\right)^2=m^2\left(\frac{1}{m}\right)^2=1.\end{aligned}\right\} \tag{19.3}$$

The inequality in (19.3) will be strict unless both ν and ℓ are constants with probability 1, and this is ruled out by the assumption that G is not a lattice distribution. This gives the desired result. □

The mean convergence of W_t to W implies also that the distribution of W_t converges to that of W. Levinson (1960) has shown that even if $h''(1)=\infty$, the distribution of W_t converges to the distribution of a non-constant random variable, provided h satisfies the weaker condition of Sec. I.8.1, Remark 3, and provided G is absolutely continuous. It is not known whether in this case there is convergence in probability of W_t to a random variable W.

20. Functional equation for the moment-generating function of W

Definitions 20.1. Let

$$\varphi(s,t)=\mathscr{E}e^{-sW_t}=F(e^{-s/n_1e^{\alpha t}},t),\qquad Re(s)\geqq 0,$$

$$\varphi(s)=\mathscr{E}e^{-sW},\qquad Re(s)\geqq 0.$$

Replacing s by $e^{-s/n_1e^{\alpha t}}$ in (7.3), we obtain

$$\varphi(s,t)=e^{-s/n_1e^{\alpha t}}(1-G(t))+\int_0^t h[\varphi(se^{-\alpha u},t-u)]\,dG(u). \tag{20.1}$$

---

[1] Bellman and Harris (1948, 1952).

The mean convergence, and convergence in distribution, of W_t to W imply that $\varphi(s,t) \to \varphi(s)$. Considering (20.1) as $t \to \infty$, we thus obtain the functional equation[1]

$$\varphi(s) = \int_0^\infty h[\varphi(se^{-\alpha u})]\, dG(u), \qquad Re(s) \geqq 0, \tag{20.2}$$

which holds under the conditions of Theorem 19.1.

Remark 1. It is obvious that $Z(t) \to 0$ implies $W=0$. We want to show that the converse is true with probability 1. (See Remark 1, Sec. I.8.1.) We can do this by showing that $P\{W=0\}=q$, the probability of extinction (see Theorem 5.2, Corollary). Let $q_1 = P\{W=0\}$. Then $q_1 = \lim_{s\to\infty} \varphi(s)$. Letting $s \to \infty$ in (20.2), we obtain the equality $q_1 = h(q_1)$. Since the variance of W is positive, we cannot have $q_1 = 1$, and hence we must have $q_1 = q$.

Remark 2. Levinson (1960) has shown that under the conditions mentioned at the end of Sec. 19, φ is the only solution of (20.2) defined for $0 \leqq s < \infty$ such that $\varphi(0)=1$, $\varphi'(0)=-1$, $|\varphi(s)| \leqq 1$.

It can be shown[2] that φ is analytic in a neighborhood of $s=0$ in the case $h(s)=s^2$. It seems likely that φ has this property whenever h is analytic at $s=1$; this is true for the Galton-Watson process.

21. Probability 1 convergence of $Z(t)/n_1 e^{\alpha t}$

We next want to show that if $m>1$, then $Z(t)$ is asymptotically exponential, that is, $Z(t) \sim n_1 W e^{\alpha t}$, $t \to \infty$, where W is the random variable of the preceding two sections. We shall see that this is true provided G satisfies a certain condition that is reasonable for most applications.

Theorem 21.1.[3] *Suppose $m>1$, $h''(1) < \infty$, and G is not a lattice distribution. Suppose in addition that*

$$\int_0^\infty \mathscr{E}(W_t - W)^2 dt < \infty. \tag{21.1}$$

Then $Z(t)/n_1 e^{\alpha t}$ converges with probability 1 to the random variable W as $t \to \infty$.

Corollary. *If $m>1$, $h''(1) < \infty$, and G has a density $G'(t) = g(t)$, such that $\int_0^\infty (g(t))^p\, dt < \infty$ for some $p>1$, then we have convergence with probability 1.*

The corollary is deduced from Theorem 21.1 by using Lemma 18.1. This implies that $\mathscr{E}(W_t - W)^2 = O(e^{-\varepsilon_1 t})$.

---

[1] Bellman and Harris (1948, 1952).

[2] Bellman and Harris (1952).

[3] See note to Theorem 25.1.

Proof of Theorem 21.1. We first give the proof for the case $p_0=0$, considering only family histories ω such that $Z(t,\omega)$ is a finite non-decreasing step function of t. According to Sec. 14, this excludes only a set of family histories whose total probability is 0.

From (21.1) it follows that the integral $\int_0^\infty (W_t - W)^2\,dt$ is finite with probability 1.

Since Z is nondecreasing in t, we have

$$W_{t+\tau} = \frac{Z(t+\tau)}{n_1 e^{\alpha t} e^{\alpha\tau}} \geqq e^{-\alpha\tau} W_t, \qquad \tau \geqq 0. \tag{21.2}$$

Now, putting $W_t = W_t(\omega)$ and $W = W(\omega)$ for explicitness, suppose that there is an ω such that the statement $\lim_{t\to\infty} W_t(\omega) = W(\omega)$ is not true. Since W is positive with probability 1 (Sec. 20, Remark 1), we shall suppose $W(\omega) > 0$. Let us suppose first that $\limsup_{t\to\infty} W_t(\omega) > W(\omega)$. Then there is a $\delta > 0$ and a sequence $t_1 < t_2 < \ldots$ such that $t_{i+1} - t_i > \delta/\alpha(1+\delta)$, and such that $W_{t_i}(\omega) > W(\omega)(1+\delta)$, $i = 1, 2, \ldots$; δ and the t_i may depend on ω.

Using (21.2) and the relation $e^{-\alpha\tau} \geqq 1 - \alpha\tau$, we have, for the same ω,

$$W_{t_i+\tau} \geqq (1-\alpha\tau)(1+\delta) W, \qquad \tau \geqq 0,$$

and hence, putting $t = t_i + \tau$ in the second integral in (21.3),

$$\left.\begin{aligned} &\int_{t_i}^{t_{i+1}} [W_t(\omega) - W(\omega)]^2\,dt \geqq \int_{t_i}^{t_i+\delta/\alpha(1+\delta)} [W_t - W]^2\,dt \\ &\geqq [W(\omega)]^2 \int_0^{\delta/\alpha(1+\delta)} [\delta - \alpha(1+\delta)\tau]^2\,d\tau = \frac{W^2\delta^3}{3\alpha(1+\delta)}\,. \end{aligned}\right\} \tag{21.3}$$

But from (21.3) we see that $\int_0^\infty [W_t(\omega) - W(\omega)]^2\,dt$ is infinite. Hence the inequality $\limsup W_t(\omega) > W(\omega)$ cannot hold for a set of ω's of positive probability.

Now suppose $\liminf_{t\to\infty} W_t(\omega) < W(\omega)$ for some ω with $W(\omega) > 0$. Then there is a sequence $t_1 < t_2 < \ldots$, $t_{i+1} - t_i > 1/\alpha$, and a number δ, $0 < \delta < 1$, such that $W_{t_i} < (1-\delta) W$. From (21.2)

$$W_{t_i-\tau} \leqq (1-\delta) W e^{\alpha\tau} \leqq \frac{(1-\delta) W}{1-\alpha\tau}, \qquad 0 \leqq \tau < \frac{1}{\alpha}. \tag{21.4}$$

It follows from (21.4) that $\int_{t_i-\delta/\alpha}^{t_i} (W - W_t)^2\,dt$ is bounded below by a positive number independent of i, and hence $\int_0^\infty (W - W_t)^2\,dt = \infty$. Hence $\liminf W_t(\omega) \geqq W(\omega)$, with probability 1. Hence if $p_0 = 0$, we must have $P\{\lim_{t\to\infty} W_t = W\} = 1$.

If $p_0 > 0$, then we use the decomposition already used in Sec. 14; this decomposition has been used for other purposes in unpublished work of R. N. SNOW. We write

$$Z(t) = \overline{Z}(t) - Z_D(t),$$

where $\overline{Z}(t)$ and $Z_D(t)$ are, respectively, the total number of objects born at or before t (counting the initial object) and the total number dying at or before t. The generating function of $\overline{Z}(t)$ satisfies an equation similar to (7.3), whose derivation we leave to the reader. There is likewise a counterpart to (10.1). We can then show that $\overline{Z}(t)/n_1 e^{\alpha t}$ converges in mean square to a random variable $\overline{W}$, and since $\overline{Z}$ is monotone we can argue as above to show that there is convergence with probability 1. Similarly $Z_D(t)/n_1 e^{\alpha t}$ converges with probability 1 to a random variable W_D. Hence $Z(t)/n_1 e^{\alpha t}$ converges to W with probability 1. □

The argument used above, going from the finiteness of $\int_0^\infty (W_t - W)^2 \, dt$ to the existence of a limit for W_t is similar to that employed to study the behavior of the solutions of certain differential equations; see BELLMAN (1953, pp. 155–156).

22. The distribution of W

The distribution of W can be studied by means of the functional equation (20.2) satisfied by its moment-generating function φ. For example, the second and higher moments of W are uniquely determined by differentiation of (20.2) at $s=0$. It can be shown that if the k-th derivative $h^{(k)}(1)$ is finite then $\mathscr{E} W^k$ is finite, $k = 3, 4, \ldots$, and $\mathscr{E}[Z(t)]^k \sim n_1^k e^{k\alpha t} \mathscr{E} W^k$, $t \to \infty$, provided G is not a lattice distribution.

We can also make deductions about W by studying the behavior of $\varphi(it)$ as $t \to \pm\infty$. This was done for the case $h(s) = s^2$ by BELLMAN and HARRIS (1948, 1952), and we state the more general results here without proof since the methods are similar, after some manipulation of (20.2)[1].

Theorem 22.1. *Suppose $m > 1$, $h''(1) < \infty$, and G is not a lattice distribution. Then the distribution $K(u) = P\{W \leq u\}$ is continuous except for a jump of magnitude q at $u = 0$, where q is the smallest nonnegative root of $s = h(s)$. If in addition G satisfies $1 - G(t) = O(e^{-ct})$, $t \to \infty$, for some $c > 0$, then K is absolutely continuous for $u > 0$.*

---

[1] In BELLMAN and HARRIS (1952) the words "for t sufficiently large" should be added to formula (12), p. 291. On the following page, add to the end of the line following (19) the words "(12) holds for $t \geq t_0$ and so that...."

23. Application to colonies of bacteria

The most obvious application of age-dependent branching processes is to the growth of a colony of bacteria, or other cases in which reproduction occurs only once during the life of the organism. POWELL (1955) has discussed the realism of the assumption that the life-lengths of all bacteria in a colony are independent, with the same distribution. For the bacteria considered by POWELL, there seems to be a positive correlation between the life-lengths of sister cells, although not between mothers and daughters. We shall indicate in a later section how extensions of our model can take account of certain kinds of dependence between life-lengths. Perhaps a more serious defect of our model is that inhomogeneities develop in the actual culture, and different subfamilies develop at different rates.

Nevertheless, experiments do show that the growth of a colony is exponential and almost deterministic during a period known as the *logarithmic phase.* This period begins after the early stages, and lasts until the bacteria are so concentrated that they interfere with one another. The reader can find diagrams illustrating the logarithmic phase in HINSHELWOOD (1946, Chapters II and III). These diagrams show that $\log Z(t)$ is approximately linear during the logarithmic phase.

To the extent that our model is valid, repeated experiments with the same type of bacteria should show a period of logarithmically linear growth in each case, with the same slope α in each case but different intercepts, the variation in intercepts representing the variability of W. The author does not know of any experiments of this kind. In any case the results would have to be treated cautiously, since a culture of real bacteria will sometimes undergo a lag before multiplication begins, and the variability of this lag would be confounded with the variability of W.

For the binary case, (19.2) reduces to

$$\text{Variance } W = \frac{8 \text{ Variance } e^{-\alpha \ell}}{1 - 4 \text{ Variance } e^{-\alpha \ell}}, \tag{23.1}$$

where ℓ is the life-length of a single object. Conversely, if the variance of W could be determined from experiments, then (23.1) would enable us to determine the variance of $e^{-\alpha \ell}$.

Data of KELLY and RAHN (1932; see also KENDALL 1948b) suggest that we can approximate the life-length density for bacteria of a certain type by

$$G'(t) = \frac{b^k}{\Gamma(k)} t^{k-1} e^{-bt}, \qquad k > 0,\ b > 0,$$

where $k = 20$ and $b = 40$ when time is measured in hours. In this case $\int_0^\infty e^{-st} dG(t) = (1 + s/b)^{-k}$, $\alpha = b(2^{1/k} - 1)$, $\mathscr{E}\ell = k/b = 1/2$, $\operatorname{Var} \ell = k/b^2 = 1/80$, and $\operatorname{Var}(e^{-\alpha\ell}) = (1+2\alpha/b)^{-k} - \frac{1}{4} = 0.00587$. Then from (23.1) we have $\operatorname{Var} W = 0.048$ and $\sigma(W) = \sqrt{\operatorname{Var} W} = 0.22$. It is interesting to note that $\sigma(\ell)/\mathscr{E}\ell = 1/\sqrt{k} = 0.22$, and $\sigma(W)/\mathscr{E}W = \sigma(W) = 0.22$. That is, *the coefficient of variation of W is approximately equal to the coefficient of variation of* ℓ. This relationship was first pointed out by D. G. KENDALL (1948b). WAUGH (1955) has studied the coefficient of variation for models of the type mentioned in Sec. 28.1.

24. The age distribution

Consider a simple renewal or replacement process; this is an age-dependent branching process of a degenerate sort, with $h(s) \equiv s$. If $X(t)$ is the age of the object present at time t, it is a well-known result of renewal theory that under suitable conditions on G

$$\lim_{t\to\infty} \operatorname{Prob}\left(X(t) \leqq x\right) = \frac{\int_0^x [1-G(t)]\,dt}{\int_0^\infty [1-G(t)]\,dt}. \tag{24.1}$$

Notice that the mean life-length $\int_0^\infty t\,dG(t)$ is also equal to $\int_0^\infty [1-G(t)]\,dt$.

In a branching process, one might expect the ages of the objects present after a long time to be distributed, at least in an average sense, according to (24.1), since any one line of descent represents a renewal process. However, the multiplication of objects changes the situation. For example, if $m > 1$ then the proportion of younger objects is greater than in (24.1), since each old object that dies is replaced, on the average, by more than one new one.

We shall see that in an age-dependent branching process with the Malthusian parameter α, we have instead of (24.1) the limiting distribution

$$A(x) = \frac{\int_0^x e^{-\alpha t}[1-G(t)]\,dt}{\int_0^\infty e^{-\alpha t}[1-G(t)]\,dt}, \tag{24.2}$$

under suitable conditions on G. If $\alpha \neq 0$ (i.e., $m \neq 1$), we have the alternative expression

$$A(x) = \frac{\alpha \int_0^x e^{-\alpha t}[1-G(t)]\,dt}{1-1/m}. \tag{24.2a}$$

The form of the distribution (24.2) has been known at least since 1907; see LOTKA (1922) for references.

Definition 24.1. We shall call $A(x)$ the *limiting age distribution.*

We shall see that under appropriate conditions the expected number of objects of age $\leqq x$ is approximately proportional to $A(x)$ when t is large. *Moreover, if* $m>1$, *then the* actual *fraction of the population of age* $\leq x$ *converges to* $A(x)$ *when* $Z(t)$ *does not go to* 0. *That is,* A *is a true limiting distribution of ages whenever the population does not die out.*

We make the following definition, first recalling that $Z(x,t)$ is the number of objects at time t of age $\leq x$.

Definition 24.2. Let $M(x,t)=\mathcal{E}Z(x,t)$.

Before proving the above statements, let us consider an elementary argument given by R. A. Fisher (1930a, Chapter II) that enables us to deduce the form of $A(x)$ and also the value of the Malthusian parameter α.

If a population has a stationary age distribution $A(x)$ (see Sec. 24.2), it is reasonable to suppose that the expected population size is proportional to an exponential, say $e^{\alpha t}$. Then the birth rate is also proportional to $e^{\alpha t}$. Take the birth rate at time 0 to be 1; then the rate at time $-x<0$ is $e^{-\alpha x}$. Now the mean number of births between times 0 and dx due to a parent born at $-x$ is $mg(x)\,dx$, where $g(x)=G'(x)$ and m is the mean number of objects per birth. Hence we have the equation $1=m\int_0^\infty e^{-\alpha x}g(x)\,dx$, which determines α.

In order to determine $A(x)$, we observe that an object is of age $\leq x$ at time 0 if and only if it was born at time $-t\geqq -x$ and survived at least to the age t. Since the birth rate t units prior to 0 is $e^{-\alpha t}$ and the probability of survival is $1-G(t)$, $A(x)$ must be proportional to $\int_0 e^{-\alpha t}[1-G(t)]\,dt$. Thus we obtain the form (24.2) for $A(x)$.

24.1. The mean age distribution. If $F(x,s,t)$ is the generating function of $Z(x,t)$, then F satisfies the equation[1]

$$\left.\begin{aligned}F(x,s,t)=[1-G(t)]\,[sJ(x-t)+1-J(x-t)]\\ +\int_0^t h[F(x,s,t-y)]\,dG(y),\end{aligned}\right\}\tag{24.3}$$

where $J(t)=1$, $t\geqq 0$, and $J(t)=0$, $t<0$. Equation (24.3) is deduced almost exactly as (7.3). From (24.3) we then deduce the equation

$$M(x,t)=[1-G(t)]\,J(x-t)+m\int_0^t M(x,t-y)\,dG(y).\tag{24.4}$$

Applying Lemma 2 of the Appendix, we then have the following result for the mean age distribution.

Theorem 24.1. *If* G *is not a lattice distribution, if there exists a real* α *satisfying* (17.1), *and if* $\int_0^\infty te^{-\alpha t}\,dG(t)<\infty$, *then for each* x, $0\leqq x<\infty$, *we have*

$$M(x,t)\sim n_1 e^{\alpha t}A(x),\qquad t\to\infty,\tag{24.5}$$

[1] This equation appeared with part of the right side missing in Harris (1951).

where n_1 *and* α *are defined by* (17.3) *and* (17.1), *except that if* $m=1$, *then* $n_1=1$. (*If* $m>1$, *then* α *exists and is positive and the condition*

$$\int_0^\infty t e^{-\alpha t} dG(t) < \infty$$

is satisfied automatically. Notice that the moment condition on G *is weaker than in Theorem* 17.3.)

24.2. Stationarity of the limiting age distribution. As we should expect, the limiting age distribution has the property of *stationarity*. That is, if at time 0 the expected number of objects of age $\leq x$ is proportional to $A(x)$ for each $x \geq 0$, then at all later times the same is true.

In order to make this statement more precise we must extend the notion of an age-dependent branching process to include the case in which the initial object has an arbitrary age x. The generating function and the moments then satisfy integral equations similar to those for the case in which $x=0$.

Let $M(x, y, t)$ be the expected number of objects of age $\leq y$ at time t, if there is one object of age x at time 0. Suppose $N(x)$ is the number of objects of age $\leq x$ in some population at time 0. Then the expected number of objects of age $\leq x$ at time t is $M_t N(x)$, where the linear operator M_t is defined by

$$M_t N(x) = \int_0^\infty M(z, x, t)\, dN(z).$$

According to Lotka (1922) we have

$$\int_0^\infty M(z, x, t)\, dA(z) = e^{\alpha t} A(x), \qquad x \geq 0, \tag{24.6}$$

and it is in this sense that A is stationary. It is not clear under just what conditions Lotka's arguments can be applied. However, the author believes (24.6) to be true whenever the conditions of Theorem 24.1 hold. The simplest way to prove (24.6) seems to be to prove the corresponding relation for the Laplace transforms on t.

24.3. The reproductive value. We may regard the distribution $A(x)$ as a left eigenfunction of the operator M_t, corresponding to the eigenvalue $e^{\alpha t}$. There is also a right eigenfunction $V(x)$, not a distribution, defined only for nonnegative values of x such that $G(x)<1$, by

$$V(x) = \frac{e^{\alpha x}}{1-G(x)} \int_x^\infty e^{-\alpha u}\, dG(u),$$

and satisfying the equation

$$\int_0^\infty V(y)\, d_y M(x, y, t) = e^{\alpha t} V(x). \tag{24.7}$$

The function V has the following interpretation. Suppose that at time 0 we conduct a V-census. That is, each object contributes an amount depending on its age, an object of age x contributing the amount $V(x)$. If the ages of the objects present at time 0 are $x_1, x_2, \ldots, x_n$, then the value of the V-census is $V(x_1) + \cdots + V(x_n)$; according to (24.7), the expected value of the V-census at time t is $e^{\alpha t}(V(x_1) + \cdots + V(x_n))$. Thus the expected growth of a V-census is exactly in accordance with the Malthusian rate, while an ordinary census may change at a different rate if the age distribution of the population is different from the stationary distribution A. This feature of the V-census was pointed out by R. A. FISHER (1930a, Chapter II), who called $V(x)$ (or a value proportional to it) the *reproductive value* of an object of age x. The author believes (24.7) to hold whenever the conditions of Theorem 24.1 are satisfied.

If V_t is the value of a V-census taken at time t, then from (24.7) we deduce, at least heuristically, that $e^{-\alpha t} V_t$ is a martingale (compare Sec. II.12.1).

25. Convergence of the actual age distribution

It is reasonable to suppose that the distribution of ages in the population converges to the distribution A if the population does not die out. In this connection we have the following result. Recall that $Z(x, t)$ is the number of objects of age $\leqq x$ at time t. We are again supposing the initial object to have age 0 at $t=0$.

Theorem 25.1.[1] *Suppose $m>1$, $h''(1)<\infty$, and G has a density $G'(t)=g(t)$, where $\int_0^\infty (g(t))^p\,dt<\infty$ for some $p>1$. Then*

$$P\left\{\lim_{t\to\infty} \frac{Z(x,t)}{n_1 e^{\alpha t}} = A(x)\,W \quad \textit{for each} \quad x\geqq 0\right\}=1. \tag{25.1}$$

Note. The result is probably true under only the conditions of Theorem 21.1.

Corollary. *Under the condition that $Z(t)$ does not go to 0 as $t\to\infty$, the ratio $Z(x,t)/Z(t)$ converges to $A(x)$ for each x, with probability 1.*

We shall only briefly indicate the principal steps of the proof.

If f is any nonnegative bounded function, we put

$$Z_f(t) = \int_0^\infty f(x)\,d_x Z(x,t).$$

By methods similar to those of the earlier sections of this chapter we can deduce integral equations for the generating function and the

[1] Announced by HARRIS (1960c). A similar result for mean square convergence appeared in HARRIS (1951).

moments of Z_f. It can then be shown that $Z_f(t)/n_1 e^{\alpha t}$ converges in mean square as $t \to \infty$ to $(n_f/n_1) W$, where n_1 is defined by (17.3),

$$\frac{n_f}{n_1} = \frac{\alpha m}{m-1} \int_0^\infty f(t)\,(1-G(t))\, e^{-\alpha t}\, dt,$$

and W is the random variable of Theorems 21.1 and 19.1. (This holds under the milder restrictions of Theorem 19.1.)

If $f_x(y)=1$ for $0 \leqq y \leqq x$ and $f_x(y)=0$ otherwise, then $Z_{f_x}(t)=Z(x,t)$, and it would be natural to prove Theorem 25.1 by deducing the probability 1 convergence of $Z(x,t)/n_1 e^{\alpha t}$ from its mean square convergence, as in Theorem 21.1. The argument used there cannot apply without modification, however, since even if $h(0)=0$, the quantity $Z(x,t)$ need not be an increasing function of t.

We carry out the argument by considering first a special class C of functions $f(y)=f_{a,b}(y)$, $0 \leqq a \leqq \infty$, $0 \leqq b < \infty$, where

$$\begin{aligned} f_{a,b}(y) &= 1, & 0 \leqq y < a, \\ f_{a,b}(y) &= e^{-b(y-a)}, & y \geqq a. \end{aligned}$$

These functions have the property

$$f_{a,b}(y_2) \geqq f_{a,b}(y_1)\, e^{-b(y_2-y_1)}, \qquad y_2 \geqq y_1. \tag{25.2}$$

If $p_0=0$, so that we may suppose $Z(t)$ to be nondecreasing in t, then (25.2) implies that for $f=f_{a,b}$ we have, for any $\tau \geqq 0$,

$$e^{-\alpha(t+\tau)} Z_f(t+\tau) \geqq e^{-(b+\alpha)\tau}\, e^{-\alpha t} Z_f(t). \tag{25.3}$$

We can then use (25.3), as (21.2) is used in the proof of Theorem 21.1, to show that $Z_f(t)/n_1 e^{\alpha t}$ converges with probability 1 to $n_f W/n_1$ whenever f is a function of class C and $p_0=0$.

In order to extend the result to $Z(x,t)$, supposing first that $p_0=0$, we observe that if $0 \leqq a < x$, then

$$\int_0^\infty f_{a,b}(y)\, d_y Z(y,t) = \int_0^x + \int_x^\infty \leqq Z(x,t) + e^{-b(x-a)} Z(t).$$

Hence

$$e^{-\alpha t} Z_{f_{a,b}}(t) - e^{-b(x-a)}\, e^{-\alpha t} Z(t) \leqq e^{-\alpha t} Z(x,t) \leqq e^{-\alpha t} Z_{f_{x,b}}(t). \tag{25.4}$$

Letting $t \to \infty$ in (25.4), we have, with probability 1,

$$\left.\begin{aligned} & n_{f_{a,b}} W - e^{-b(x-a)}\, n_1 W \leqq \liminf_{t\to\infty} e^{-\alpha t} Z(x,t) \\ & \leqq \limsup_{t\to\infty} e^{-\alpha t} Z(x,t) \leqq n_{f_{x,b}} W. \end{aligned}\right\} \tag{25.5}$$

Letting $b \to \infty$ in (25.5) and recalling the definition of f_x, we obtain

$$n_{f_a} W \leqq \liminf \leqq \limsup \leqq n_{f_x} W. \tag{25.6}$$

If we then let $a \to x$ in (25.6) and note that $n_{f_x} = n_1 A(x)$, we obtain the desired result when $p_0 = 0$. If $p_0 > 0$, we obtain the result by treating births and deaths separately, somewhat as in the proof of Theorem 21.1.

26. Applications of the age distribution

In this section we make the assumptions of Theorem 19.1. Note that ℓ_1 and ℓ_2 do not have the same meaning as in Def. 2.2.

26.1. The mitotic index. Let us consider a family of cells (e.g., bacteria) multiplying by fission. Let m denote the mean number per fission (usually two). Before dividing, a cell undergoes a period of preparation called *mitosis*. The fraction of cells undergoing mitosis at a given time is called the *mitotic index*.

We shall suppose that the life-length of a cell is $\ell = \ell_1 + \ell_2$, where ℓ_1 represents the premitotic and ℓ_2 the mitotic period. In a large population where each cell, at death, is replaced by an average of just one new cell, the mitotic index should be equal to $\mathcal{E}\ell_2/\mathcal{E}(\ell_1+\ell_2)$. However, we have already pointed out in Sec. 24 that in an increasing population the proportion of younger cells is increased, and hence the observed mitotic index should be less than the value just suggested.

Suppose that a living cell of age y has a probability $p(y)$ of being in mitosis. Then the mitotic index in a population whose age distribution is approximately the limiting distribution A defined by (24.2) is approximately $\int_0^\infty p(y)\, dA(y)$. We shall call this quantity the *limiting mitotic index*. A simple calculation shows that, whether or not ℓ_1 and ℓ_2 are independent, $p(y) = [G_1(y) - G(y)]/[1 - G(y)]$, where G_1 is the distribution of ℓ_1. Hence, using (24.2),

$$\text{limiting mitotic index} = \frac{\int_0^\infty [G_1(y) - G(y)]\, e^{-\alpha y}\, dy}{\int_0^\infty [1 - G(y)]\, e^{-\alpha y}\, dy}. \tag{26.1}$$

If ψ_1 and ψ are the moment-generating functions of ℓ_1 and ℓ, then (26.1) becomes[1]

$$\frac{m\,\psi_1(\alpha) - 1}{m - 1}, \tag{26.2}$$

where we have used the relation $\int_0^\infty e^{-\alpha y} G(y)\, dy = \psi(\alpha)/\alpha = 1/m\alpha$ (see (17.1)).

Formula (26.2) becomes very simple if we can make the following assumption, whose plausibility will not be defended.

Assumption. The premitotic and mitotic periods are random quantities that are independent of one another. Moreover, the premitotic period is

[1] Harris (1959b).

distributed as the sum of n independent periods, each distributed like the mitotic period. Here n is an arbitrary positive integer.

Roughly speaking, the premitotic period is assumed to be n times as long as the mitotic period and to have n times as much variance.

Let ψ_2 be the moment-generating function of ℓ_2.

Under the above assumption, we have $\psi_1(\alpha)=(\psi_2(\alpha))^n$, and

$$1/m=\psi(\alpha)=\psi_1(\alpha)\,\psi_2(\alpha)=(\psi_1(\alpha))^{(n+1)/n}.$$

Then (26.2) becomes $(m^{1/(n+1)}-1)/(m-1)$. In the case of binary fission the limiting mitotic index is then $2^{1/(n+1)}-1$, which is approximately $\log 2/(n+1)=0.693/(n+1)$ when n is large. The factor 0.693 thus represents the effect of the expansion of the population, since in a population of steady size, the limiting mitotic index would be $1/(n+1)$.

The mitotic index is important in some biological investigations; see, e.g., WALKER (1954).

HOFFMAN, METROPOLIS, and GARDINER (1956) studied the mitotic index using random sampling experiments. They give the formula $0.693\,\mathcal{E}\ell_2/\mathcal{E}\ell_1$ for the limiting mitotic index, under certain conditions. (See footnote, p. 177 of their paper.)

26.2. The distribution of life fractions. If a cell at a certain instant of time t has the age x, and if its total life span, past and future, is ℓ, we shall call the ratio x/ℓ the *life fraction* of the cell at t, denoting this ratio by r. We could conceivably relate r to some measurable characteristic of the cell, such as its content of the basic substance DNA; see, for example, WALKER (1954) for a discussion of one theory of such a relation, and some of its shortcomings.

Thus it is of some interest to be able to calculate the distribution of r in a population of cells. Let $R(\varrho)$ be the expected proportion of cells for which the life fraction r is less than ϱ, if the population has the limiting age distribution A. We leave it to the reader to show that

$$R(\varrho)=\frac{m}{m-1}\left[1-\int_0^\infty e^{-\alpha\varrho t}\,dG(t)\right],\qquad 0\leqq\varrho\leqq 1. \tag{26.3}$$

27. Age-dependent branching processes in the extended sense

We recall the definition $Z(x,t,\omega)=$ number of objects of age $\leqq x$ at t for the family history ω. For each t and ω, we may consider $Z(\cdot,t,\omega)$ as a point-distribution with objects of types (ages) $x_1, x_2, \ldots$, where $x_1, x_2, \ldots$ are the jumps of Z. Hence we have a random function of t whose values are point-distributions.

It is intuitively plausible that this is an example (in continuous time) of the general branching processes discussed in Chapter III. Moreover, the expectations $M(x, y, t)$ of Sec. 24.2 should have the semigroup property

$$M(x, y, t_1 + t_2) = \int_0^\infty M(z, y, t_2)\, d_z M(x, z, t_1). \tag{27.1}$$

We shall not prove or use the above assertions. The reader might try constructing a proof of (27.1) as an exercise in the use of conditional probabilities.

28. Generalizations of the mathematical model

Our model for age-dependent branching processes can be generalized in a number of ways, of which we shall mention three briefly in this section. In the following section another model will be discussed at somewhat greater length.

28.1. Transformation probabilities dependent on age. We have supposed that when an object is transformed, the probabilities p_n of transformation into n objects are independent of the age of the object. WAUGH (1955, 1961) has generalized the model to allow the p_n to depend on the age at the time the transformation occurs and has studied certain cases extensively.

28.2. Correlation between sister cells. We have already pointed out that data of POWELL (1955) suggest that in the multiplication of bacteria, the life-lengths of sister cells are correlated, while the life-lengths of mother and daughter are not. Let us formulate a model that enables us to treat some kinds of dependencies in the case of binary fission. Suppose that the life-length of a cell has the mean value μ and the variance σ^2, and that ϱ is the correlation between the life-lengths of sister cells, $0 \leqq \varrho < 1$. We can attain these parametric values by assuming that if ℓ_1 and ℓ_2 are the life-lengths of sister cells, then $\ell_1 = X + Y_1$, $\ell_2 = X + Y_2$, where X, Y_1, and Y_2 are independent, $\mathscr{E}X + \mathscr{E}Y_1 = \mathscr{E}X + \mathscr{E}Y_2 = \mu$, Variance $X = \varrho\sigma^2$, Variance Y_1 = Variance $Y_2 = (1-\varrho)\sigma^2$.

Suppose further that Y_1 and Y_2 have the same probability distribution. Let $F_x(s, t)$ be the generating function of the number of objects at time t if initially there is one object of age 0, given that X for the initial object has the value x. The reader should now be able to write an integral equation for F_x, somewhat more complicated than (7.3), and to treat the moments by means of renewal-type equations. The reader might prove, as a research exercise, that the mean number of objects is the same as it would be for the same distribution of life-lengths with independence, while the variance is at least as large (and presumably larger).

28.3. Multiple types. Suppose there are k types of objects, the i-th type having a life-length distribution G_i and a generating function $h_i(s_1, s_2, \ldots, s_k)$ for the new objects when a transformation occurs. We can then write down a system of k equations for the generating functions $F_i(s_1, \ldots, s_k, t)$, where F_i corresponds to an initial object of type i. Such systems have been studied in detail by R. N. SNOW (1959a, 1959b). Perhaps they can be useful for treating the case of correlations between mother-and-daughter objects or sister objects. The moments can be studied by means of systems of renewal-type equations; see BELLMAN and COOKE (1963), pp. 257—264.

An important case of more than one type occurs when there are *mutations*. The well-known experiments and theoretical discussion of LURIA and DELBRÜCK (1943) are a most interesting example of inductive reasoning about the nature of certain mutations. We shall not discuss this topic here, but refer the reader to papers of ARMITAGE (1952), HARRIS (1951), and particularly KENDALL (1952) for further results, discussion, and references to additional work.

29. Age-dependent birth-and-death processes

In this section we discuss, without attempting rigor, a process that is closely related to age-dependent branching processes. Suppose that a living object of age x has a probability $\lambda(x)\,dx$ of giving birth to another object before it reaches age $x+dx$, this probability being independent of the past history of the object. Multiple births are excluded. Moreover, an object of age x has a probability $\mu(x)\,dx$ of dying before age $x+dx$, independently of past history. What complicates the new model is that when a new object has just been born, we have both an object of age 0 and one of age x, which may still produce further objects; whereas in the earlier model the objects present after the first birth are all of age 0.

The above model seems to have been first considered (in a probabilistic, as opposed to a mean-value, sense) by D. G. KENDALL (1949); see also KENDALL (1952) and BARTLETT (1955).

Age-dependent birth-and-death processes are possible models for the multiplication of animals or plants that reproduce more than once in a lifetime. Just as in the case of other branching processes, we must suppose the environment adequate so that the creatures in the process behave independently of one another. If there are two sexes, we should presumably follow only one.

We shall suppose throughout that λ and μ are continuous for $0 \leq x < \infty$.

Let ℓ be the life-length of the initial object and let $N(t)$ be the random number of children born to the initial object in the time interval $(0, t)$. Let $F(s, t)$ be the generating function of the number of objects at time t, defining $F(s, t) = 1$ for $t < 0$. Let $J(t) = 1$ or 0 according as $t \geqq 0$ or $t < 0$.

If the initial object has ν children, born at times $t_1 < t_2 < \cdots < t_\nu$, then the generating function at time t for the family of the i-th child,

$i \leqq \nu$, is $F(s, t-t_i)$; the generating function for the initial object, if l is fixed, is $J(t-\ell)+[1-J(t-\ell)]s$. Hence we have, putting $J(t-\ell)=J$,

$$\left.\begin{aligned} F(s,t) &= \mathscr{E}\{F(s,t-t_1)\dots F(s,t-t_\nu)\,[J+(1-J)s]\} \\ &= \mathscr{E}\left\{e^{\int_0^t \log F(s,t-\tau)\,dN(\tau)}[J+(1-J)s]\right\}. \end{aligned}\right\} \tag{29.1}$$

If we differentiate (29.1) at $s=1$ and put $M(t)=\partial F(1,t)/\partial s=\mathscr{E}Z(t)$, we obtain

$$\left.\begin{aligned} M(t) &= \mathscr{E}\left\{1-J(t-\ell)+\int_0^t M(t-\tau)\,dN(\tau)\right\} \\ &= e^{-\int_0^t \mu(x)\,dx}+\mathscr{E}\int_0^t M(t-\tau)\,dN(\tau) \\ &= e^{-\int_0^t \mu(x)\,dx}+\int_0^t M(t-\tau)\,d\mathscr{E}N(\tau). \end{aligned}\right\} \tag{29.2}$$

In order to evaluate the right side of (29.2), note that the probability that a birth occurs to the initial object in the time interval $(\tau, \tau+d\tau)$ is the probability that the object survives to τ times the probability that a birth occurs if the object is alive at τ, i.e.,

$$e^{-\int_0^\tau \mu(x)\,dx}\lambda(\tau)\,d\tau. \tag{29.3}$$

Hence we replace $d\mathscr{E}N(\tau)$ in (29.2) by the expression in (29.3), obtaining

$$M(t)=e^{-\int_0^t \mu(x)\,dx}+\int_0^t M(t-\tau)\,e^{-\int_0^\tau \mu(x)\,dx}\lambda(\tau)\,d\tau. \tag{29.4}$$

This is an equation of the renewal type, to which the results of the Appendix can be applied provided $\int_0^\infty e^{-\int_0^t \mu(x)\,dx}\lambda(t)\,dt<\infty$.

Let $m^{(2)}(t)=\mathscr{E}\{Z(t)(Z(t)-1)\}$. A second differentiation of (29.1), followed by arguments similar to those above but somewhat more involved, yields the following equation for $m^{(2)}$. Note that it is also of the renewal type, if we assume M to be known:

$$\begin{aligned} m^{(2)}(t) &= \int_0^t m^{(2)}(t-\tau)\,e^{-\int_0^\tau \mu(u)\,du}\lambda(\tau)\,d\tau \\ &\quad + 2e^{-\int \mu(u)\,du}\int_0^t M(t-u)\,\lambda(u)\,du \\ &\quad + 2\iint\limits_{0\leqq\sigma\leqq\tau\leqq t} M(t-\tau)\,M(t-\sigma)\,e^{-\int_0^\tau \mu(u)\,du}\lambda(\tau)\,\lambda(\sigma)\,d\tau\,d\sigma. \end{aligned}$$

Kendall (1949) treated not only the total number of objects but also the first and second moments of the age distribution. For the case in which λ and μ are independent of age, KENDALL (1949, 1952) has found explicitly the moment-generating functional $\Phi(s,t)=\mathscr{E} e^{-\int s(x)\,dZ(x,t)}$, where $Z(x,t)$ is the number of objects of age $\leqq x$ at time t.

Appendix

Results on the renewal equation. In this appendix we collect certain known results about the integral equation of renewal theory.

Let G be a distribution function on $(0, \infty)$. We shall assume $G(0)=G(0+)=0$. Consider the equation

$$K(t)=f(t)+m\int_0^t K(t-u)\,dG(u), \tag{1}$$

where K is the unknown function, m is a positive constant, and f is a known Borel measurable function that is bounded on every finite interval. We adopt the convention that G is continuous to the right. The integral in (1) is interpreted as $\int_0^{t+0}$ and is a Lebesgue-Stieltjes integral. The total variation of any function g on the interval $[t_1, t_2]$ is denoted by $\int_{t_1}^{t_2}|dg(t)|$. The convolutions of G are defined by $G_1=G$, $G_n(t)=\int_0^t G_{n-1}(t-u)\,dG(u)$. The function $H(m,t)$, which we shall see below is finite, is defined by

$$H(m,t)=\sum_{n=1}^{\infty} m^{n-1} G_n(t). \tag{2}$$

Lemma 1. *Equation* (1) *has one and only one solution that is bounded on every finite interval. This solution has the form*

$$K(t)=f(t)+m\int_0^t f(t-u)\,dH(m,u). \tag{3}$$

In particular, if $f(t)=G(t)$, *then* $K(t)=H(m,t)<\infty$.

This result is essentially due to FELLER (1941); it is not necessary to require that f be nonnegative.

Now define the real constant α, when it exists, by

$$m\int_0^{\infty} e^{-\alpha t}\,dG(t)=1. \tag{4}$$

There is never more than one real α satisfying (4). If $m>1$, there is a positive root; if $m=1$, the root is 0; if $m<1$, there may or may not be a real root, and if there is, it is negative.

Lemma 2. *Suppose that G is not a lattice distribution and that α exists. Suppose also that either of the two following conditions holds:* (a) *$f(t)e^{-\alpha t}$ is the difference of two nonincreasing functions, each of which is integrable on $(0,\infty)$, and $\int_0^{\infty} te^{-\alpha t}\,dG(t)<\infty$; or* (b) *$f(t)e^{-\alpha t}$ approaches 0 as $t\to\infty$ and is bounded and integrable on $(0,\infty)$, and $\int_0^{\infty} t^2 e^{-\alpha t}\,dG(t)<\infty$. Then $K(t)\sim n_f e^{\alpha t}$ as $t\to\infty$, where*

$$n_f=\frac{\int_0^{\infty} f(t)\,e^{-\alpha t}\,dt}{m\int_0^{\infty} te^{-\alpha t}\,dG(t)}. \tag{5}$$

(Here n_f is not the same as in Sec. 25.)

Proof. Multiply both sides of (3) by $e^{-\alpha t}$ and let $t \to \infty$. Since $f(t)e^{-\alpha t} \to 0$, the limit of the right side, if it exists, is (*) $\lim \int_0^t f(t-u)\, e^{-\alpha(t-u)}\, dH_1(m, u)$, where $dH_1(m, u) = m e^{-\alpha u}\, dH(m, u)$. Notice that if we define a distribution $\overline{G}(t)$, $t \geqq 0$, by $d\overline{G}(t) = m e^{-\alpha t}\, dG(t)$, then the n-th convolution of $\overline{G}$ has the differential $d\overline{G}_n(t) = m^n e^{-\alpha t}\, dG_n(t)$. This follows immediately by comparison of the Laplace transforms. Now suppose (a) above is satisfied. Referring to SMITH (1958), let $\overline{G}$ correspond to the distribution of SMITH's random variable X_1, let $f(t)\, e^{-\alpha t}$ correspond to SMITH's $Q(t)$, and let H_1 correspond to SMITH's H. Then, using SMITH's formula (1.3), we see that the limit (*) exists and is equal to n_f as given by (5). If (b) rather than (a) is satisfied, we can use SMITH's formula (1.4) to evaluate the limit (*). $\square$

An analogous result holds if G is a lattice distribution.

Lemma 3. *Suppose that $m > 1$ and that G has a density g satisfying $\int_0^\infty (g(t))^p\, dt < \infty$ for some $p > 1$. Suppose that f is continuous and $\int_0^\infty |df(t)| < \infty$. Then*

$$K(t) = n_f e^{\alpha t}\, [1 + O(e^{-\varepsilon t})] \tag{6}$$

for some $\varepsilon > 0$ (which may depend on f), where n_f is defined by (5).

We sketch the proof of (6). Let $\tilde{K}(s) = \int_0^\infty e^{-st} K(t)\, dt$, $\tilde{f}(s) = \int_0^\infty f(t)\, e^{-st}\, dt$, $\tilde{g}(s) = \int_0^\infty e^{-st} g(t)\, dt$. The conditions on f imply that $\tilde{f}(s)$ is analytic for $Re(s) > 0$. Moreover, $K(t) = O(e^{\alpha t})$ so that $\tilde{K}(s)$ is analytic for $Re(s) > \alpha$. Hence we may take the Laplace transform of both sides of (1), obtaining

$$\tilde{K}(s) = \frac{\tilde{f}(s)}{1 - m\tilde{g}(s)}, \qquad Re(s) > \alpha.$$

Moreover, the continuity of $G(t)$ implies that $H(m, t)$ is continuous in t. Hence from (3) we see that K is a continuous function of bounded variation on every finite interval. Using the inversion formula for the Laplace transform, we obtain

$$K(t) = \lim_{T \to \infty} \frac{1}{2\pi} \int_{-T}^{T} e^{(x+iy)t}\, \frac{\tilde{f}(x+iy)}{1 - m\tilde{g}(x+iy)}\, dy, \qquad x > \alpha. \tag{7}$$

The denominator in (7) has a simple zero when $x + iy = \alpha$, but has no other zeros in a sufficiently narrow strip $|x - \alpha| < \delta$. We take $\alpha - \delta > 0$ so that $\tilde{f}$ and $\tilde{g}$ are both defined in the strip. From the theory of residues we can write

$$K(t) = n_f e^{\alpha t} + \lim_{T \to \infty} \frac{1}{2\pi} \int_{-T}^{T} \frac{e^{(x_1+iy)t}\, \tilde{f}(x_1 + iy)\, dy}{1 - m\tilde{g}(x_1 + iy)}, \tag{8}$$

for some $x_1 < \alpha$ with $|x_1 - \alpha| < \delta$, where n_f is given by (5). We can write the integral in (8) as

$$\frac{1}{2\pi} \int_{-T}^{T} e^{(x_1+iy)t}\, \tilde{f}(x_1 + iy)\, dy + \frac{m}{2\pi} \int_{-T}^{T} \frac{e^{(x_1+iy)t}\, \tilde{g}(x_1 + iy)\, \tilde{f}(x_1 + iy)\, dy}{1 - m\tilde{g}(x_1 + iy)}. \tag{9}$$

The first term in (9) approaches $f(t)$ as $T \to \infty$. The second term is bounded by

$$\text{Constant} \cdot e^{x_1 t} \int_{-T}^{T} |\tilde{g}(x_1 + iy)\, \tilde{f}(x_1 + iy)|\, dy, \tag{10}$$

since $|1-m\tilde{g}(x_1+iy)|$ has a positive lower bound as y varies. Since $\int_0^\infty |df| < \infty$, we have $|\tilde{f}| = O\left(\frac{1}{1+|y|}\right)$, $|y| \to \infty$.

If $\int_0^\infty (g)^p\, dt < \infty$ is true for some $p > 1$, it is also true for some p, $1 < p \leqq 2$, since $\int_0^\infty g(t)\, dt = 1$. (This remark should have been inserted on p. 285 of BELLMAN and HARRIS (1952).) Hence assume $1 < p \leqq 2$, and define p' by

$$\frac{1}{p} + \frac{1}{p'} = 1. \tag{11}$$

Then $\int_{-\infty}^\infty |\tilde{g}(x_1+iy)|^{p'} dy \leqq C_p \left[\int_0^\infty e^{-px_1 t}(g(t))^p dt\right]^{\frac{1}{p-1}}$, where C_p depends only on p. (See TITCHMARSH (1937, 4.1.2); the integral on the right should be raised to the power $1/(p-1)$.) Using HÖLDER's inequality, we see that the integral in (10) is bounded by

$$\left(\int_{-\infty}^\infty |\tilde{g}|^{p'} dy\right)^{1/p'} \left(\int_{-\infty}^\infty |\tilde{f}|^p dy\right)^{1/p} < \infty.$$

Thus the remainder in (8) is $O(e^{x_1 t})$, $x_1 < \alpha$. □

Lemma 4. *Suppose $m < 1$ and $\lim_{t\to\infty} f(t) = c$. (G may or may not be a lattice distribution.) Then $K(t) \to c/(1-m)$, and*

$$\left.\begin{aligned} \left|K(t) - \frac{c}{1-m}\right| &\leqq \frac{1}{1-m} \sup_{y \geqq t/2} \left|f(y) - c - \frac{mc}{1-m}(1-G(y))\right| \\ &+ m\left[H(m,t) - H\left(m, \frac{t}{2}\right)\right] \sup_{y\geqq 0} \left|f(y) - c - \frac{mc}{1-m}(1-G(y))\right|. \end{aligned}\right\} \tag{12}$$

Proof of Lemma 4. We see from (2) that if $m < 1$, then $\lim_{t\to\infty} H(m,t) = 1/(1-m)$. Now put $K^*(t) = K(t) - c/(1-m)$ and $f^*(t) = f(t) - c - \frac{mc}{1-m}(1-G(t))$. Then (1) is satisfied if we replace K and f by K^* and f^*. From (3) the solution is $K^*(t) = f^*(t) + m\int_0^t f^*(t-u)\, dH(m,u)$. If we write the last integral as $\int_0^{t/2} + \int_{t/2}^t$ we obtain (12). □

Lemma 5. *Suppose the conditions of Theorem 18.1 hold. If $\overline{H}(\overline{m}, t)$ and $\overline{G}(t)$ are defined as in (18.3), then $\overline{H}(\overline{m}, t) \uparrow (1-\overline{m})^{-1}$ and $(1-\overline{m})^{-1} - \overline{H}(\overline{m}, t) \to 0$ exponentially as $t \to \infty$.*

Proof. Since $\overline{H}$ satisfies the equation $\overline{H}(\overline{m}, t) = \overline{G}(t) + \overline{m}\int_0^t \overline{H}(\overline{m}, t-u)\, d\overline{G}(u)$, we have, putting $u(t) = (1-\overline{m})^{-1} - \overline{H}(\overline{m}, t)$,

$$u(t) = \frac{1-\overline{G}(t)}{1-\overline{m}} + \overline{m}\int_0^t u(t-y)\, d\overline{G}(y). \tag{13}$$

From the definition of $\overline{G}$, we see that there is a positive number δ such that $\int_0^\infty e^{\delta y}\, d\overline{G}(y) < \infty$. Putting $dL(y) = e^{\delta y}\, d\overline{G}(y) \Big/ \int_0^\infty e^{\delta y}\, d\overline{G}(y)$, we can write (13) as

$$v(t) = \left(\frac{1-\overline{G}(t)}{1-\overline{m}}\right) e^{\delta t} + m' \int_0^t v(t-y)\, dL(y), \tag{14}$$

where $v(t) = u(t)\, e^{\delta t}$ and $m' = \overline{m}\int_0^\infty e^{\delta y}\, d\overline{G}(y)$. If δ is small enough then $m' < 1$. Hence, applying Lemma 4 to (14), and noting that $(1-\overline{G}(t))\, e^{\delta t} \to 0$ if δ is small enough, we see that $v(t) \to 0$ if δ is small. This gives the desired result. □

Chapter VII

Branching processes in the theory of cosmic rays (electron-photon cascades)

1. Introduction

The atmosphere of the earth is continually penetrated by *cosmic rays*, radiation of exceedingly high energy originating chiefly outside the solar system. It has been established experimentally that the rays reaching the top of the earth's atmosphere, the *primary rays*, consist mainly of protons, with some helium nuclei and smaller amounts of nuclei of heavier elements. When these rays pass through the atmosphere, they collide with atmospheric atoms, initiating very complicated events that result in the production of showers of additional protons, as well as neutrons, electrons, photons (gamma rays), and particles called *mesons*, of which there are several types.

If the number of cosmic-ray particles is measured as a function of height above the ground, it is observed that there is a maximum at a height of about 16 kilometers, with lesser amounts above and below. Early researchers felt the need for a theory to explain this phenomenon.

Independently, and almost simultaneously, Bhabha and Heitler (1937) and Carlson and Oppenheimer (1937) formulated a cascade theory for the electrons and photons in cosmic-ray showers. (The electrons and photons together are sometimes called the *soft component*.) According to the theory, a downward-moving energetic photon, high in the atmosphere, will disappear after traveling a random distance, being replaced by two electrons, one of the ordinary sort with a negative charge, the other positively charged, that is, a "positron". The energy of the photon is shared in a random fashion between the two electrons. Each electron proceeds downward, continually radiating photons, and losing energy to them, in a process called *Bremsstrahlung* (literally, "braking radiation"). The electrons also lose energy in other ways. The new photons, unless their energies are too small, produce additional electrons in the same manner as the original one, and the process continues as a branching process.

This theory, which will be discussed in detail below, gives a satisfactory explanation of the behavior of actual electron-photon cascades under some experimental conditions, for example, when a very energetic electron initiates a shower in lead. The multiplication of electrons and photons postulated by the mathematical model explains the initial increase of the number of particles as the depth increases, while the eventual decrease comes because the energy per particle is eventually so small that particles are not registered on a detection device.

It was once thought that the theory explained the intensity pattern of electrons and photons in the atmosphere. However, it is now realized that the situation is more complicated than was once believed, since showers are initiated at different heights by photons with different energies. (See the paper by PUPPI and DALLAPORTA in WILSON (1952), particularly pp. 350–351.) The theory may still be considered of basic importance, although, according to RAMAKRISHNAN (1962, p. 425), "the literature ... has grown out of proportion to ... [its] use ... in explaining the increasing mass of experimental data".

Another component of cosmic radiation, which has received extensive treatment but will not be discussed here, is the *nucleon cascade* (a nucleon is a proton or a neutron). When a proton of the primary radiation strikes a nucleus, it knocks loose one or more nucleons, as well as mesons that have no part in the production of further nucleons. The new nucleons then produce further nucleons by a similar process, and so on. For a discussion of nucleon cascades, we refer the reader to the article by MESSEL in WILSON (1954); see also Sec. III.8.1.

In both the electron-photon cascade and the nucleon cascade we study random functions such as $N(E, t)$, the number of particles of energy greater than E that have traversed a distance or thickness t. We shall treat t just as if it were a time parameter. Physically this seems justified because we assume that the particles all move in one direction, without backward or even lateral motion. (A small proportion of the literature does consider lateral motion.) Moreover, the photons all move with the same speed, i.e., the speed of light, and the electrons whose energies are in the range of interest to us move almost with the speed of light, so that the particles in a shower reach a given distance from the point of origin almost simultaneously. (See ROSSI (1959).)

If E is fixed, the random function $N(E, t)$ is not a Markov process, and this causes difficulties in the treatment. Many of the early theoretical studies avoided such difficulties by using various Markov processes as approximations. The most extensive treatment of this kind was given by ARLEY (1943). ARLEY allowed for the loss of energy of the particles in a cascade by using Markov processes with transition probabilities that depend on time. His mathematical model for a cascade with one type of particle is a birth-and-death process of the sort discussed in Sec. V.7 with $\lambda(t)$ constant but $\mu(t)$ proportional to t. The mean number of particles in such a cascade rises to a maximum and then decreases. ARLEY then extends this idea to deal with two kinds of particles.

Since the publication of ARLEY's book, the tendency has been toward studying the original cascade model rather than Markov-chain approximations. Analytic and numerical results have been obtained for the first and second moments, and, to a much lesser extent, for the probability distributions involved. Numerous functional equations have been developed for the moments, probabilities, or probability-generating functions. However, except for the first and second moments, and the probabilities for showers of small size, the computational difficulties have not been overcome, and only a few limit theorems have been obtained (see Secs. 14.1 and 16).

In the present chapter no attempt will be made to summarize all the lines of attack that have been proposed for electron-photon cascades. Rather, we shall try to obtain a careful formulation, from the point of view of probability theory, of Approximation A, one of the most important models. After showing how one kind (out of a number of possible kinds) of functional equation can be derived for the generating function of $N(E, t)$, the number of electrons of energy $>E$ at thickness t (Sec. 11), we derive familiar equations for the expected values (Sec. 12). Next (Secs. 13 and 14), we define a certain Markov process, the *expectation process*, which is useful for deriving and interpreting some of the properties of the expected values. In Sec. 15 we study the total energy in the electrons. Sec. 16 has some results and conjectures about the asymptotic distributions for large thicknesses. In Sec. 17 a representation is given for the energy of a single electron under Approximation B, which is one step more complex than Approximation A. In Sec. 18 we deal very briefly with electron-photon cascades under Approximation B.

Surveys and bibliographies of mathematical work on the problem have been given by RAMAKRISHNAN and MATHEWS (1954), BHARUCHA-REID (1960, Chapter 5), and RAMAKRISHNAN (1962, Chapter XIII). On the physical side, a very readable account of the present state of knowledge about cosmic rays is contained in the article by K. GREISEN in recent (e.g., 1958) editions of the *Encyclopedia Britannica*. A popular account of some recent discoveries is given by ROSSI (1959). More technical accounts of recent developments are given in the series of books *Progress in Cosmic Ray Physics* (1952, 1954, 1956), edited by J. G. WILSON, and *Progress in Elementary Particle and Cosmic Ray Physics* (1958), edited by J. G. WILSON and S. A. WOUTHUYSEN. The brief account of cascade theory in HEISENBERG (1943) is useful.

2. Assumptions concerning the electron-photon cascade

We envisage the particles in a cascade as moving in one direction, always in the same sense. When one particle generates another, the second particle continues in the same direction as the parent particle. The number of particles that eventually reach a given distance t from the origin of the shower will be called *the number of particles at the thickness t*, and we shall treat t exactly as a time parameter. It will also be convenient to use words such as "before" and "after" when we refer to events ordered by t.

The word *particle* will refer to either a photon or an electron, and the word *electron* will refer to either a negative electron or a positive electron (positron).

We shall first describe informally what is usually called Approximation A for the electron-photon cascade (JÁNOSSY (1950a)). A precise mathematical description of the process will be given in Sec. 7.

2.1. Approximation *A*. A photon of energy E, moving through a thickness dt of homogeneous material, has a probability $\lambda dt + o(dt)$ of being transformed into two electrons, one positive and one negative; here λ is a positive constant depending on the nature of the material but, under our assumptions, not depending on E or t. The thickness ℓ traveled by a photon before it is transformed thus has the probability density function $\lambda e^{-\lambda\ell}$. The transformation of a photon into two electrons is called *pair production*.

The energy of a photon does not change during its lifetime. When it is finally transformed, its energy is shared by the two new electrons. The positive electron receives the energy Eu, where u has the probability density $q(u)$, and the negative electron receives the energy $E(1-u)$; the function q is assumed not to depend on E or t.

An electron (positive or negative) always preserves its identity but radiates photons as it travels, in a process called *Bremsstrahlung*. (See Sec. 1 above.) As usually described (neglecting higher infinitesimals), an electron of energy E has a probability $k(u)\,du\,dt$ of emitting a photon of energy between Eu and $Eu + E\,du$ in the thickness interval dt, where $k(u)$ is a function that is assumed not to depend on E or t. When an emission occurs, the energy that goes to the photon is subtracted from the energy of the electron.

The photons radiated by an electron undergo pair production, and the resulting electrons radiate photons, etc., resulting in a branching process.

Since k and q do not depend on E or t, the proportion of its energy that a particle loses in any interval does not depend on the energy at the beginning of the interval, nor on the location of the interval.

2.2. Approximation *B*. This is different from Approximation A only in the following respect: An electron having a positive energy loses the nonrandom amount of energy $\beta\,dt$, β constant, in the thickness interval dt, by a process called *ionization loss*. In addition, the electron still loses energy by Bremsstrahlung. Thus the energy of an electron will eventually reach 0, and at this time it is considered to be removed from the cascade.

When an electron has a high energy, its energy loss by ionization is small compared with the loss from Bremsstrahlung, which is proportional to the energy. Hence Approximation A is suitable for high energies.

We shall deal chiefly with Approximation A, giving a few results for Approximation B. Even the latter is far simpler than the real situation (see JÁNOSSY (1950a)), but the two approximations are considered reasonable for high and moderate energies, respectively.

3. Mathematical assumptions about the functions q and k

We shall impose the following conditions on the functions q and k of Sec. 2. The degree of generality is sufficient to cover the assumptions usually made by physicists.

Conditions 3.1. The function q is symmetric about $\frac{1}{2}$, vanishes outside the interval $[0, 1]$, and has continuous first and second derivatives in $[0, 1]$; at 0 or 1 this refers to the right-hand and left-hand derivatives, respectively. Of course, $q \geqq 0$ and $\int_0^1 q(u)\, du = 1$.

Conditions 3.2. The function k has the form

$$k(u) = \frac{\mu}{u} + k_0(u), \qquad 0 < u \leqq 1, \tag{3.1}$$

where μ is a nonnegative constant, which will be assumed to be strictly positive unless the contrary is stated; k_0 is continuous on $[0, 1]$; and k_0 has a continuous derivative on $(0, 1)$ satisfying

$$|k_0'(u)| \leqq a(1-u)^{-b} \tag{3.2}$$

for some $a > 0$, $0 \leqq b < 2$. The function k_0 may have negative values, although, of course, k does not.

We allow an unbounded derivative for k_0 in order to cover the form $k(u) = -1/\log(1-u)$, which has been used by some authors because it simplifies the treatment. (See Example 4.1.)

We shall next give the particular forms usually assumed by physicists for k and q, followed by some discussion of their validity and meaning.

3.1. Numerical values for k, q, and λ; units. The thickness t is usually measured in terms of a *cascade unit*, which varies from one material to another. In air it is about 0.3 kilometers at sea level (references vary) and is greater at higher altitudes; in lead it is about $\frac{1}{2}$ centimeter. The unit of energy is the *electron volt*, or more usually a million electron volts (MEV). The constant β (Sec. 2.2) is about 100 MEV per unit of length for air and about 7 for lead. According to ROSSI (1959), a few primaries with energies as high as 10^{18} electron volts have been detected indirectly.

If t is measured in cascade units, then k, q, and λ may be approximated for various different materials by the forms

$$\left.\begin{aligned} k(u) &= \frac{1.36}{u} - 1.36 + u\,, \\ q(u) &= \frac{1-1.36\,(u-u^2)}{0.77333\ldots}\,, \\ \lambda &= 0.77333\ldots. \end{aligned}\right\} \tag{3.3}$$

These forms may be found in JÁNOSSY (1950a, p. 206) or RAMAKRISHNAN (1962, p. 432). They come from expressions deduced by BETHE and HEITLER (1934).

The expected energy lost by an electron of energy E in dt from Bremsstrahlung is $E\left(\int_0^1 u\,k(u)\,du\right)dt$ and the amount lost by ionization is $\beta\,dt$. Since $\int_0^1 u\,k(u)\,du$ is nearly 1, we see that Approximation A can be valid only if E/β is large. In fact E/β is very large for many of the electrons observed in cascades.

3.2. Discussion of the cross sections. The functions k and q are known as the *cross sections* for Bremsstrahlung and pair production, respectively. It must be stressed that they are only approximations to the real cross sections, which are more complicated. In particular, the assumption that k and q do not depend on the absolute energy level E is not strictly true.

From the mathematical point of view, the most interesting property of the form (3.3) usually assumed for k is that $\int_0^1 k(u)\,du = \infty$. We call this the case of an *infinite cross section*, corresponding to a strictly positive value of μ in (3.1). In the contrary case, where $\mu = 0$, we shall speak of a *finite cross section*.

We shall see that the assumption $\mu > 0$ corresponds to a cascade in which every electron produces infinitely many photons in every interval of thickness. Physicists have called this situation the *infrared catastrophe*. Actually, although many photons of low energy are produced, the number is not infinite. In fact, TER-MIKAELYAN (1955) has shown that if the effect of the medium is properly accounted for, then under certain conditions the expression for $k(u)$ in (3.3) should be multiplied by $(1+\omega^2/u^2)^{-1}$, where ω is a constant depending on the material; it is about 1.9×10^{-4} for lead and 7.5×10^{-5} for air. Numerical computations by TIMOFEEV (1961) indicate that introduction of TER-MIKAELYAN'S factor makes little practical difference. However, it eliminates the infrared catastrophe, since the new expression for k behaves like $1.36\,u/\omega^2$ when u is small compared with ω.

It appears to the author that there is some interest in studying a branching process in which infinitely many objects can exist. Accordingly, we shall follow the usual practice and assume $\mu>0$ in (3.1) unless the contrary is stated. However, the theorems in this chapter will remain true if $\mu=0$, unless modifications are explicitly stated. Statements other than theorems will either remain true when $\mu=0$ or will be true with obvious modifications unless the contrary is stated. The main complication introduced by the presence of an infinite cross section is in the derivation of the basic equations (11.14). Accordingly, we shall point out in Sec. 9.3 how the derivation may be carried out more simply in the case of a finite cross section.

4. The energy of a single electron (Approximation *A*)

We next give a mathematical description of the manner in which a single electron loses energy.

Definitions 4.1. Let $\varepsilon(t)$ be the energy at thickness t of an electron having energy 1 when $t=0$. Let $X(t)=-\log\varepsilon(t)$, where log means the natural logarithm.

Consideration of Sec. 2.1 suggests that $\varepsilon(t)$ and $X(t)$ should have the following properties.

$\varepsilon(t)$

(a) If $t_1<t_2<\cdots<t_n$, the ratios $\varepsilon(t_2)/\varepsilon(t_1)$, $\varepsilon(t_3)/\varepsilon(t_2)$, ..., $\varepsilon(t_n)/\varepsilon(t_{n-1})$ are independent random variables.

(b) The distribution of the ratio $\varepsilon(t+h)/\varepsilon(t)$, $h>0$, is independent of t.

(c) The random function $\varepsilon(t)$ changes only by negative jumps as t varies, each jump representing a loss of energy by the electron to the photon that is emitted when the jump occurs.

(d) The expected number of jumps in the interval $t_1\leqq t\leqq t_2$ at which ε is multiplied by a factor between $1-u_1$ and $1-u_2$ is

$$(t_2-t_1)\int_{u_1}^{u_2} k(u)\,du\,,\quad 0<u_1<u_2<1\,.$$

$X(t)$

(a) The increments $X(t_2)-X(t_1)$, ..., $X(t_n)-X(t_{n-1})$ are independent.

(b) The distribution of the increment $X(t+h)-X(t)$ is independent of t.

(c) $X(t)$ changes only by positive jumps.

(d) The expected number of jumps in $t_1\leqq t\leqq t_2$ of magnitude between x_1 and x_2 is

$$(t_2-t_1)\int_{x_1}^{x_2} k(1-e^{-x})\,e^{-x}\,dx\,,\quad 0<x_1<x_2<\infty\,.$$

Properties (a) and (b) imply that $X(t)$ is a *temporally homogeneous additive process* (process with stationary and independent increments), and that the probability distribution of $X(t)$ is one of the *infinitely divisible* laws. (See, e.g., LÉVY (1948, Chapter V, especially pp. 154ff.); DOOB (1953, Chapter VIII, especially pp. 423—424); GNEDENKO (1962, Chapter 9 and Chapter 10, Sec. 56).) Properties (c) and (d), in the light of these references, then determine the probability distribution of $X(t)$ through its characteristic function (or Fourier transform), and we have the following result[1].

Theorem 4.1. *Let the function $k(u)$ satisfy Conditions 3.2. Then the characteristic function of $X(t)=-\log\varepsilon(t)$ is given by*

$$\mathscr{E} e^{i\tau X(t)}=\theta_t(\tau)=\exp\left\{t\int_0^\infty (e^{i\tau x}-1)\,k(1-e^{-x})\,e^{-x}\,dx\right\}. \tag{4.1}$$

The mean and variance of $X(t)$ are finite and are given by

$$\left.\begin{aligned}\mathscr{E} X(t)&=t\int_0^\infty x\,k(1-e^{-x})\,e^{-x}\,dx,\\ \operatorname{Var} X(t)&=t\int_0^\infty x^2 k(1-e^{-x})\,e^{-x}\,dx.\end{aligned}\right\} \tag{4.2}$$

If $\mu>0$, then ε and X have, with probability 1, infinitely many jumps on every t-interval. We shall examine this jump behavior more closely in the next section.

The term *straggling* is sometimes used to denote, in a general way, the random character of $\varepsilon(t)$.

Example 4.1. Suppose $k(u)$ has the form $-1/\log(1-u)$, used by BHABHA and HEITLER (1937) in some of their calculations. Then $\theta_t(\tau)=(1-i\tau)^{-t}$, and, as these authors showed, $X(t)$ has the probability density function

$$h_t(x)=\frac{x^{t-1}e^{-x}}{\Gamma(t)}. \tag{4.3}$$

5. Explicit representation of $\epsilon(t)$ in terms of jumps

Having decided that $-\log\varepsilon(t)$ should be an additive process that changes only by jumps, with the characteristic function given by (4.1), we now change our point of view. We want to define our cascade process in terms of certain random variables that can be interpreted in terms of a family-tree structure, as in the case of age-dependent branching processes. We begin by defining the random function $\varepsilon(t)$ in terms of certain random variables $t_1, u_1, t_2, u_2, \ldots$. (Def. 5.2.) However, in order to motivate the definition, we shall first discuss informally some properties of the jumps in an additive random process.

---

[1] The distribution of $X(t)$ was found by BHABHA and HEITLER (1937) for the case of example 4.1, and for certain other forms of $k(u)$ by FURRY (1937).

For convenience and simplicity in the treatment, *we shall hereafter suppose that all values of t considered are less than some large positive number T.* The functional equations to be derived, involving the parameter t, will not depend on T provided $t<T$. There will be no difficulty in providing a rigorous interpretation of the meaning in those few instances where we shall allow t to go to ∞. *If we say that a function depending on t is bounded, or that some relation holds for all t, it is understood that this is for $0 \leqq t < T$.*

Definitions 5.1. Let T be a positive number that will remain fixed in the rest of the chapter. It is understood that $0 \leqq t < T$ always. Let V be the region $0<t<T$, $0<u<1$ in the Cartesian (t, u) plane.

A jump of ε can be represented by a random point (t', u') in V, where $1-u'$ is the ratio $\varepsilon(t'+0)/\varepsilon(t'-0)$; that is, ε loses the proportion u' at t'. If (t_1, u_1), (t_2, u_2), ... are the jumps of ε, where the system of enumeration will be discussed below, then the number of points (t_i, u_i) in a region A in V is known to have a Poisson distribution with mean $\iint_A k(u)\, du\, dt$, provided the integral is finite, and the numbers of points in disjoint regions are independent random variables (see, e.g., Lévy (1948, pp. 154ff.); or Doob (1953, pp. 423—424)). The total number of points (t_i, u_i) in V is infinite with probability 1 if $\mu>0$.

As the parameter u decreases from 1 to 0, the number of points (t_i, u_i) with $u_i > u$ is a random function of u of the generalized Poisson type; that is, the intensity for an "event" between u and $u+du$ (i.e., a point (t_i, u_i) with $u<u_i<u+du$) is $Tk(u)\,du$. Thus we can enumerate the jumps of ε in the order of decreasing u-coordinates: (t_1, u_1), (t_2, u_2), ..., with $1>u_1>u_2>\cdots>0$. From the Poissonian nature of the distribution of points (t_i, u_i) in V, we can determine the probability distribution of the t_i and the u_i. *The t's are uniformly and independently distributed on the interval $(0, T)$. The u's are independent of the t's; the conditional probability density of u_{n+1}, given $u_n, u_{n-1}, \ldots$, with $1>u_1>u_2>\cdots>u_n>0$, is*

$$\left.\begin{aligned} \varphi(u_{n+1}\,|\,u_n) &= T k(u_{n+1}) \exp\left[-T \int_{u_{n+1}}^{u_n} k(u)\, du\right], \\ 0 &< u_{n+1} < u_n;\quad n=0, 1, 2, \ldots, \end{aligned}\right\} \tag{5.1}$$

where for convenience we have put $u_0=1$. Formula (5.1) is simply the analogue, for a Poisson process with the variable rate $Tk(u)$, of the fact that in a Poisson process with a constant rate the times between events have an exponential distribution.

The divergence of the integral $\int_0^1 k(u)\, du$ implies that $\int_0^{u_n} \varphi(u\,|\,u_n)\, du = 1$. There is no difficulty in modifying the treatment in case $\mu=0$. We adopt the convention that if $u_n > 0$, then u_{n+1} has probability

$$1-\int_0^{u_n}\varphi(u|u_n)\,du$$

of being 0, $n=0, 1, 2, \ldots$, and if $u_n=0$ then $u_{n+1}=0$ with probability 1. It can then be seen that with probability 1 some one of the u_i and all its successors will be 0.

The following lemma will be useful. (It is true for $\mu \geqq 0$.)

Lemma 5.1. *If $u_1, u_2, \ldots$ have the probability densities (5.1), then*

$$\mathscr{E}(u_{n+1}|u_n, u_{n-1}, \ldots, u_1) \leqq \left(\frac{cT}{1+cT}\right)u_n, \qquad n=0, 1, \ldots, \tag{5.2}$$

where $c=\max\limits_{0\leqq x\leqq 1} x k(x)$.

Proof. If we are given $u_n=u$, $0<u\leqq 1$, then the conditional expectation of u_{n+1}/u can be written, after an integration by parts, as

$$1-\frac{1}{u}\int_0^u \exp\left\{-T\int_x^u k(y)\,dy\right\}dx. \tag{5.3}$$

Defining c as in the statement of the lemma, we have $\int_x^u k(y)\,dy \leqq c\log(u/x)$, whence the expression (5.3) has the upper bound $cT/(1+cT)$. ☐

As a corollary to (5.2) we see by induction on n that

$$\mathscr{E}u_n \leqq \left(\frac{cT}{1+cT}\right)^n, \qquad n=1, 2, \ldots, \tag{5.4}$$

and hence

$$\sum_{i=1}^{\infty}\mathscr{E}u_i = \mathscr{E}\sum_{i=1}^{\infty}u_i < \infty, \tag{5.5}$$

where we can interchange $\mathscr{E}$ and $\sum$ in (5.5) because the u_i are nonnegative. Hence the sum $\sum u_i$ converges with probability 1, and hence from (5.5) we see that with probability 1 we have

$$\prod_{i=1}^{\infty}(1-u_i) > 0. \tag{5.6}$$

We are now ready to define $\varepsilon(t)=\varepsilon(t; t_1, u_1, t_2, u_2, \ldots)$ in terms of the jumps (t_i, u_i). Let $t_1, t_2, \ldots$ be independent random variables uniformly distributed on $(0, T)$ and let $u_1, u_2, \ldots$ be random variables independent of the t's, having the probability distributions determined by (5.1). We shall also suppose that

$$1 > u_1 > u_2 > \cdots, \ \sum u_i < \infty; \tag{5.7}$$

$$t_i \neq t_j \quad \text{if} \quad i \neq j; \tag{5.8}$$

$$\text{there are infinitely many } t_i \text{ on every } t\text{-interval in } (0, T). \tag{5.9}$$

We can make these assumptions, since sequences not satisfying (5.7)—(5.9) have probability 0.

Definition 5.2. Define the random function $\varepsilon(t)$ by

$$\varepsilon(t)=\varepsilon(t;\ t_1, u_1, t_2, u_2, \ldots)=\prod_{i:t_i\leqq t}(1-u_i), \qquad 0\leqq t<T. \tag{5.10}$$

Note that $\varepsilon(0)=1$, since $t_i>0$ for each i.

We shall take it as a known consequence of the theory of additive processes that *if* $\varepsilon(t)$ *is defined by* (5.10) *then* $X(t)=-\log\varepsilon(t)$ *is an additive random process with the characteristic function* (4.1). We see from (5.6) that with probability 1, $\varepsilon(t)$ is positive for every t.

It is also a consequence of the definition of ε that with probability 1 the random function $\varepsilon(t)$ is continuous to the right for each t. Thus $\varepsilon(t)=\varepsilon(t+0)$, but $\varepsilon(t-0)$ will be greater than $\varepsilon(t)$ if t is a point of discontinuity.

5.1. Another expression for $\varepsilon(t)$. Let us look at the jump-like character of ε in more detail. If $w_1, w_2, \ldots, w_k$ are any numbers between 0 and 1, the following identity can be proved by mathematical induction:

$$\left.\begin{aligned}&(1-w_1)(1-w_2)\ldots(1-w_k)\\&=1-w_1-w_2(1-w_1)-w_3(1-w_1)(1-w_2)\\&\quad-\cdots-w_k(1-w_1)(1-w_2)\ldots(1-w_{k-1}).\end{aligned}\right\} \tag{5.11}$$

Let (t_1, u_1), (t_2, u_2), ... be a sequence with $0<u_1, u_2, u_3, \ldots<1$ and $\sum u_i<\infty$, the t_i being any real numbers, no two of which are the same. Notice that there need not be a smallest number among the t_i. For each positive integer k, let us rearrange the numbers u_i, $1\leqq i\leqq k$ in the order of increasing t_i, denoting the new ordering by $u_{k1}, u_{k2}, \ldots, u_{kk}$. If we now apply (5.11) with $w_i=u_{ki}$, and write the result in terms of $u_1, u_2, \ldots, u_k$, we have

$$\prod_{i=1}^{k}(1-u_i)=1-\sum_{i=1}^{k} v_i^k u_i, \tag{5.12}$$

where v_i^k is the product of all factors $(1-u_j)$ such that $1\leqq j\leqq k$ and $t_j<t_i$; take $v_i^k=1$ if there are no such factors.

If we let $k\to\infty$ in (5.12), we obtain the following result.

Lemma 5.2. *Let* (t_1, u_1), (t_2, u_2), (t_3, u_3), ... *be a sequence with* $0<u_1, u_2, \ldots<1$ *and* $\sum u_i<\infty$, *the* t_i *being any distinct real numbers. Then*

$$\prod_{i=1}^{\infty}(1-u_i)=1-\sum_{i=1}^{\infty} v_i u_i, \tag{5.13}$$

where v_i *is the product of all factors* $(1-u_j)$ *such that* $t_j<t_i$. *Take* $v_i=1$ *if there are no such factors.*

Now let us apply (5.13) to the expression for $\varepsilon(t)$ in (5.10), including in (5.13) only those u_i for which $t_i\leqq t$. Notice that v_i will then be $\varepsilon(t_i-0)$.

Hence we have

$$\varepsilon(t)=1-\sum_{t_i\leq t} u_i\,\varepsilon(t_i-0)\,, \tag{5.14}$$

where $u_i\,\varepsilon(t_i-0)$ represents the amount of energy lost at t_i.

6. Distribution of $X(t)=-\log\varepsilon(t)$ when t is small

For later use we must find the form of the distribution of $\varepsilon(t)$ or $X(t)$ when t is small. In the language of the theory of semigroups, we consider the semigroup of transformations U^t, where $U^t F(x)=\mathscr{E}F\bigl(x+X(t)\bigr)$. If t is small, we should expect to have $U^t F(x)=F(x)+t\,U_0F(x)+o(t)$, where U_0 is an operator called the *infinitesimal generator* of the semigroup U^t. The form of U_0 is known; see MASLOV and POVZNER (1958, Eq. (22)); however, the factor $y^2/(1+y^2)$ in (22) should be replaced by its reciprocal. The constant γ in (22), for our case, is such that there is no term involving $F'(x)$.

For our purposes it is convenient to give a direct derivation of the form of U_0 for certain functions F. This derivation will provide a needed bound for the term $o(t)$.

We begin by showing that $X(t)$ has a probability density $h_t(x)$ given by either of the expressions on the right side of (6.1) below[1]. (In case $\mu=0$, the necessary modifications are discussed in Remark 2 following Theorem 6.3.)

Theorem 6.1. *If $t>0$ then $X(t)$ has a probability density $h_t(x)$, continuous for $x>0$, satisfying*

$$\left.\begin{aligned} h_t(x)=\frac{1}{2\pi i x}\int_{-\infty}^{\infty} e^{-i\tau x}\left(\frac{d\theta_t(\tau)}{d\tau}\right)d\tau=\lim_{T_1\to\infty}\frac{1}{2\pi}\int_{-T_1}^{T_1} e^{-i\tau x}\,\theta_t(\tau)\,d\tau,\\ x>0,\end{aligned}\right\} \tag{6.1}$$

where the first integral in (6.1) *is absolutely convergent, and $\theta_t(\tau)$ is given by* (4.1).

Proof. Referring to (4.1), we may write

$$\left.\begin{aligned} \frac{\log\theta_t(\tau)}{t} &= \int_0^\infty e^{i\tau x}\left(k(1-e^{-x})-\frac{\mu}{x}\right)e^{-x}\,dx\\ &\quad-\int_0^\infty\left(k(1-e^{-x})-\frac{\mu}{x}\right)e^{-x}\,dx+\mu\int_0^\infty (e^{i\tau x}-1)\left(\frac{e^{-x}}{x}\right)dx. \end{aligned}\right\} \tag{6.2}$$

[1] The existence of a density for $X(t)$ follows from a result of TUCKER (1962) and also from a result of FISZ and VARADARAJAN (1963).

The last integral in (6.2) is $-\mu\log(1-i\tau)$. After integrating the first integral by parts we obtain

$$\frac{\log\theta_t(\tau)}{t}=-\mu\log(1-i\tau)-A+\frac{g(0)}{1-i\tau}+\frac{\xi(\tau)}{1-i\tau}, \tag{6.3}$$

where

$$\left.\begin{aligned}&g(x)=k(1-e^{-x})-\frac{\mu}{x}, \qquad \xi(\tau)=\int_0^\infty e^{i\tau x}e^{-x}g'(x)\,dx,\\ &A=\xi(0)+g(0)=\int_0^\infty e^{-x}g(x)\,dx.\end{aligned}\right\} \tag{6.4}$$

Conditions 3.2 insure that $\xi(\tau)$, $\xi'(\tau)$, and $\xi''(\tau)$ are bounded continuous functions of τ. We also note that $g(0)=\frac{1}{2}\mu+k_0(0)$.

From (6.3) we see that for each fixed $t>0$, we have $|\theta_t(\tau)|=O(|\tau|^{-\mu t})$ and $|d\theta_t(\tau)/d\tau|=O(|\tau|^{-\mu t-1})$ as $\tau\to\pm\infty$. Hence

$$\int_{-\infty}^{\infty}|d\theta_t(\tau)/d\tau|\,d\tau<\infty.$$

This implies that $X(t)$ has a probability density $h_t(x)$, continuous for $x>0$, given by the first integral on the right side of (6.1). (See, for example, the argument on p. 480 of HARRIS (1948).) If we write this integral as $\lim_{T_1\to\infty}\int_{-T_1}^{T_1}$, integrate by parts, and let $T_1\to\infty$, we obtain the second integral on the right side of (6.1). Thus the usual Fourier inversion formula (i.e., the second integral) holds even though we have not proved that $h_t(x)$ is of bounded variation in any x-interval. □

By using (6.1) and (6.3) we can determine the behavior of $h_t(x)$ and of expectations $\int_0^\infty h_t(x)F(x)\,dx$ for certain functions F when t is small. The results are given in the following two theorems. However, since the proofs are somewhat lengthy and involve numerous routine estimates, we defer them to Appendix 1.

Theorem 6.2. *The probability density $h_t(x)$ satisfies*

$$\left.\begin{aligned}h_t(x)=e^{-At-x}&\left(\frac{x^{\mu t-1}}{\Gamma(\mu t)}+\frac{t g(0)\,x^{\mu t}}{\Gamma(\mu t+1)}\right)\\ &+te^{-x}\big(g(x)-g(0)\big)+t^2R(t,x),\end{aligned}\right\} \tag{6.5}$$

where A and g are defined by (6.4), *and $R(t,x)$ and $\int_0^\infty|R(t,x)|\,dx$ are bounded for $0\leq t<T$ and $x\geq 0$.*

Theorem 6.3. *Let $F(x)$ be a bounded function such that*

$$\int_0^c|F(x)-F(0)|\cdot\left|\frac{\log x}{x}\right|dx<\infty$$

for any $c>0$, *and define* I_F *(which is then finite) by*

$$I_F=\sup_{x\geqq 0}|F(x)-F(0)| + \int_0^\infty x^{-1}|F(x)-F(0)|\,(1+x^{\mu T})\,(1+|\log x|)\,(1+x)\,e^{-x}\,dx.$$

Then

$$\left.\begin{aligned}&\int_0^\infty h_t(x)F(x)\,dx\\ &=F(0)+t\int_0^\infty (F(x)-F(0))\,k(1-e^{-x})\,e^{-x}\,dx+I_F t^2 R_1(t),\end{aligned}\right\}\qquad(6.6)$$

where $|R_1|$ *has a finite upper bound that is independent of the form of* F.

Remark 1. Formula (6.6) has the following interpretation. The energy of an electron decreases by jumps, and we may write $\varepsilon(t)=(1-U_1(t))\times(1-U_2(t))\ldots$, where $U_1(t), U_2(t), \ldots$ are the largest, next largest, ... of the u_i for which $0<t_i\leqq t$. (See Sec. 5.) Now from (5.1) $U_1(t)$ has the probability density

$$t\,k(u)\exp\left[-t\int_u^1 k(v)\,dv\right], \qquad (6.7)$$

and it can be verified that, neglecting terms in t^2, the expression (6.6) is the same as would be obtained by taking $X(t)=-\log(1-U_1(t))$. In other words, *the only jump that is important, if t is small, is the largest one* (where "large" is used in a proportional rather than in an absolute sense).

Remark 2. If $\mu=0$, then $X(t)$ has the probability

$$\exp\left(-t\int_0^1 k(u)\,du\right)=e^{-At}$$

of being 0, but its distribution has a density at all positive values when $t>0$; we still denote this density by $h_t(x)$. Theorems 6.1 to 6.3 require only minor modifications. The first expression for $h_t(x)$ in (6.1) remains true, while the second is true if $\theta_t(\tau)$ is replaced by $\theta_t(\tau)-e^{-At}$. In Theorem 6.2, we have the simpler expression

$$h_t(x)=t\,e^{-x}k(1-e^{-x})+t^2R_0(t,x), \qquad (6.8)$$

where R_0 has the properties describing the function R in Theorem 6.2. Theorem 6.3 remains true if we add $e^{-At}F(0)$ to the left side of (6.6). We can now replace I_F by $\sup_{x\geqq 0}|F(x)-F(0)|$, and we need require of F only that it be a bounded Borel-measurable function.

7. Definition of the electron-photon cascade and of the random variable $N(E,t)$ (Approximation A)

We are now ready to define the electron-photon cascade. We shall make the definition in detail only for the case in which the initial particle is an electron, the case of an initial photon being similar. Moreover, it suffices to consider an initial particle of energy 1, other cases involving only simple changes.

7.1. Indexing of the particles. The electrons and photons are indexed, somewhat as in the scheme of Chapter VI, by a symbol ι, which is 0 or a finite sequence of positive integers $i_1 j_1 i_2 j_2 \ldots i_k j_k$ or $i_1 j_1 \ldots i_k$. The i's may take any positive integer values, but each j is either 1 or 2. The object corresponding to the index ι is denoted by $\langle\iota\rangle$. The symbol $\langle 0\rangle$ denotes the original electron; next, $\langle 1\rangle, \langle 2\rangle, \ldots$ are the first, second, ... photons produced by $\langle 0\rangle$, where the terms "first", "second", etc., are explained below. The symbols $\langle i1\rangle$ and $\langle i2\rangle$ denote, respectively, the first and second electrons produced by $\langle i\rangle$ when it disappears. For definiteness we may suppose that the first electron is the positive one and the second is the negative one. Next, $\langle i_1 j_1 i_2\rangle$ denotes the i_2-th photon produced by $\langle i_1 j_1\rangle$. It should now be clear how the enumeration proceeds.

An electron $\langle i_1 j_1 \ldots i_k j_k\rangle$ is in the k-th electron generation; $\langle 0\rangle$ is the 0-th electron generation; a photon $\langle i_1 j_1 \ldots i_k\rangle$ is in the k-th photon generation.

In this section we agree that if $\langle\iota\rangle$ is a particle, then $\langle\iota'\rangle$ is the parent of $\langle\iota\rangle$. For example, if $\iota = i_1 j_1 i_2$, then $\iota' = i_1 j_1$.

A symbol such as ιi means the sequence ι with i added at the right end; if $\iota = 0$, then ιi is simply i.

7.2. Histories of lives and energies. The history of a cascade is given in terms of certain numbers and random functions, which we now list. (See Fig. 6.)

For each electron or photon $\langle\iota\rangle$ we put

τ_ι = time at which $\langle\iota\rangle$ is born;

$\varepsilon_{(\iota)}$ = energy of $\langle\iota\rangle$ at birth;

$\varepsilon_\iota(t)$ = energy of $\langle\iota\rangle$ at time t (write $\varepsilon(t)$ for $\varepsilon_0(t)$);

u_ι = fraction received by $\langle\iota\rangle$ of parental energy at time $\langle\iota\rangle$ is born; that is, $\varepsilon_{(\iota)} = \varepsilon_{\iota'}(\tau_\iota - 0)\, u_\iota$, where $\langle\iota'\rangle$ is the parent of $\langle\iota\rangle$.

In addition, if $\langle\iota\rangle$ is a photon, we let ℓ_ι denote its life-length.

We agree that $\varepsilon_\iota(t) = 0$ if $t < \tau_\iota$. In the case of a photon, $\varepsilon_\iota(t) = 0$ if $t \geqq \tau_\iota + \ell_\iota$.

We must also introduce numbers $t_{\iota 1}, t_{\iota 2}, \ldots$, defined whenever $\langle\iota\rangle$ is an electron. Here $\tau_\iota + t_{\iota i}$ represents the time of birth of the photon $\langle\iota i\rangle$.

The numbers t_ι, u_ι, and ℓ_ι are positive and are subject to the following additional conditions. For each pair of sister electrons, say $\langle\iota 1\rangle$ and $\langle\iota 2\rangle$, we have

$$0 < u_{\iota 1},\ u_{\iota 2} < 1, \quad \text{and} \quad u_{\iota 1} + u_{\iota 2} = 1. \tag{7.1}$$

Moreover, if $\langle\iota\rangle$ is an electron, then

$$0 < t_{\iota i} < T, \quad i = 1, 2, \ldots, \quad \text{and} \quad t_{\iota i} \neq t_{\iota j} \quad \text{if} \quad i \neq j, \tag{7.2}$$

and

$$\text{there are infinitely many } t_{\iota i} \text{ on every interval in } (0, T). \tag{7.3}$$

Furthermore, if $\langle\iota\rangle$ is an electron, then

$$1 > u_{\iota 1} > u_{\iota 2} > \cdots > 0 \quad \text{and} \quad \sum_{i=1}^{\infty} u_{\iota i} < \infty. \tag{7.4}$$

If we are given numbers t_ι, u_ι, and ℓ_ι, then we define the other quantities for the cascade in the following manner, remembering that $\langle\iota'\rangle$ is the parent of $\langle\iota\rangle$.

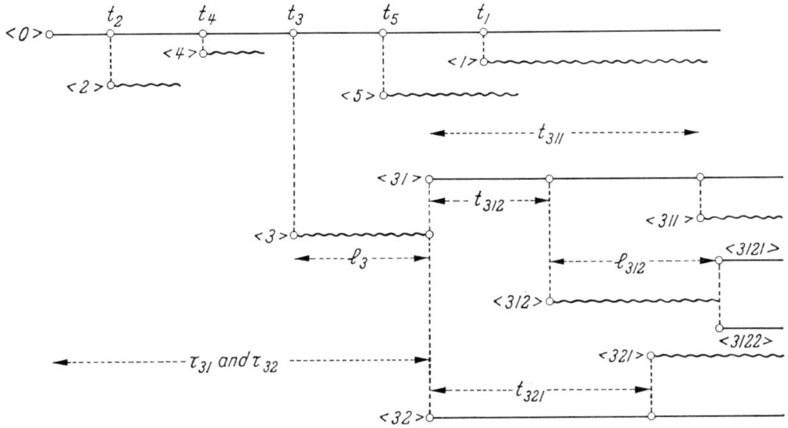

Fig. 6. In the figure wavy lines represent life spans of photons; solid lines, electrons. The photon $\langle 3\rangle$ is born at t_3 and lives until $t_3 + \ell_3$, when it is replaced by the electrons $\langle 31\rangle$ and $\langle 32\rangle$. The photon $\langle 312\rangle$ is born at $t_3 + \ell_3 + t_{312}$ and lives until $t_3 + \ell_3 + t_{312} + \ell_{312}$, etc.

For the electron $\langle\iota\rangle$ we have

$$\left.\begin{aligned} &\tau_\iota = \tau_{\iota'} + \ell_{\iota'}, \qquad \iota \neq 0; \qquad \tau_0 = 0; \\ &\varepsilon_{(\iota)} = u_\iota \varepsilon_{(\iota')}, \qquad \iota \neq 0; \qquad \varepsilon_{(0)} = 1; \\ &\varepsilon_\iota(t) = \varepsilon_{(\iota)} \varepsilon(t - \tau_\iota;\ t_{\iota 1}, u_{\iota 1}, t_{\iota 2}, u_{\iota 2}, \ldots), \\ &\quad t \geqq \tau_\iota, \end{aligned}\right\} \tag{7.5}$$

where ε is the function defined by (5.10).

For the photon $\langle\iota\rangle$ we have

$$\left.\begin{aligned} &\tau_\iota = \tau_{\iota'} + t_\iota, \\ &\varepsilon_{(\iota)} = u_\iota \varepsilon_{\iota'}(\tau_\iota - 0). \end{aligned}\right\} \tag{7.6}$$

7.3. Probabilities in the cascade; definition of Ω. We must now prescribe joint probability distributions for the numbers t_ι, u_ι, ℓ_ι. In this manner we determine a probability measure P on the space Ω of sequences $\omega=(t_1, u_1, \ell_1, u_{11}, \ldots)$, where the coordinates in ω are the t_ι, u_ι, and ℓ_ι, enumerated in some definite order. The coordinates are nonnegative, and we can assume that Ω contains only sequences satisfying (7.1)—(7.4), since all other sequences will have total probability 0.

We assume that the t's are uniformly distributed on $(0, T)$ and are mutually independent and independent of everything else. The ℓ's have the density $\lambda e^{-\lambda \ell}$, $\ell>0$, and have the same independence properties as the t's. The u's have the following dependency properties. If $\langle\iota\rangle$ is an electron, then $u_{\iota 1}, u_{\iota 2}, \ldots$ have the same joint probability law as do $u_1, u_2, \ldots$ in (5.1). If $\langle\iota\rangle$ is a photon, then $u_{\iota 1}+u_{\iota 2}=1$; $u_{\iota 1}$ has the probability density q, and from symmetry so does $u_{\iota 2}$ (see Conditions 3.1). Otherwise, the u's are independent of one another and of everything else.

In case $\mu=0$, then for each electron $\langle\iota\rangle$ we have $u_{\iota i}=0$ for all sufficiently large i, with probability 1. Thus many particles in the cascade are born with energy 0, but this causes no difficulty.

For the nucleon cascade, NEY (1961) has used a different scheme to establish a probability measure.

7.4. Definition of $N(E, t)$.

Definition 7.1. Let $N(E, t)=N(E, t, \omega)$ be the number of electrons whose energy at t is $>E$, where $E>0$.

Remark. Strictly speaking, we should define two functions in Def. 7.1, one corresponding to an initial electron and one to an initial photon. However, when confusion will not arise, we use the same notation N for each.

8. Conservation of energy (Approximation A)

In this section we assume that initially there is one electron with energy 1.

According to Approximation A (Sec. 2.1), loss of energy from a particle is accompanied by an equal transmission of energy to other particles. Accordingly, we should expect the total energy in all particles at any thickness t to equal 1, the original energy. It is not quite obvious that this is true, and the reader might for an exercise construct a cascade (i.e., a single ω) for which it is not true. FILIPPOV (1961), in discussing a related process with one type of particle, gives examples where (in our terminology) the form of k depends on E, and where there is a loss of total energy, although each individual transformation conserves it. However, in our case, such exceptions have total probability 0, as the following result shows.

Theorem 8.1. *Let $\varepsilon_\iota(t)$ be the energy of $\langle\iota\rangle$ at t, with $\varepsilon_0(t)=\varepsilon(t)$. Then with probability 1 we have*

$$\sum_\iota \varepsilon_\iota(t)=1, \quad \textit{all} \quad t. \tag{8.1}$$

Proof. Recall the definition of ω (Sec. 7.3). Let $\varepsilon_\iota^*(t)=\varepsilon_\iota^*(t,\omega)$ be defined as $\varepsilon_{(\iota)}$, the birth energy of $\langle\iota\rangle$, if $\langle\iota\rangle$ is born at or before t; otherwise $\varepsilon_\iota^*(t)=0$. The first step is to show that the relation

$$\varepsilon(t)+\sum_{i=1}^{\infty}\varepsilon_i^*(t)=1 \tag{8.2}$$

holds for every t and every ω. Now if $t_\iota>t$, then $\varepsilon_i^*(t)=0$, while if $t_i\leqq t$ then $\varepsilon_i^*(t)=\varepsilon_{(i)}=u_i\varepsilon(t_i-0;t_1,u_1,t_2,u_2,\ldots)$ from (7.6) and (7.5). Using (5.14), we then obtain (8.2).

Next, since the photon $\langle i\rangle$ shares its energy when it dies between the electrons $\langle i1\rangle$ and $\langle i2\rangle$, we have $\varepsilon_i^*(t)=\varepsilon_i(t)+\sum_{j=1}^{2}\varepsilon_{ij}^*(t)$. Then, arguing as for (8.2), we obtain $\varepsilon_{ij}^*(t)=\varepsilon_{ij}(t)+\sum_{r=1}^{\infty}\varepsilon_{ijr}^*(t)$. Continuing this process, we have the following identity holding for every $t\geqq 0$ and every ω in Ω:

$$\left.\begin{aligned}1&=\varepsilon(t)+\sum_{i_1}\varepsilon_{i_1}(t)+\sum_{i_1j_1}\varepsilon_{i_1j_1}(t)+\cdots\\&\quad+\sum_{i_1j_1\ldots j_{k-1}i_k}\varepsilon_{i_1j_1\ldots j_{k-1}i_k}(t)\\&\quad+\sum_{i_1j_1\ldots i_kj_k}\varepsilon^*_{i_1j_1\ldots i_kj_k}(t)\\&=R_k(t,\omega)+S_k(t,\omega),\end{aligned}\right\} \tag{8.3}$$

where $S_k(t)$ is the last summation in (8.3), and $R_k(t)$ is the energy at t of all photons in generations 1 through k of photons and of all electrons in generations 0 through $k-1$ of electrons (see Sec. 7.1).

Let us show that $\lim_{k\to\infty}\mathscr{E}S_k=0$. Putting $i_1j_1\ldots j_k=\iota$, we may write a typical term in $S_k(t)$ as $\varepsilon_{(\iota)}J(t-\tau_\iota)$, where $\varepsilon_{(\iota)}$ is the birth energy of $\langle\iota\rangle$, τ_ι is the time of birth of $\langle\iota\rangle$, and $J(x)=0$ for $x<0$ and 1 for $x\geqq 0$. Now $\tau_\iota>\ell_{i_1}+\ell_{i_1j_1i_2}+\ell_{i_1j_1i_2j_2i_3}+\cdots+\ell_{i_1j_1\ldots j_{k-1}i_k}$, the sum being distributed as the sum of k independent random variables, each having the density $\lambda e^{\lambda\ell}$. Hence

$$\left.\begin{aligned}P\{J(t-\tau_\iota)>0\}&\leqq\frac{\lambda^k}{(k-1)!}\int_0^t x^{k-1}e^{-\lambda x}\,dx\\&\leqq\frac{\lambda^k}{(k-1)!}\int_0^t x^{k-1}\,dx=\frac{(\lambda t)^k}{k!}.\end{aligned}\right\} \tag{8.4}$$

Next, we see from the definition of $\varepsilon_{(\iota)}$ that

$$\varepsilon_{(\iota)}\leqq u_{i_1}u_{i_1j_1}u_{i_1j_1i_2}\cdots u_{i_1j_1i_2\ldots j_k}. \tag{8.5}$$

Now from (5.2) we see that $\mathscr{E} u_{i_1} \leqq \left(\frac{cT}{1+cT}\right)^{i_1}$, $\mathscr{E} u_{i_1 j_1 i_2} \leqq \left(\frac{cT}{1+cT}\right)^{i_2}$, etc., while $\mathscr{E} u_{i_1 j_1} = \mathscr{E} u_{i_1 j_1 i_2 j_2} = \cdots = \frac{1}{2}$ because of the symmetry of q (Conditions 3.1). Since the u's in (8.5) are independent, we see that

$$\mathscr{E} \varepsilon_{(\iota)} \leqq \left(\frac{1}{2}\right)^k \left(\frac{cT}{1+cT}\right)^{i_1+i_2+\cdots+i_k}.$$

Moreover, $\varepsilon_{(\iota)}$ and $J(t-\tau_\iota)$ are independent. Hence

$$\mathscr{E} S_k \leqq \sum_{i_1,\ldots,i_k=1}^{\infty} \sum_{j_1,\ldots,j_k=1}^{2} \frac{(\lambda t)^k}{k!} \left(\frac{1}{2}\right)^k \left(\frac{cT}{1+cT}\right)^{i_1+\cdots+i_k} = \frac{(c\lambda t T)^k}{k!}. \tag{8.6}$$

Hence for any t, $\lim_{k\to\infty} \mathscr{E} S_k(t) = 0$.

Now the random function $R_k(t, \omega)$ in (8.3) is a nondecreasing function of k for each t and ω, and hence $S_k(t, \omega)$ is nonincreasing in k. Since $\mathscr{E} S_k \to 0$, this implies that for each fixed t we have $P\{\lim_{k\to\infty} S_k(t) = 0\} = 1$.

On the other hand, since $S_k(t, \omega)$ is the total initial energy received at or before t by a certain set of electrons, it must be nondecreasing in t for each k and ω. Hence, if for some ω and t we have $\lim_{k\to\infty} S_k(t, \omega) = 0$, then $S_k(t', \omega) \to 0$ for all $t' < t$. This suffices to prove the theorem[1]. □

9. Functional equations

9.1. Introduction. During most of the rest of the chapter we shall be investigating the random variable $N(E, t)$, the number of electrons at t of energy $> E$, under the respective hypotheses

H_1: one initial photon of energy 1,

H_2: one initial electron of energy 1.

Expectations under the two hypotheses will be denoted by $\mathscr{E}_1$ and $\mathscr{E}_2$, respectively. Although we shall deal for the most part only with the number of electrons, entirely similar equations can be obtained for photons, or for photons and electrons jointly.

Several different kinds of functional equations can be used for the investigation. At the present time no deep investigation of the relationships among the various approaches has been made, and convenience has been the main criterion as to which one should be used.

There is first a choice of what functions of the probabilities should be considered — moments, product densities, generating functions, characteristic functionals, etc. Then the type of functional equation must be selected: (a) We may consider the cascade at times 0, Δ, and t, letting

[1] It has long been recognized that the functional equations for cascades are consistent with Theorem 8.1, but the author is not aware of a prior proof of the theorem.

$\Delta \downarrow 0$. This corresponds to the approach used in getting the "backward equations" for Markov processes, in particular to (V.3.4). (b) We may consider the cascade at 0, t, and $t+\Delta$, letting $\Delta \downarrow 0$. This corresponds to the "forward equations" for Markov processes, in particular to (V.3.3). (c) We may consider the system at a point of regeneration, as in the case of the age-dependent processes of Chapter VI.

We shall use the approach (a) in conjunction with the generating functions

$$f_i(s, E, t) = \mathscr{E}_i s^{N(E,t)}, \qquad i = 1, 2. \tag{9.1}$$

This approach leads to the basic equations (11.14), which have the form $\partial f_i/\partial t = \Lambda_i(f_1, f_2)$, $i = 1, 2$, where Λ_1 and Λ_2 are nonlinear integral operators. Differentiation of these equations with respect to s at $s=1$ leads to linear equations in the moments. Equations similar to (11.14) can be obtained, at least formally, for the moment-generating functionals

$$\Phi_i(S_1, S_2, t) = \mathscr{E}_i \exp\{-\int [S_1(E)\, d_E N(E,t) + S_2(E)\, d_E N_0(E,t)]\},$$

where S_1 and S_2 are nonnegative functions of E, and N_0 is the number of *photons*.

If we try the approach (b), we can obtain forward equations for Φ_1 and Φ_2, but *not* for f_1 and f_2. The forward equations for Φ_1 and Φ_2 are linear, as we should expect from consideration of the analogous equations (V.4.2), while the nonlinearity of the backward equations for f_1 and f_2 is to be expected from the nonlinearity of (V.4.3).

The point-of-regeneration method (c) can be used when there is one initial photon, the point of regeneration being the transformation of the initial photon. If there is initially one electron, the method at first sight seems unusable, since there is no *first* production of a photon. However, a modification of the method can be used, following a procedure employed in a somewhat different connection by MOYAL (1956). We shall not discuss this possibility. (If $\mu = 0$, the method presents no difficulty.)

A few of the more important references are CARLSON and OPPENHEIMER (1937), approach (b) for the first moments; BHABHA and HEITLER (1937), direct treatment of the first moments by study of successive generations; NORDSIECK, LAMB, and UHLENBECK (1940), approach (b) for the probabilities; JÁNOSSY (1950b), approach (c) for the generating functions, leading indirectly to the differential equations (JÁNOSSY's "G-equations") that arise through approach (a); and BHABHA (1950), RAMAKRISHNAN (1950), and BHABHA and RAMAKRISHNAN (1950), product densities and approach (b). There have been numerous papers by H. MESSEL and collaborators, some of them listed in the article by MESSEL in WILSON (1954). RAMAKRISHNAN and SRINIVASAN (1956) have studied the total number of particles of specified energies produced between 0 and t, rather than the number at t.

9.2. An integral equation. We shall establish an integral equation that will be used in Sec. 10 to establish certain properties of f_1, f_2, $\mathscr{E}_1 N(E,t)$, and $\mathscr{E}_2 N(E,t)$. These properties, besides being important in themselves, are required for the derivation of the functional equations of Theorem 11.1.

In the case of an initial electron of energy 1, the total energy in all particles is always 1 (Theorem 8.1). A similar result holds for an initial photon. Hence in either case we have $N(E,t) < 1/E$, and hence f_1 and f_2 are, for fixed E and t, polynomials in s.

Theorem 9.1. *The generating functions f_1 and f_2 are related by the equation, holding for each complex s,*

$$\left.\begin{aligned} &f_1(s,E,t+\Delta) = e^{-\lambda\Delta} f_1(s,E,t)\\ &+\lambda\int_0^\Delta e^{-\lambda\tau}\,d\tau\int_0^1 f_2\left(s,\frac{E}{u},t+\Delta-\tau\right) f_2\left(s,\frac{E}{1-u},t+\Delta-\tau\right) q(u)\,du,\\ &\qquad\qquad \Delta\geqq 0,\ t\geqq 0,\ 1\geqq E>0.\end{aligned}\right\}\qquad(9.2)$$

Equation (9.2), similar to one obtained by JÁNOSSY (1950b), has an obvious interpretation. If the initial photon is still present at Δ, then the situation is the same at Δ as it was at 0. However, if the initial photon is transformed at $\tau < \Delta$, then two electrons of energy u and $1-u$ will be created, which then produce independent families. From the definition of our process, an electron of energy u has the same probability distribution for progeny of energy $> E$ as does an electron of energy 1 for progeny of energy $> E/u$. A similar consideration applies to an electron of energy $1-u$. Thus we explain the product of two generating functions on the right side of (9.2). We omit the detailed proof, which presents no difficulties and is similar to the proof of Theorem VI.7.1.

9.3. Derivation of the basic equations (11.14) in case $\mu = 0$. In case $\mu = 0$, the probability that an initial electron produces no photons in the interval $(0,t)$ is e^{-At}, where $A = \int_0^1 k(u)\,du$. (Compare (6.4).) Considerations similar to those leading to (9.2) then yield the integral equation, holding for all complex s,

$$\left.\begin{aligned} &f_2(s,E,t+\Delta) = e^{-A\Delta} f_2(s,E,t)\\ &+\int_0^\Delta e^{-A\tau}\,d\tau\int_0^1 f_1\left(s,\frac{E}{u},t+\Delta-\tau\right) f_2\left(s,\frac{E}{1-u},t+\Delta-\tau\right) k(u)\,du,\\ &\qquad\qquad \Delta\geqq 0,\ t\geqq 0,\ E>0.\end{aligned}\right\}\qquad(9.3)$$

From (9.2) and (9.3) we see that f_1 and f_2 are continuous in t uniformly for $t \geqq 0$, $1 \geqq E > 0$, and $|s| \leqq 1$. Letting $\Delta \to 0$ with t fixed, we then see that equations (11.14a) and (11.14b) are satisfied if the derivatives are interpreted as right-hand derivatives. Similarly, considering $t+\Delta$ fixed and letting $t\to t+\Delta$, we see that (11.14a) and (11.14b) are satisfied for the left-hand derivatives, and hence for the usual derivatives. Thus in case $\mu=0$, the arguments in Sec. 11 are unnecessary.

10. Some properties of the generating functions and first moments

We define the first moments as follows.

Definitions 10.1. Let $M_1(E,t)=\partial f_1(1,E,t)/\partial s=\mathscr{E}_1 N(E,t)$, $M_2(E,t)=\partial f_2(1,E,t)/\partial s=\mathscr{E}_2 N(E,t)$.

Note that $M_1(E,t)=M_2(E,t)=0$ for $E\geqq 1$, and $M_1(E,0)=0$. Moreover, since $N(E,t)<1/E$, we have

$$M_1(E,t)<1/E,\ M_2(E,t)<1/E,\qquad E>0.$$

We shall now demonstrate some properties of the M_i and the f_i that imply that these functions are differentiable with respect to E and t. At the same time we obtain useful bounds for the difference quotients.

Remarks on notation. In Secs. 10 and 11 we shall use c to denote a constant, not necessarily the same each time. This constant is independent of t, E, Δ, s, or other quantities unless some limitation is stated. For example, Theorem 10.3 means that given any $\delta>0$ we can find a constant c_δ such that (10.3) is true with $c=c_\delta$ if $|s|$, $1/|s|$, and E_1 are $>\delta$, if $E_1\leqq E_2<\infty$, and if $0\leqq t<T$. We continue to use T as in Def. 5.1, and always $0\leqq t<T$.

Theorem 10.1. *The expectations M_1 and M_2 satisfy the equation*

$$M_1(E,t)=2\lambda\int_E^1 q(u)\left\{\int_0^t e^{-\lambda\tau} M_2\left(\frac{E}{u},t-\tau\right)d\tau\right\}du,\qquad 0<E\leqq 1. \tag{10.1}$$

Proof. Set $t=0$ in (9.2), replace Δ by t, and differentiate with respect to s at $s=1$. Using the symmetry of q (Conditions 3.1) and the relations $M_2(E,t)=0$ for $E\geqq 1$, $M_1(E,0)=0$, we arrive at (10.1). □

Theorem 10.2. *The expectation $M_1(E,t)$ satisfies*

$$|M_1(E_2,t)-M_1(E_1,t)|\leqq\frac{ct|E_2-E_1|}{E_1^2},\qquad 0<E_1\leqq E_2<\infty. \tag{10.2}$$

Proof. Put $E=e^{-y}$, $u=e^{-z}$ in (10.1), $0\leqq y,\ z<\infty$. Then (10.1) can be written

$$M_1(E,t)=2\lambda I(y,t)=2\lambda\int_0^y \varphi(y-z)\psi(z,t)\,dz,$$

where $\varphi(y)=e^{-y}q(e^{-y})$ and $\psi(z,t)=\int_0^t e^{-\lambda\tau}M_2(e^{-z},t-\tau)\,d\tau$. Using the inequality $M_2(e^{-z}, t) \leqq e^z$ and the boundedness of q and its derivative, we obtain

$$|I(y+\Delta, t) - I(y, t)| \leqq cte^y |\Delta + e^\Delta - 1|, \qquad 0 \leqq y \leqq y + \Delta.$$

Putting $e^{-y} = E_2$, $e^{-(y+\Delta)} = E_1$, $\Delta = \log(E_2/E_1) \leqq (E_2 - E_1)/E_1$, we obtain (10.2) for $0 < E_1 \leqq E_2 \leqq 1$. Since $M_1(E, t) = 0$ for $E \geqq 1$, (10.2) holds for the range of values $0 < E_1 \leqq E_2 < \infty$. □

Theorem 10.3. *The generating function f_1 satisfies*

$$|f_1(s, E_1, t) - f_1(s, E_2, t)| \leqq ct|E_1 - E_2|, \qquad 0 < E_1 \leqq E_2 < \infty, \tag{10.3}$$

provided E_1, $|s|$, and $1/|s|$ are bounded away from 0.

Proof. Let $N(E_i, t) = N_i$, $i = 1, 2$. Then, if $0 < E_1 \leqq E_2$, we have $1/E_1 > N_1 \geqq N_2$, and

$$\left.\begin{aligned} &|f_1(s, E_1, t) - f_1(s, E_2, t)| = |\mathscr{E}_1(s^{N_1} - s^{N_2})| \\ &\leqq \mathscr{E}_1\{|s|^{N_2}|s^{N_1-N_2} - 1|\} = \mathscr{E}_1\{|s|^{N_2}|e^{(N_1-N_2)\log s} - 1|\}, \end{aligned}\right\} \tag{10.4}$$

where we take $0 \leqq$ argument $\log s < 2\pi$.

Under the conditions of the theorem, $\log s$ and $(N_1 - N_2)\log s$ are bounded and hence $|s^{N_1-N_2} - 1| \leqq c(N_1 - N_2)$. Hence (10.4) leads to

$$\left.\begin{aligned} &|f_1(s, E_1, t) - f_1(s, E_2, t)| \leqq c\,\mathscr{E}_1(N_1 - N_2) \\ &= c\big(M_1(E_1, t) - M_1(E_2, t)\big). \end{aligned}\right\} \tag{10.5}$$

Theorem 10.3 now follows from (10.5) and Theorem 10.2. □

Theorem 10.4. *The generating function f_1 satisfies*

$$|f_1(s, E, t+\Delta) - f_1(s, E, t)| \leqq c\Delta, \qquad E > 0, \ \Delta > 0,$$

provided E and $1/|s|$ are bounded away from 0.

This result follows from (9.2), if we observe that f_1 and f_2 are bounded by $|s|^{1/E} + 1$.

Theorem 10.5. *The function M_2 satisfies the inequality*

$$|M_2(E_1, t) - M_2(E_2, t)| \leqq ct|E_1 - E_2|, \qquad 0 < E_1 \leqq E_2 < 1,$$

provided E_1 and E_2 are bounded away from 0 and 1.

Proof. If we take expectations $\mathscr{E}_2$ of both sides of equation (2) of Appendix 2, we obtain

$$M_2(E, t) = P\big(\varepsilon(t) > E\big) + \mathscr{E}_2 \sum_{i=1}^{\infty} M_1\left(\frac{E}{\varepsilon(i)}, t - t_i\right), \tag{10.6}$$

where $\varepsilon_{(i)}$ and t_i are the energy at birth and time of birth of the photon $\langle i \rangle$, $\varepsilon(t)$ is the energy of $\langle 0 \rangle$ at t, and M_1 is 0 if $t - t_i < 0$. From Theorem

6.2 it follows that the contribution to the difference $|M_2(E_1, t) - M_2(E_2, t)|$ by the first term on the right side of (10.6) is bounded by

$$ct|E_2 - E_1|, \qquad 0 < E_1 \leq E_2 < 1, \tag{10.7}$$

provided E_1 and E_2 are bounded away from 0 and 1. From Theorem 10.2 the difference $M_1(E_2/\varepsilon_{(i)}, t - t_i) - M_1(E_1/\varepsilon_{(i)}, t - t_i)$ is bounded in magnitude by

$$ct\left|\frac{E_2}{\varepsilon_{(i)}} - \frac{E_1}{\varepsilon_{(i)}}\right| \bigg/ \left(\frac{E_1}{\varepsilon_{(i)}}\right)^2 = \frac{ct\,\varepsilon_{(i)}|E_2 - E_1|}{E_1^2}. \tag{10.8}$$

We see from (8.2) that $\sum \varepsilon_{(i)} \leq 1$, since $\varepsilon_i^*(t) = \varepsilon_{(i)}$ for $t \geq t_i$. We then obtain the desired result by applying (10.7) and (10.8) to (10.6). □

Theorem 10.6. *The generating function f_2 satisfies the inequality*

$$|f_2(s, E_1, t) - f_2(s, E_2, t)| \leq ct|E_1 - E_2|, \qquad 0 < E_1 \leq E_2 < 1, \tag{10.9}$$

provided $|s|$, $1/|s|$, and E_1 are bounded away from 0 and provided E_2 is bounded away from 1.

Proof. This follows from Theorem 10.5 almost as Theorem 10.3 follows from Theorem 10.2. □

11. Derivation of functional equations for f_1 and f_2

In this section we shall frequently use the following conditions on the variables s, t, E, and Δ. The definition of T is in Def. 5.1. We continue to use c to mean a constant, not necessarily the same each time.

Conditions 11.1. The quantities $|s|$ and $1/|s|$ are bounded away from 0; E lies between 0 and 1 and is bounded away from 0 and 1; $0 < t < T$; $0 < \Delta < T - t$. The meaning of "bounded away from 0" and similar expressions should be clear from the remarks on notation near the beginning of Sec. 10.

Special meaning of the O-notation. In this section a statement such as $F = O(\Delta^2)$ will mean $|F| \leq c\Delta^2$, where the constant c can be chosen independently of s, E, t, and Δ *provided* those variables satisfy Conditions 11.1.

We are now ready to derive the basic functional equations for f_1 and f_2. The derivation is rather lengthy, largely because of the presence of the "infrared catastrophe" (see Sec. 3). Accordingly, we shall first sketch the main ideas. We have seen, in Sec. 9.3, that the procedure is much simpler in the case of a finite cross section.

We have already established the integral equation (9.2). The meaning of this equation is given just below it. We then obtain one of our basic equations, (11.14a), by letting $\Delta \to 0$ in (9.2).

In order to establish (11.14b), the first step is to establish (11.1) for the case of an initial electron of energy 1. The quantities t_i' and u_i' appearing in (11.1) are defined above it, and the meaning of the equation is explained just below it.

Equation (11.1) is still too complicated to use directly. However, it turns out that we can discard all but the first factor in the infinite product in (11.1), with an error that is $O(\Delta^2)$ as $\Delta \to 0$. In other words, *if Δ is small, nearly all the energy lost by the electron $\langle 0 \rangle$ in the interval* $(0, \Delta)$ *goes to one photon.* Moreover, we can replace the argument $t+\Delta-t_1'$ in (11.1) by t, again with an error that is $O(\Delta^2)$, rather than $O(\Delta)$ as might be expected at first sight. Thus we arrive at equation (11.4).

In order to take the final step of letting $\Delta \downarrow 0$ in (11.4), we use Theorem 6.3, which enables us to evaluate the expectation of functions of $X(t)$ or $\varepsilon(t)$ as t goes to 0. In this way we arrive at the equation (11.14b).

We now carry out the derivation.

11.1. Singling out of photons born before Δ. We defined (Def. 5.2) the energy of an electron in terms of a sequence $t_1, u_1, t_2, u_2, \ldots$, where $0<t_i<T$ and $1>u_1>u_2>\cdots>0$, each pair (t_i, u_i) corresponding to the emission of a photon. Now let Δ be a fixed positive number $<T$, and let us single out those points (t_i, u_i) for which $0<t_i<\Delta$. We shall denote them, ordered according to decreasing u-coordinates, by (t_1', u_1'), $(t_2', u_2'), \ldots$. The sequence, of course, depends on Δ, and according to (7.3) there are infinitely many points (t_i', u_i'). We shall suppose that none of the t_i is equal to Δ, the contrary event having probability 0. The t_i' and the u_i' have the joint distributions of the t_i and the u_i prescribed in Sec. 5, except that Δ replaces T. (See Appendix 2.) Referring to Def. 5.2, we see that if $t \leqq \Delta$, then $\varepsilon(t; t_1, u_1, \ldots) = \varepsilon(t; t_1', u_1', \ldots)$, since only points (t_i, u_i) with $t_i < \Delta$ are involved.

Our first result is the following lemma.

Lemma 11.1. *The generating function f_2 satisfies the relation*

$$\left.\begin{aligned} & f_2(s, E, t+\Delta) \\ & = \mathscr{E}_2\left\{f_2\left(s, \frac{E}{\varepsilon(\Delta; t_1', u_1', \ldots)}, t\right)\right. \\ & \quad \left. \times \prod_{i=1}^{\infty} f_1\left(s, \frac{E}{u_i'\varepsilon(t_i'-0; t_1', u_1', \ldots)}, t+\Delta-t_i'\right)\right\}, \qquad E>0. \end{aligned}\right\} \quad (11.1)$$

The proof of Lemma 11.1 will be indicated in Appendix 2. The interpretation of (11.1) is as follows. The original electron $\langle 0 \rangle$ will have the energy $\varepsilon(\Delta; t_1, u_1, \ldots) = \varepsilon(\Delta; t_1', u_1', \ldots)$ at Δ; the history of this electron and its descendants born after Δ follows the same probability law as the original process except that all energies are multiplied by $\varepsilon(\Delta)$. This part of the process accounts for the factor f_2 on the right side of (11.1). Any first-generation photon born before Δ corresponds to a point (t_i', u_i'), and the energy at birth of the photon is $u_i' \varepsilon(t_i'-0; t_1', u_1', t_2', u_2', \ldots)$. The electrons descended from this photon account for the i-th factor of the product on the right side of (11.1).

Since $\lim_{i\to\infty} u_i = 0$, and since $f_1(s, E, t) = 1$ when $E \geqq 1$, only finitely many terms in the infinite product differ from 1.

11.2. Simplification of equation (11.1). The first step is the replacement of $\prod_{i=1}^{\infty}$ in (11.1) by the single factor corresponding to $i=1$, with an error that is $O(\Delta^2)$. That is, only the "biggest" first-generation photon born in $(0, \Delta)$ is important.

Lemma 11.2. *Under Conditions* 11.1 *and the additional requirement* $|s| \leqq 1$, *the function* f_2 *satisfies*

$$\left.\begin{aligned} &f_2(s, E, t+\Delta) \\ &= \mathscr{E}_2\left\{f_2\left(s, \frac{E}{\varepsilon(\Delta)}, t\right) f_1\left(s, \frac{E}{u_1'\varepsilon(t_1'-0)}, t+\Delta-t_1'\right)\right\} + O(\Delta^2). \end{aligned}\right\} \tag{11.2}$$

Proof. From (5.4) we have $\mathscr{E}_2\left(\sum_{i=2}^{\infty} u_i'\right) = O(\Delta^2)$. Now if the numbers $y, z_1, z_2, \ldots, z_n$ are each bounded in absolute value by 1, we have the algebraic inequality

$$|y z_1 z_2 \ldots z_n - y z_1| \leqq \sum_{i=2}^{n} |1 - z_i|, \tag{11.3}$$

whose proof for $n=3$, for example, is suggested by writing $y z_1 z_2 z_3 - y z_1 = (y z_1 z_2 z_3 - y z_1 z_2) + (y z_1 z_2 - y z_1)$. Hence if we replace the product $\prod_{i=1}^{\infty}$ in (11.1) by the first factor of the product, the error is bounded by

$$\mathscr{E}_2 \sum_{i=2}^{\infty} \left| f_1\left(s, \frac{E}{u_i'\varepsilon(t_i'-0)}, t+\Delta-t_i'\right) - 1\right|.$$

Since the i-th term of the sum is bounded by 1 and is 0 if $u_i' < E$, the error is bounded by $\sum_{i=2}^{\infty} P(u_i' \geqq E) \leqq (1/E) \sum_{i=2}^{\infty} \mathscr{E}_2(u_i') = O(\Delta^2)$. Hence the lemma is proved. □

Our next step is to show that the f_1-term in (11.2) can be replaced by $f_1\left(s, \frac{E}{1-\varepsilon(\Delta)}, t\right)$, again producing an error that is $O(\Delta^2)$.

Lemma 11.3. *Under Conditions* 11.1, *the function* f_2 *satisfies*

$$\left.\begin{aligned} &f_2(s, E, t+\Delta) \\ &= \mathscr{E}_2\left\{f_2\left(s, \frac{E}{\varepsilon(\Delta)}, t\right) f_1\left(s, \frac{E}{1-\varepsilon(\Delta)}, t\right)\right\} + O(\Delta^2), \qquad |s| \leqq 1. \end{aligned}\right\} \tag{11.4}$$

Proof. Refer to (8.2). If the photon $\langle i\rangle$ is born before Δ, i.e., if $t_i<\Delta$, then t_i must appear as one of the t''s, say t'_j, and $\varepsilon_i^*(\Delta)=u'_j\,\varepsilon(t'_j-0)$. Hence, putting $t=\Delta$ in (8.2), we obtain

$$1-\varepsilon(\Delta)=\sum_{i=1}^{\infty}u'_i\,\varepsilon(t'_i-0). \tag{11.5}$$

We have pointed out above that $\mathscr{E}_2\sum_{i=2}^{\infty}u'_i=O(\Delta^2)$. Hence, from (11.5),

$$\mathscr{E}_2[1-\varepsilon(\Delta)-u'_1\,\varepsilon(t'_1-0)]=O(\Delta^2). \tag{11.6}$$

Put $\varphi_E(x)=1$, $0<x\leqq E$; $\varphi_E(x)=E/x$, $E\leqq x\leqq 1$. Then from Theorem 10.3 we have, for $0<x,\ y\leqq 1$,

$$|f_1(s,\varphi_E(x),t)-f_1(s,\varphi_E(y),t)|\leqq c\,|\varphi_E(x)-\varphi_E(y)|. \tag{11.7}$$

Since $|\varphi'_E(x)|\leqq 1/E$, the right side of (11.7) is bounded by $c|x-y|/E$. Since $f_1(s,u,t)=1$ if $u\geqq 1$, we can write $f_1\left(s,\frac{E}{x},t\right)=f_1(s,\varphi_E(x),t)$ for $0<x\leqq 1$. Hence,

$$\left.\begin{aligned}\left|f_1\left(s,\frac{E}{x},t\right)-f_1\left(s,\frac{E}{y},t\right)\right|\leqq c\,|\varphi_E(x)-\varphi_E(y)|\leqq\frac{c|x-y|}{E},\\ 0<x,\ y\leqq 1.\end{aligned}\right\} \tag{11.8}$$

Applying (11.6) and (11.8) with $x=u'_1\,\varepsilon(t'_1-0)$ and $y=1-\varepsilon(\Delta)$, we see that

$$\left.\begin{aligned}&f_1\left(s,\frac{E}{u'_1\,\varepsilon(t'_1-0)},t+\Delta-t'_1\right)\\ &=f_1\left(s,\frac{E}{1-\varepsilon(\Delta)},t+\Delta-t'_1\right)+R,\qquad \mathscr{E}_2R=O(\Delta^2).\end{aligned}\right\} \tag{11.9}$$

In order to conclude the proof of Lemma 11.3, we must be able to replace $t+\Delta-t$ by t with a suitably small error. Consider the difference

$$R_1=f_1\left(s,\frac{E}{1-\varepsilon(\Delta)},t+\Delta-t'_1\right)-f_1\left(s,\frac{E}{1-\varepsilon(\Delta)},t\right). \tag{11.10}$$

If $\varepsilon(\Delta)\geqq 1-E$, then $R_1=1-1=0$. Moreover, $\varepsilon(\Delta)=(1-u'_1)(1-u'_2)\ldots$ depends only on the u'_i and hence is independent of t'_1. Hence, using Theorem 10.4, we have

$$\mathscr{E}_2|R_1|\leqq c\,P[\varepsilon(\Delta)<1-E]\,\mathscr{E}_2(\Delta-t'_1).$$

Now $\mathscr{E}_2(\Delta-t'_1)=\Delta/2$, and (see Def. 4.1 and equation (4.2))

$$P(\varepsilon(\Delta)<1-E)=P[X(\Delta)>-\log(1-E)]\leqq\frac{\mathscr{E}X(\Delta)}{-\log(1-E)}\leqq\frac{c\Delta}{-\log(1-E)}.$$

Hence

$$\mathscr{E}_2|R_1|=O(\Delta^2). \tag{11.11}$$

Lemma 11.3 now follows from (11.2), (11.9), (11.10), and (11.11). □

11.3. Limiting form of $f_2(s,E,t+\Delta)$ as $\Delta\downarrow 0$. We are now ready to apply Theorem 6.3 to determine the limiting form of $f_2(s,E,t+\Delta)$ as

$\Delta \downarrow 0$, using (11.4). Recalling that $\varepsilon(\Delta)=\exp(-X(\Delta))$, we take the function $F(x)$ of Theorem 6.3 to be

$$F(x)=f_1\left(s, \frac{E}{1-e^{-x}}, t\right) f_2(s, E e^x, t), \tag{11.12}$$

and we consider s, E, and t fixed, $0<E<1$, $0<|s|\leqq 1$. From Theorem 10.6 we can verify that F satisfies the conditions of Theorem 6.3, noting that $f_1=1$ when x is near 0.

If we apply Theorem 6.3 to the right side of (11.4), we obtain, for $0<|s|\leqq 1$ and $0<E<1$,

$$\left.\begin{aligned} &f_2(s, E, t+\Delta)=f_2(s, E, t)\\ &\qquad+\Delta\int_0^\infty\left[f_1\left(s, \frac{E}{1-e^{-x}}, t\right) f_2(s, E e^x, t)-f_2(s, E, t)\right]\\ &\qquad\times k(1-e^{-x})\, e^{-x}\, dx+I_F R_2 \Delta^2+O(\Delta^2), \end{aligned}\right\} \tag{11.13}$$

where I_F is defined in Theorem 6.3, R_2 has a bound that is independent of s, E, and t, and it can be verified that I_F is bounded provided Conditions 11.1 hold.

Assuming $0<E<1$ and $0<|s|\leqq 1$, we see that f_2 has a right-hand derivative in t, namely, the coefficient of Δ in (11.13), and also that f_2 is a continuous function of t. By a routine argument we can show that the coefficient of Δ on the right side of (11.13) is a continuous function of t. Hence f_2 has a continuous right-hand t-derivative. Using a result of the theory of real variables (HOBSON (3d ed., 1927, Vol. 1, p. 365)), we see that f_2 has a t-derivative for each t and is an absolutely continuous function of t.

Since f_2 is a power series in s (in fact, a polynomial), we can express the coefficients of the powers of s by means of CAUCHY's integral formula. In this way we can see that the coefficients are themselves differentiable in t.

Theorem 11.1.[1] *The generating functions f_1 and f_2 satisfy the equations*

$$\left.\begin{aligned} &\frac{\partial f_1(s, E, t)}{\partial t}=-\lambda f_1(s, E, t)\\ &+\lambda\int_0^1 f_2\left(s, \frac{E}{u}, t\right) f_2\left(s, \frac{E}{1-u}, t\right) q(u)\, du, \qquad E>0,\ t>0, \end{aligned}\right\} \tag{11.14a}$$

$$\left.\begin{aligned} &\frac{\partial f_2(s, E, t)}{\partial t}\\ &=\int_0^1\left\{f_1\left(s, \frac{E}{u}, t\right) f_2\left(s, \frac{E}{1-u}, t\right)-f_2(s, E, t)\right\} k(u)\, du,\\ &\qquad\qquad E>0,\ t>0, \end{aligned}\right\} \tag{11.14b}$$

[1] These equations are comparable to equations (15) of JÁNOSSY (1950b), the "G-equations". See also HARRIS (1957b).

for all complex s, with the conditions

$$\left.\begin{array}{lll} f_1(s,E,0)=1, & E>0;\; f_2(s,E,0)=s, & 0<E<1; \\ & f_2(s,E,0)=1, & E\geqq 1. \end{array}\right\} \qquad (11.14\text{c})$$

Proof. First suppose $0<E<1$, $0<|s|\leqq 1$. If we let $\Delta\downarrow 0$ in (11.13), we see that the right-hand t-derivative of f_2 is equal to the right side of (11.14b), after making the change of variable $u=1-e^{-x}$. We have already seen that f_2 has a t-derivative; hence (11.14b) holds if $0<E<1$, $0<|s|\leqq 1$. We may then extend the range of E to all $E>0$, since both sides of (11.14b) are 0 if $E\geqq 1$. Finally, since f_1 and f_2 are both polynomials in s, the equation must hold for all values of s.

We obtain (11.14a) in a similar manner by letting $\Delta\downarrow 0$ in (9.2).

The truth of the conditions (11.14c) is obvious from the definition of the process. □

We shall state a uniqueness condition for the solution of (11.14) after we have discussed the moments.

12. Moments of $N(E,t)$

Since $N(E,t)<1/E$, it is evident that all the moments of N are finite if $E>0$. The factorial moments (and hence the moments) can be obtained by solving the linear equations that arise when we differentiate (11.14) with respect to s and set $s=1$.

12.1. First moments. Refer to Def. 10.1. Differentiation of (11.14) at $s=1$ yields the following pair of equations (we use the relations $q(1-u)=q(u)$ and $M_i(E,t)=0$ for $E\geqq 1$):

$$\frac{\partial M_1(E,t)}{\partial t}=-\lambda M_1(E,t)+2\lambda\int_E^1 M_2\left(\frac{E}{u},t\right)q(u)\,du, \qquad (12.1\text{a})$$

$$\frac{\partial M_2(E,t)}{\partial t}=\int_0^1\left\{M_1\left(\frac{E}{u},t\right)+M_2\left(\frac{E}{1-u},t\right)-M_2(E,t)\right\}k(u)\,du, \qquad (12.1\text{b})$$

with the initial conditions

$$\left.\begin{array}{lll} M_1(E,0)=0, & E>0;\; M_2(E,0)=1, & 0<E<1; \\ & M_2(E,0)=0, & E\geqq 1. \end{array}\right\} \qquad (12.1\text{c})$$

Carlson and Oppenheimer (1937) gave forward equations for the first-moment densities. Equations (12.1), which are backward equations, correspond to equation (24) of Jánossy (1950b).

In order to solve (12.1), we follow the customary procedure, introducing the Mellin transforms

$$\hat{M}_i(s,t)=\int_0^1 E^{s-1}M_i(E,t)\,dE, \qquad i=1,2, \qquad (12.2)$$

which are defined for $Re(s) > 1$, since $M_i(E, t) < 1/E$. If we multiply both sides of (12.1a) and (12.1b) by E^{s-1} and integrate on E from 0 to 1, we obtain

$$\frac{\partial \hat{M}_1(s,t)}{\partial t} = -\lambda \hat{M}_1(s,t) + 2\lambda \left[\int_0^1 u^s q(u)\, du\right] \hat{M}_2(s,t), \tag{12.3a}$$

$$\left.\begin{aligned} \frac{\partial \hat{M}_2(s,t)}{\partial t} &= \left[\int_0^1 u^s k(u)\, du\right] \hat{M}_1(s,t) \\ &\quad - \left[\int_0^1 (1-u^s) k(1-u)\, du\right] \hat{M}_2(s,t), \end{aligned}\right\} \tag{12.3b}$$

$$\hat{M}_1(s,0) = 0, \quad \hat{M}_2(s,0) = \frac{1}{s}. \tag{12.3c}$$

The justification of the operations leading to (12.3) is routine in view of the properties of M_1 and M_2 deduced in Sec. 10.

For fixed s, (12.3) is a linear system of differential equations in t, which can be solved explicitly. The Mellin transforms can then be inverted numerically to yield values for $M_1(E, t)$ and $M_2(E, t)$. Such calculations have been carried out by a number of authors. Values for M_2 may be found in RAMAKRISHNAN (1962, p. 538), the parameter y in the table being the negative of the natural logarithm of our parameter E. The calculations indicate that if E is fixed, then M_1 and M_2 first increase and then decrease as t increases, as suggested in Sec. 1. In Sec. 14, we make a closer study of the first moments.

12.2. Second and higher moments. Calculations for the second moments have been carried out by JÁNOSSY and MESSEL (1950), RAMAKRISHNAN and MATHEWS (1954), RAMAKRISHNAN and SRINIVASAN (1955), and others. See the tables on pp. 539–543 of RAMAKRISHNAN (1962). Calculation of higher moments would be complicated.

12.3. Probabilities. The probability distribution of $N(E, t)$ can be obtained from (11.14) as follows. For $E > 0$, f_1 and f_2 are polynomials of degree $< 1/E$. Hence the probability distribution of $N(E, t)$, for any particular E, is uniquely determined by the factorial moments of order $< 1/E$. These moments can be obtained, in principle, by differentiating (11.14) the required number of times at $s = 1$, using the Mellin transform, and solving the resulting linear equations. See MESSEL (1956). This method is, of course, not practical if E is small, and indeed no practical methods have been devised other than perhaps artificial sampling (or Monte Carlo) methods.

12.4. Uniqueness of the solution of (11.14). The Mellin transforms $\hat{M}_1$ and $\hat{M}_2$ are uniquely determined by equations (12.3), and the first moments M_1 and M_2 are then uniquely determined because of the uniqueness property of the Mellin transform. Similar considerations show that all the higher moments of $N(E, t)$ are uniquely determined through (11.14). Thus the pair (f_1, f_2) is the only solution of (11.14) consisting of power series in s whose coefficients are regular enough so that (11.14) can be differentiated arbitrarily often at $s=1$. However, we shall not investigate the extent of this class.

13. The expectation process[1]

In order to motivate what follows, let us consider a process with only one type of particle rather than two. We suppose that the total energy of all particles is constant, and that the proportion of its energy that a particle loses in a collision is independent of the magnitude of the energy. Let $m(E, t)\,dE$ be the expected number of particles with energies in $(E, E+dE)$ at thickness t, if initially there is one particle of energy 1. If necessary, the density m may have a δ-function component at $E=1$. Then the above assumptions lead to the equations

$$\int_0^1 E\, m(E, t)\, dE = 1, \tag{13.1}$$

$$m(E, t_1+t_2) = \int_E^1 m(u, t_1)\, m\left(\frac{E}{u}, t_2\right) \frac{du}{u}. \tag{13.2}$$

If we make the change of variable $E=e^{-z}$, $z \geqq 0$, and if we put $p(z, t) = e^{-2z} m(e^{-z}, t)$, then equations (13.1) and (13.2) become, respectively,

$$\int_0^\infty p(z, t)\, dz = 1, \tag{13.3}$$

$$p(z, t_1+t_2) = \int_0^z p(z-y, t_1)\, p(y, t_2)\, dy. \tag{13.4}$$

In other words, *we can regard $p(z, t)$ as the probability density of a non-negative additive random process $Z(t)$*. The statement that $Z(t)$ has approximately a Gaussian distribution when t is large can be translated in terms of the density $m(E, t)$, or its integral over certain ranges. However, we must be careful about the range of validity of the approximation, as will be explained in more detail below, and the situation is not as satisfactory as we should like.

Physicists have obtained approximations to the moments in electron-photon cascades by the use of saddle-point methods; see, e.g., JÁNOSSY (1950a) and the references of Sec. 12.2. In principle, we are rather following KHINCHIN (1949), who used the central limit theorem extensively in statistical mechanics as a substitute for saddle-point methods.

[1] The expectation process for electron-photon cascades was introduced by HARRIS (1957b). A similar process for cascades with one type of particle has been used by FILIPPOV (1961).

KHINCHIN used the device of a "conjugate distribution" to shift the region of the approximation to the position of most interest. An analogue to this device can be defined for the more complex problem to be treated below, but at present it is impossible to judge whether it would be useful to do this, and we shall not discuss it.

The treatment below is only an outline and is heuristic rather than rigorous.

13.1. The probabilities for the expectation process. We continue to use the index i or $j=1$ for photons and 2 for electrons.

In order to define the expectation process, we must consider the case in which there is initially one photon of energy 1 as well as the case of one initial electron, and we must consider the number of photons at t as well as the number of electrons.

Definitions 13.1. Let $M_{ij}(E, t)$ be the expected number of particles of type j and energy $>E$ at t if there is initially one particle of type i and energy 1. Let $\hat{M}_{ij}(s, t)=\int_0^1 E^{s-1} M_{ij}(E, t)\, dE$, and let $\hat{\boldsymbol{M}}$ be the matrix $(\hat{M}_{ij})$.

Of course, M_{12} and M_{22} are identical with M_1 and M_2 of Def. 10.1. Moreover, equations (12.1a) and (12.1b) are satisfied if we replace M_1 and M_2 by M_{11} and M_{21}, respectively, with the initial conditions

$$\left.\begin{aligned} M_{11}(E, 0)=1, \quad 0<E<1;\; M_{11}(E, 0)&=0, \quad E\geqq 1; \\ M_{21}(E, 0)&=0, \quad E>0. \end{aligned}\right\} \tag{13.5}$$

The argument is similar to that used in deriving (12.1).

The integral $\int_{E_0}^1 -E\, d_E M_{ij}(E, t)$ is the expected amount of energy at t in particles of type j with energies $>E_0$, if initially there is one particle of type i and energy 1. Hence, from Theorem 8.1 and its counterpart when the initial particle is a photon, we have

$$\sum_{j=1}^{2} \int_0^1 -E\, d_E M_{ij}(E, t)=1, \qquad i=1, 2. \tag{13.6}$$

This suggests that it may be useful to define $-E\,dM_{ij}(E, t)$ as the differential element of a probability. However, the treatment is simpler if we make the transformation $E=e^{-z}$. Hence we define

$$d P_{ij}(z, t)=e^{-z} d_z M_{ij}(e^{-z}, t), \qquad z\geqq 0, \tag{13.7}$$

putting $P_{ij}(0-, t)=0$. Note that $P_{11}(0+, t)=M_{11}(1-0, t)=e^{-\lambda t}$, while $P_{ij}(0+, t)=M_{ij}(1-0, t)=0$ for $t>0$ and $(i, j)\neq(1, 1)$. We take $P_{ij}(z, t)$ to be continuous to the right in z.

Then (13.6) becomes

$$\sum_{j=1}^{2} \int_{0-}^{\infty} d P_{ij}(z, t) = \lim_{z\to\infty} \sum_{j=1}^{2} P_{ij}(z, t) = 1. \tag{13.8}$$

Now define $\hat{P}_{ij}(s, t)$ and the matrix $\hat{\boldsymbol{P}}(s, t)$ by

$$\left.\begin{aligned} \hat{P}_{ij}(s, t) &= \int_{0-}^{\infty} e^{-sz}\, d P_{ij}(z, t), \\ \hat{\boldsymbol{P}}(s, t) &= \big(\hat{P}_{ij}(s, t)\big). \end{aligned}\right\} \tag{13.9}$$

Because of (13.8), the functions $\hat{P}_{ij}$ are defined at least for $Re(s) \geqq 0$. Let us note the relation

$$\hat{M}_{ij}(s, t) = \frac{1}{s} \hat{P}_{ij}(s-1, t),$$

which follows from (13.7) and integration by parts.

The matrix $\hat{\boldsymbol{P}}(s, t)$ satisfies the following equation, whose derivation is similar to that of (12.3):

$$\frac{\partial \hat{\boldsymbol{P}}(s, t)}{\partial t} = \boldsymbol{B}(s)\, \hat{\boldsymbol{P}}(s, t), \qquad \hat{\boldsymbol{P}}(s, 0) = \boldsymbol{I}; \ Re(s) \geqq 0, \tag{13.10}$$

where $\boldsymbol{I}$ is the identity matrix and $\boldsymbol{B}$ is the matrix

$$\boldsymbol{B}(s) = \begin{pmatrix} -\lambda & 2\lambda \int_0^1 u^{s+1} q(u)\, du \\ \int_0^1 u^{s+1} k(u)\, du & -\int_0^1 (1-u^{s+1})\, k(1-u)\, du \end{pmatrix}. \tag{13.11}$$

The solution of (13.10) is

$$\hat{\boldsymbol{P}}(s, t) = e^{t\boldsymbol{B}(s)}, \tag{13.12}$$

which satisfies the functional equation

$$\hat{\boldsymbol{P}}(s, t_1 + t_2) = \hat{\boldsymbol{P}}(s, t_1)\, \hat{\boldsymbol{P}}(s, t_2), \qquad t_1, t_2 \geqq 0. \tag{13.13}$$

The elements of the matrix product on the right side of (13.13) are sums of products of Laplace transforms, corresponding to sums of convolutions. Hence expressing (13.13) in terms of the P_{ij}, we have

$$P_{ij}(z, t_1 + t_2) = \sum_{k=1}^{2} \int_{0-}^{z} P_{kj}(z - y, t_2)\, d P_{ik}(y, t_1), \qquad z, t_1, t_2 \geqq 0. \tag{13.14}$$

Let us interpret (13.8) and (13.14).

13.2. Description of the expectation process. Let us picture a vector Markovian process $\big(I(t), Z(t)\big)$, $t \geqq 0$, where $I(t) = 1$ or 2 and $Z(t) \geqq 0$. We suppose that the process is additive in the Z-component; in other words, $\text{Prob}\{I(t_2) = j, Z(t_2) \leqq z_2 \,|\, I(t_1) = i, Z(t_1) = z_1\}$ depends on $i, j, z_2 - z_1$, and $t_2 - t_1$.

If we interpret the P_{ij} as

$$P_{ij}(z,t) = \text{Prob}\{I(t)=j,\ Z(t) \leqq z \mid I(0)=i,\ Z(0)=0\}, \qquad (13.15)$$

then equations (13.14) are simply the Chapman-Kolmogorov equations for the process $(I(t), Z(t))$, while (13.8) insures that the probabilities add to 1.

In order to deduce the nature of our Markov process, we may study its behavior in a small t-interval. This can be done most conveniently by considering the behavior of $\int_0^\infty H(z)\, dP_{ij}(z,t)$ for small t, where H is an arbitrary polynomial in e^{-z}. This behavior can be deduced from the relation (obtained from (13.12)) $\int_0^\infty H(z)\, d\boldsymbol{P}(z,t) = H(0)\boldsymbol{I} + t\boldsymbol{D} + O(t^2)$, where $\boldsymbol{P} = (P_{ij})$ and

$$\boldsymbol{D} = \begin{pmatrix} -\lambda H(0) & 2\lambda \int_0^\infty q(e^{-z}) H(z) e^{-2z} dz \\ \int_0^\infty e^{-2z} k(e^{-z}) H(z)\, dz & \int_0^\infty \left(H(z) - e^z H(0)\right) k(1-e^{-z}) e^{-2z} dz \end{pmatrix}.$$

It is then possible to deduce the following behavior for the expectation process $(I(t), Z(t))$, although we omit the argument. The component $I(t)$ takes the values 1 or 2 and is a jump-type Markov process with infinitesimal transition probabilities $\text{Prob}\{I(t+dt)=2 \mid I(t)=1\} = \lambda\, dt$, $\text{Prob}\{I(t+dt)=1 \mid I(t)=2\} = K\, dt$, where

$$K = \int_0^1 u k(u)\, du. \qquad (13.16)$$

The Z-component behaves one way *between jumps* of I and another way *when jumps occur*. During any t-interval when I remains equal to 1, Z remains constant. During any t-interval when I remains equal to 2, Z increases like the Markov process $X(t)$ of Sec. 4, except that the function $k(u)$ is replaced by $(1-u)k(u)$. If I jumps from 1 to 2 at t, then Z has a jump at t of magnitude $z > 0$ with probability density $2e^{-2z} q(e^{-z})$. If I jumps from 2 to 1 at t, then Z has a jump at t of magnitude $z > 0$ with density $e^{-2z} k(e^{-z})/K$. We always begin with $Z(0) = 0$.

If we plot $Z(t)$ as the abscissa and $I(t)$ as the ordinate of a point in the Cartesian plane, regarding t as a time parameter, we see that the point moves to the right, jumping back and forth between the lines $I=1$ and $I=2$. Since Z behaves like a homogeneous additive process on each line, it is plausible that the distribution of $Z(t)$ is approximately Gaussian when t is large.

The expectation process is a special case of the processes treated by NEVEU (1961). However, NEVEU does not discuss the distribution problems in which we shall be interested.

14. Distribution of $Z(t)$ when t is large

The author conjectures that $Z(t)$ has an approximately Gaussian distribution when t is large. Hence we shall want to calculate the mean and variance of $Z(t)$. It is convenient to make the calculation when $I(0)$ has the stationary distribution for $I(t)$, given by the vector

$$\boldsymbol{\pi}_0 = \left(\frac{K}{\lambda+K}, \frac{\lambda}{\lambda+K}\right). \tag{14.1}$$

Presumably, the mean and variance are asymptotically the same, whatever the distribution of $I(0)$.

The expected value of $Z(t)$ is $\boldsymbol{\pi}_0 \boldsymbol{M}(t)\boldsymbol{\nu}$, where $\boldsymbol{M}(t)$ is the matrix whose elements are $\int z\, dP_{ij}(z,t) = -\frac{\partial}{\partial s}\hat{P}_{ij}(0,t)$, and $\boldsymbol{\nu}$ denotes the olumn vector $\binom{1}{1}$. Let us note the relations

$$\left.\begin{aligned} &\boldsymbol{B}(0) = \begin{pmatrix} -\lambda & \lambda \\ K & -K \end{pmatrix}, \quad \boldsymbol{\pi}_0 \boldsymbol{B}(0) = (0,0), \quad \boldsymbol{B}(0)\,\boldsymbol{\nu} = \begin{pmatrix} 0 \\ 0 \end{pmatrix}, \\ &\hat{\boldsymbol{P}}(0,t)\,\boldsymbol{\nu} = e^{\boldsymbol{B}(0)t}\,\boldsymbol{\nu} = \boldsymbol{\nu}. \end{aligned}\right\} \tag{14.2}$$

Differentiating (13.10) with respect to s at $s=0$, we obtain

$$\frac{d\boldsymbol{M}(t)}{dt} = -\boldsymbol{B}'(0)\,\hat{\boldsymbol{P}}(0,t) + \boldsymbol{B}(0)\,\boldsymbol{M}(t), \tag{14.3}$$

where the prime denotes differentiation with respect to s. If we multiply (14.3) on the left by $\boldsymbol{\pi}_0$, on the right by $\boldsymbol{\nu}$, and use (14.2), we obtain

$$\frac{d}{dt}\left(\boldsymbol{\pi}_0 \boldsymbol{M}(t)\,\boldsymbol{\nu}\right) = \frac{d}{dt}\mathscr{E} Z(t) = -\boldsymbol{\pi}_0 \boldsymbol{B}'(0)\,\boldsymbol{\nu}, \tag{14.4}$$

and since $Z(0)=0$, this means

$$\mathscr{E} Z(t) = -t\,\boldsymbol{\pi}_0 \boldsymbol{B}'(0)\,\boldsymbol{\nu}, \qquad \boldsymbol{\nu} = \binom{1}{1}, \tag{14.5}$$

where

$$\boldsymbol{B}'(0) = \begin{pmatrix} 0 & 2\lambda \int_0^1 u\,q(u)\log u\,du \\ \int_0^1 u\,k(u)\log u\,du & \int_0^1 u\,k(1-u)\log u\,du \end{pmatrix}. \tag{14.6}$$

The evaluation of the variance is somewhat more complicated, but involves only routine methods of treating linear equations. We obtain the estimate

$$\text{Variance}\, Z(t) = \boldsymbol{\pi}_0 \left[\boldsymbol{B}''(0) - \frac{2}{(\lambda+K)^2}\boldsymbol{B}'(0)\,\boldsymbol{B}(0)\,\boldsymbol{B}'(0)\right]\boldsymbol{\nu}t + O(1), \tag{14.7}$$

again assuming that initially $Z(0)=0$ and that $I(0)$ has the stationary distribution.

14.1. Numerical calculation. If we use the numerical values given in Sec. 3, we find[1]

$$\left.\begin{aligned} &\lambda=0.774,\ K=1.01, \quad \boldsymbol{\pi}_0=(0.5661,\ 0.4339), \\ &\boldsymbol{B}(0)=\begin{pmatrix}-0.774 & 0.774\\ 1.01 & -1.01\end{pmatrix}, \quad \boldsymbol{B}'(0)=\begin{pmatrix}0 & -0.3678\\ -1.131 & -0.6761\end{pmatrix}, \\ &\boldsymbol{B}''(0)=\begin{pmatrix}0 & 0.3836\\ 2.454 & 0.3858\end{pmatrix}. \end{aligned}\right\} \qquad (14.8)$$

Performing the required calculations, we obtain

$$\mathscr{E}Z(t)=0.992t, \quad \operatorname{Var} Z(t)=1.564t+O(1), \qquad t\to\infty. \qquad (14.9)$$

The standard deviation of $Z(t)$ is thus $1.25t^{\frac{1}{2}}+O(t^{-\frac{1}{2}})$.

The distribution of $Z(t)$ has a probability density, except for the small probability concentrated at 0; let $p(z,t)$ denote this density. Then, assuming that $Z(t)$ has an approximately Gaussian distribution, we have the asymptotic relation, for β_1 and β_2 fixed,

$$\int\limits_{0.992t+1.25\beta_1\sqrt{t}}^{0.992t+1.25\beta_2\sqrt{t}} p(z,t)\,dz \to \frac{1}{\sqrt{2\pi}}\int\limits_{\beta_1}^{\beta_2} e^{-\frac{1}{2}x^2}\,dx, \qquad t\to\infty. \qquad (14.10)$$

In order to interpret (14.10), let $m(E,t)\,dE$ be the expected number of particles of both types with energies between E and $E+dE$, if initially there is one photon with probability 0.5661 or one electron with probability 0.4339; let us refer to this as the *stationary case.* Then (14.10) becomes, on referring to (13.7),

$$\int\limits_{e^{-0.992t-1.25\beta_2\sqrt{t}}}^{e^{-0.992t-1.25\beta_1\sqrt{t}}} E\,m(E,t)\,dE \to \frac{1}{\sqrt{2\pi}}\int\limits_{\beta_1}^{\beta_2} e^{-\frac{1}{2}x^2}\,dx. \qquad (14.11)$$

The left side of (14.11) is the expected amount of energy, in the stationary case, contained in particles whose energies lie between the limits of integration. For example, particles with energies $<e^{-0.992t}$ should contain, on the average, half the total energy.

The author does not know whether formula (14.11) has appeared in the literature. Since we have not proved the Gaussian character of $Z(t)$, the formula must be considered a conjecture, although a very probable one. Note that (14.11) is an asymptotic expression for the *amount of energy,* not the number of particles. Asymptotic formulas for the expected number of particles above a given energy level are given in HEISENBERG (1943, pp. 18—24). See also the paper by R. W. WILLIAMS in the collection of MENZEL (1960, Vol. 2, Sec. 2.3, p. 555).

[1] As a result of some rounding-off errors, these calculations were carried out with the value $\lambda=0.774$ rather than the value 0.77333 ... of Sec. 3.1.

It is tempting to approximate $m(E, t)$ by the formula

$$m(E,t) \overset{(?)}{\sim} \frac{1}{E^2\sqrt{(2\pi)(1.564\,t)}}\, e^{-(-\log E - 0.992\,t)^2/3.128\,t}, \qquad (14.12)$$

where t and $-\log E$ go to ∞ in such a manner that $|-\log E - 0.992t|/\sqrt{t}$ remains bounded. Now the right side of (14.12) reaches its maximum, for any given t, when $E = E_t = e^{-4.12t}$, which is presumably outside the range of validity of (14.12). However, the right side of (14.12) reaches its maximum, for a given E, when $t = t_E = -\log E/0.992 + O(1)$, where $O(1)$ is bounded as $E \to 0$. This is within the range where (14.12) should be valid, and hence we expect

$$m(E,t_E) \sim \frac{0.318}{E^2\sqrt{-\log E}}, \qquad E \to 0. \qquad (14.13)$$

As a matter of fact, $m(E, t_E)$ has been calculated by physicists using saddle-point methods, and the calculations agree with (14.13). Let us refer to MENZEL (1960, Vol. 2, formula (2), and Table 2 of Sec. 2.4, pp. 555—556). According to (2), whether the initial particle is a photon of energy 1 or an electron of energy 1, the maximum densities for photons and electrons are respectively (asymptotically for small E) $0.180/E^2(-\log E)^{\frac{1}{2}}$ and $0.137/E^2(-\log E)^{\frac{1}{2}}$. The sum of 0.180 and 0.137 is 0.317, which checks with our value 0.318 in (14.13). Since the author's calculations were rather rough, the closeness of the agreement may be partly fortuitous.

15. Total energy in the electrons

Although the total energy in all particles is always equal to 1, the amount of energy in the *electrons* is a random variable, which we shall call $W(t)$. The expected value of $W(t)$, if the initial particle is of type i, is

$$\mathscr{E}_i W(t) = -\int_0^{1+} E\, dM_{i2}(E,t) - P(I(t) - 2 \,|\, I(0) - i)$$

in the notation of Sec. 13. Since $I(t)$ is a two-state Markov process with the limiting probability distribution $(K/(\lambda+K),\ \lambda/(\lambda+K))$, we see that $\mathscr{E}_i W(t) \to \lambda/(\lambda+K)$, a result given by BHABHA and CHAKRABARTY (1943).

Actually, we have the stronger result[1]

$$\lim_{t\to\infty} \mathscr{E}_i \left[W(t) - \frac{\lambda}{\lambda+K}\right]^2 = 0, \qquad i = 1, 2. \qquad (15.1)$$

In other words, for large thicknesses the fraction of energy in the electrons is with high probability close to $\lambda/(\lambda+K) = 0.4339$.

As was pointed out in Sec. 5, our probability space has not been defined so as to give a rigorous probabilistic meaning to statements

[1] HARRIS (1957b).

involving "$t \to \infty$". However, the joint distribution of $W(t')$, $W(t'')$, ... can be shown to be independent of T provided $t', t'', \ldots < T$. Since T is arbitrary, we can define a stochastic process $W(t)$, $0 \leqq t < \infty$, by means of these joint distributions; we shall then interpret (15.1) as applying to this process.

Definition 15.1. Let $\varphi_i(s, t) = \mathscr{E}_i e^{-sW(t)}$, $Re(s) \geqq 0$; $i = 1, 2$.

In essentially the same manner as equations (11.14) were derived, we can show that the following equations hold:

$$\frac{\partial \varphi_1(s,t)}{\partial t} = -\lambda \varphi_1(s,t) + \lambda \int_0^1 \varphi_2(su,t)\, \varphi_2(s(1-u),t)\, q(u)\, du, \tag{15.2a}$$

$$\frac{\partial \varphi_2(s,t)}{\partial t} = \int_0^1 [\varphi_1(su,t)\, \varphi_2(s(1-u),t) - \varphi_2(s,t)]\, k(u)\, du, \tag{15.2b}$$

$$\varphi_1(s,0) = 1, \quad \varphi_2(s,0) = e^{-s}. \tag{15.2c}$$

As usual, we obtain the moments by differentiating (15.2a) and (15.2b) at $s = 0$. The first moments are

$$\left.\begin{aligned} \mathscr{E}_1 W(t) &= \frac{\lambda}{K+\lambda}\left(1 - e^{-(K+\lambda)t}\right), \\ \mathscr{E}_2 W(t) &= \frac{\lambda}{K+\lambda} + \frac{K}{K+\lambda}\, e^{-(K+\lambda)t}. \end{aligned}\right\} \tag{15.3}$$

It can then be shown that the ratios $\mathscr{E}_1 W^2(t)/(\mathscr{E}_1 W(t))^2$ and $\mathscr{E}_2 W^2(t)/(\mathscr{E}_2 W(t))^2$ approach 1 as $t \to \infty$, implying that the variance of $W(t)$ goes to 0. Hence (15.1) holds.

15.1. Martingale property of the energy[1]. From the definition of the electron-photon cascade we see heuristically that

$$\left.\begin{aligned} &\mathscr{E}_i[W(t+h) \mid W(\tau), 0 \leqq \tau \leqq t] \\ &= W(t)\, \mathscr{E}_2 W(h) + [1 - W(t)]\, \mathscr{E}_1 W(h) \\ &= W(t)\, e^{-(K+\lambda)h} + \frac{\lambda}{K+\lambda}\left[1 - e^{-(K+\lambda)h}\right], \qquad i = 1, 2;\ h \geqq 0. \end{aligned}\right\} \tag{15.4}$$

Introduce the random function $W^*(t) = e^{(K+\lambda)t}\left[W(t) - \frac{\lambda}{K+\lambda}\right]$; then (15.4) implies that $W^*(t)$ is a martingale. However, it does not appear easy to evaluate $\mathscr{E}_i|W^*(t)|$, and it can be shown that $\mathscr{E}_i(W^*(t))^2 \to \infty$ as $t \to \infty$. Hence it is not clear whether we can apply the martingale convergence theorem given in Doob (1953, Chapter 7).

---

[1] Harris (1957b, footnote 12). A martingale property of the energy in nucleon cascades has been investigated by Peter Ney (1961), who has studied the total energy in nucleon cascades where there may be a loss of energy at each collision. Ney (1962a, b) has also studied random functions analogous to the energy in more general kinds of cascade processes.

16. Limiting distributions

As in the case of other branching processes, it is natural to consider the limiting distributions as $t\to\infty$. We shall first quote a result about the distribution of $N(E, t)$ as $t\to\infty$ when E is *fixed*. Next, we shall make some conjectures about the situation if E goes to 0 at a suitable rate as $t\to\infty$. The papers cited in Sec. 16.1 make the assumption that $\lambda<\int_0^1 k(u)\,du<\infty$ and $0<k(u)$, $q(u)$ for $0<u<1$.

16.1. Case in which $t\to\infty$, E fixed. Let $N(E, t)$ and $N_0(E, t)$ be the respective numbers of electrons and photons with energies $>E$ at t. Let us first observe that, whether the initial particle is an electron or a photon, $N(E, t)$ and $N_0(E, t)$ converge in probability to 0 as $t\to\infty$. This was proved by URBANIK (1955b). The result also is a consequence of the statement that $\lim_{t\to\infty} P_{ij}(z, t)=0$ (see Sec. 13.1), which seems obvious from the nature of the expectation process.

It is then natural to consider the joint conditional distribution of $N(E, t)$ and $N_0(E, t)$ as $t\to\infty$ with E fixed, given that their sum is not 0. A limiting distribution exists, but the situation is quite different from the cases where YAGLOM's theorem and its generalization apply (Theorems I.9.1 or II.9.1), since the limiting distribution is degenerate.

In fact, we have the following result, due to KONWENT and ŁOPUSZAŃSKI (1956). The result, which is closely related to the work of URBANIK (1955b, c), ŁOPUSZAŃSKI (1955a, b; 1956), STACHOWIAK (1956), and CZERWONKO (1957), is as follows. (The initial particle, of energy 1, may be either an electron or a photon.)

Suppose that E and E' are fixed, $0<E<E'<1$. Under the condition that $N(E, t)+N_0(E, t)\geqq 1$, the probability approaches 1 as $t\to\infty$ that $N(E, t)=0$, that $N_0(E, t)=1$, and that $N(E', t)+N_0(E', t)=0$. In other words, if there are any particles of energy $>E$ when t is large, then with high probability there is just one particle, it is a photon, and its energy is almost exactly E. (The authors disclaim complete rigor.)

It also follows from the work of the above authors that

$$\lim_{t\to\infty}\frac{P^{ij}(E, t)}{M_{ij}(E, t)}=1, \qquad 0<E<1, \qquad i, j=1, 2, \tag{16.1}$$

where M_{ij} is defined in Def. 13.1 and $P^{ij}(E, t)$ is the probability of one or more particles of type j and energy $>E$ at t, without regard to the number of particles of the other type, if initially there is one particle of type i.

It may seem surprising that photons are completely dominant among those rare particles that are still energetic at very great thicknesses, since the electrons carry about 43 per cent of the total energy (see Sec. 15).

In view of (16.1), the dominance of the photons is closely related to the assertion that $\lim_{t\to\infty} M_{i2}(E,t)/M_{i1}(E,t)=0$, $i=1,2$; $0<E<1$. Since (13.7) implies $EM_{ij}(E,t)\leq P_{ij}(-\log E,t)\leq M_{ij}(E,t)$, $0<E<1$, this assertion is equivalent to the following one about the expectation process:

$$\lim_{t\to\infty} \frac{\mathrm{Prob}\{I(t)=2\,|\,I(0)=i,\,Z(t)\leq z\}}{\mathrm{Prob}\{I(t)=1\,|\,I(0)=i,\,Z(t)\leq z\}}=0, \qquad z>0. \tag{16.2}$$

The author conjectures that (16.2) is true. It would be interesting to prove it by probabilistic arguments.

16.2. Limit theorems when $t\to\infty$ and $E\to 0$. No limit theorems of this sort have been proved for the electron-photon cascade. If we consider the results of KOLMOGOROV (1941) and FILIPPOV (1961) mentioned in Sec. III.16, in conjunction with (14.11), it appears probable that the following sort of result holds, for either an initial electron or an initial photon. We use the numerical values of Sec. 14.1.

Let $Y(\beta,t)$ be the total amount of energy at t in particles whose energies are $>\exp\{-0.992t-1.25\beta\sqrt{t}\}$. Then the ratio $Y(\beta_2,t)/Y(\beta_1,t)$ should converge in probability to $\varphi(\beta_2)/\varphi(\beta_1)$ as $t\to\infty$, where $\varphi(x)=(2\pi)^{-\frac{1}{2}}\int_{-\infty}^{x} e^{-\frac{1}{2}t^2}\,dt$. (Here β is not the parameter of Sec. 17.)

It also seems plausible that $Y(\beta,t)/\mathscr{E}\,Y(\beta,t)$ converges in probability to a random variable, for any fixed real β.

17. The energy of an electron when $\beta>0$ (Approximation B)

Let us return to the case mentioned briefly in Sec. 2.2, where an electron loses energy at the constant rate β per unit of time (ionization loss) in addition to losses from the radiation of photons. It was pointed out in Sec. 3.1 that we should take account of ionization loss unless the ratio β/E is small. Since β is about 100 million electron volts per cascade unit in air, Approximation A should not be used for electrons in air having energies of less than some hundreds of millions of electron volts.

Let us continue to denote by $\varepsilon(t)$ the energy of an electron as defined under Approximation A (that is, the random function of Def. 5.2), and let $\varepsilon_1(t)$ denote the energy of an electron under Approximation B. We wish to find an appropriate way of expressing a sample function ε_1 in terms of a sample function ε.

We shall list certain appropriate requirements for a sample function ε_1 having photon emissions at the same thicknesses as a given sample function ε but subject also to ionization loss. We shall then see that these requirements determine ε_1 uniquely.

By neglecting certain events whose total probability is 0, we can suppose that $\varepsilon(0)=1$ and that ε is continuous to the right and always positive. (See Sec. 5.)

The requirements for ε_1 are the following three.

(a) $\varepsilon_1(0)=1$; ε_1 is continuous at every continuity point of ε, is nonnegative and nonincreasing, and is continuous to the right at every point.

(b) If τ is a point of discontinuity of ε and if $\varepsilon_1(\tau-0)>0$, then

$$\frac{\varepsilon_1(\tau)}{\varepsilon_1(\tau-0)}=\frac{\varepsilon(\tau)}{\varepsilon(\tau-0)}\,. \tag{17.1}$$

Note. From (a) and (b) we can deduce that $\varepsilon_1(t)/\varepsilon(t)$ must be a continuous function of t for all $t\geqq 0$.

(c) Let t be any point such that $\varepsilon_1(t)$ is positive. Then we require the inequality

$$-\frac{\beta}{\varepsilon(t+h)}\leqq\frac{1}{h}\left[\frac{\varepsilon_1(t+h)}{\varepsilon(t+h)}-\frac{\varepsilon_1(t)}{\varepsilon(t)}\right]\leqq-\frac{\beta}{\varepsilon(t)}\,, \tag{17.2}$$

provided h is positive and sufficiently small.

Equation (17.2) has the following interpretation. Bremsstrahlung is a multiplicative operation and ionization loss is additive; since the two operations do not commute, it is not immediately evident what we should mean by simultaneous and continual application of the two. However, it is surely reasonable to require that if $h>0$ is sufficiently small, then $\varepsilon_1(t+h)$ is *less* than what we would obtain by first subtracting the ionization loss βh from $\varepsilon_1(t)$ and then multiplying by $\varepsilon(t+h)/\varepsilon(t)$, and $\varepsilon_1(t+h)$ is *greater* than what would be obtained by performing these operations in the reverse order; that is,

$$\varepsilon_1(t)\frac{\varepsilon(t+h)}{\varepsilon(t)}-\beta h\leqq\varepsilon_1(t+h)\leqq\left(\varepsilon_1(t)-\beta h\right)\frac{\varepsilon(t+h)}{\varepsilon(t)}\,. \tag{17.3}$$

But equation (17.3) is equivalent to (17.2).

From (17.2) we see that $\varepsilon_1(t)/\varepsilon(t)$ is differentiable on the right at every point t such that $\varepsilon_1(t)>0$, the right-hand derivative being equal to $-\beta/\varepsilon(t)$ whenever $\varepsilon_1(t)>0$. Since $\varepsilon_1/\varepsilon$ is also continuous, it must have a derivative, equal to $-\beta/\varepsilon(t)$, at each continuity point t of ε for which $\varepsilon_1(t)>0$ (Hobson (1927, Vol. 1, p. 365)). Since $\varepsilon_1(0)/\varepsilon(0)=1$, we have[1]

$$\varepsilon_1(t)=\varepsilon(t)\max\left[0,\,1-\beta\int_0^t\frac{ds}{\varepsilon(s)}\right]. \tag{17.4}$$

It can be shown that the function ε_1 defined by (17.4) satisfies (a), (b), and (c).

---

[1] Harris (1957b).

The form of the expression (17.4) does not depend on the Markovian nature of the process ε. However, the energy of an electron under Approximation B can be appropriately defined as a Markov process and studied by means of differential equations; see, e.g., RAMAKRISHNAN and MATHEWS (1956). An alternative approach is suggested by the expression (17.4). Put

$$Y(t) = \varepsilon(t) - \beta\,\varepsilon(t) \int_0^t \frac{ds}{\varepsilon(s)}\,. \tag{17.5}$$

Since $\varepsilon_1(t) = \max\big(0, Y(t)\big)$, the distribution of $\varepsilon_1(t)$ is determined by the distribution of $Y(t)$. This distribution is determined by its moments, which can be found by tedious but direct calculation from (4.1) and the expression (17.5).

We now show that our definition of ε_1 makes the proper disposition of energy.

Theorem 17.1. *If ε_1 is defined by (17.4), then*

$$\varepsilon_1(t) = 1 - \beta t - (\textit{sum of magnitudes of jumps of } \varepsilon_1 \textit{ preceding or at } t)\,, \tag{17.6}$$

provided $\varepsilon_1(t) > 0$.

Proof. Suppose $\varepsilon_1(t_0) > 0$. Let $t_1, t_2, \ldots$ be the location of the jumps of ε preceding or at t_0. Then, from (17.4),

$$\left.\begin{aligned} &\sum_{i=1}^{\infty} [\varepsilon_1(t_i - 0) - \varepsilon_1(t_i + 0)] \\ &= \sum [\varepsilon(t_i - 0) - \varepsilon(t_i + 0)] \left[1 - \beta \int_0^{t_i} \frac{ds}{\varepsilon(s)}\right]. \end{aligned}\right\} \tag{17.7}$$

To simplify (17.7), observe that $\sum [\varepsilon(t_i - 0) - \varepsilon(t_i + 0)] = 1 - \varepsilon(t_0)$ and that if f is any continuous function, then

$$\sum [\varepsilon(t_i - 0) - \varepsilon(t_i + 0)]\, f(t_i) = -\int_0^{t_0} f(t)\, d\varepsilon(t)\,.$$

Thus we can rewrite (17.7) as

$$1 - \varepsilon(t_0) + \beta \int_0^{t_0} \left[\int_0^t \frac{ds}{\varepsilon(s)}\right] d\varepsilon(t) = 1 - \varepsilon(t_0) + \beta \int_0^{t_0} \left[\int_s^{t_0} d\varepsilon(t)\right] \frac{ds}{\varepsilon(s)}$$

$$= 1 - \varepsilon(t_0) + \beta \int_0^{t_0} \frac{\varepsilon(t_0) - \varepsilon(s)}{\varepsilon(s)}\, ds = 1 - \beta t_0 - \varepsilon_1(t_0)\,,$$

which proves the theorem. □

Formula (17.4) is useful for production of ε_1 sample functions by artificial sampling (Monte Carlo), since a single ε function can be used to produce numerous ε_1 functions corresponding to different values of β; or, if β is fixed, for many different values of the initial energy.

18. The electron-photon cascade (Approximation *B*)

When electrons lose energy according to the law described in Sec. 17, the treatment of electron-photon cascades becomes much more complicated than for Approximation A, and many authors have given analytical and numerical treatments. Various types of equations for the first and second moments have been derived; see BHARUCHA-REID (1960, Chapter 5) and RAMAKRISHNAN (1962, Chapter XIII). The first moments have been tabulated (see CHAKRABARTY and GUPTA (1956) and RAMAKRISHNAN (1962, p. 537)), but tabulations of the second moments do not seem to have been performed.

The following heuristic discussion shows some rather surprising properties of the cascade process under Approximation B if $\mu > 0$. The reader must keep in mind that here we are speaking only of the *mathematical model*. The physical significance, if any, of these remarks must be very limited.

In what follows, we suppose that there is one initial electron with energy 1.

We let $N(t)$ be the number of electrons with positive energies at t. Since infinitely many photons are created (and hence infinitely many electrons), we might expect $N(t)$ to be infinite. However, we must keep in mind that electrons disappear when they reach 0 energy, and hence it is not immediately obvious whether N is finite or infinite.

From Theorem 17.1 we see that the total energy in the electrons and photons decreases at the rate $\beta N(t)$. Since the total energy lost cannot exceed 1, we have

$$\beta \int_0^\infty N(t)\, dt \leqq 1, \tag{18.1}$$

with probability 1, and accordingly,

$$\beta \int_0^\infty \mathscr{E} N(t)\, dt \leqq 1. \tag{18.2}$$

This means that $\mathscr{E} N(t)$ must be finite for almost all t. Indeed, it was shown by BHABHA and CHAKRABARTY (1943) that $\mathscr{E} N(t) < \infty$ for each t. We shall not prove this statement here. It can be shown that likewise $\mathscr{E} N^2(t) < \infty$ for each t.

Although the relation $\mathscr{E} N(t) < \infty$ implies that for each particular t the random variable $N(t)$ is finite with probability 1, the random function N has the following peculiar property[1]. *In each t-interval the random function N is infinite for infinitely many values of t with probability* 1.

The following argument should convince the reader that this statement is true. Suppose that initially there is one electron, and let $\langle 1 \rangle$,

[1] HARRIS (1957b).

$\langle 2\rangle, \ldots$ be the first-generation photons born from the initial electron before its energy reaches 0. Let $I=[t_1, t_2]$ be any fixed closed interval. With probability 1, one of the photons, say $\langle n_1\rangle$, gives birth to an electron that lives through some random closed interval $I_1 \subset I$. With probability 1 there is a photon $\langle n_2\rangle$, $n_2 > n_1$, such that $\langle n_2\rangle$ gives birth to an electron that lives through some random closed interval $I_2 \subset I_1$. Carrying on this argument, we see that there is a sequence of photons $\langle n_1\rangle, \langle n_2\rangle, \ldots$ such that $\langle n_i\rangle$ gives birth to an electron that lives through a closed interval I_i, where $I \supset I_1 \supset I_2 \supset \ldots$. Since the intervals I_i are closed, they must have a common point $\bar{t} \in I$, and obviously $N(\bar{t})=\infty$.

Another property of N is that it must vanish somewhere in every t-integral of length greater than $1/\beta$. This follows from (18.1) and the fact that $N(t)$ is an integer.

Appendix 1

Proofs of Theorems 6.2 and 6.3. The constants $c, c_1, c_2, \ldots$ are independent of x, t, and τ. The statement that a function is bounded applies to the ranges $0 \leqq t < T$, $-\infty < \tau < \infty$, $0 \leqq x < \infty$, whichever are applicable.

From Conditions 3.2 we see that $e^{-x} g'(x) = O(e^{-cx})$, $x \to \infty$, for some $c > 0$. Hence $\int_{-\infty}^{\infty} |e^{-x} g'(x)|^2 \, dx < \infty$, and hence, from the Parseval equality, $\int_{-\infty}^{\infty} |\xi(\tau)|^2 \, d\tau < \infty$. Then from the Schwarz inequality we see that $\xi(\tau)/(1-i\tau)$ is absolutely integrable. From (6.3) we have

$$\left.\begin{aligned} \theta_t(\tau) = {} & \frac{e^{-At}}{(1-i\tau)^{\mu t}} \left\{1 + \frac{t g(0)}{1-i\tau} + \frac{t\xi(\tau)}{1-i\tau}\right\} \\ & + \frac{e^{-At}}{(1-i\tau)^{\mu t}} \, \frac{t^2 [g(0)+\xi(\tau)]^2}{(1-i\tau)^2} \, \psi\left(\frac{t[g(0)+\xi(\tau)]}{1-i\tau}\right), \end{aligned}\right\} \tag{1}$$

where $\psi(z) = (e^z - 1 - z)/z^2$.

Let us write the right side of (1) as $g_1 + t g_2 + t g_3 + t^2 g_4$, where $g_1 = e^{-At}(1-i\tau)^{-\mu t}$, $g_2 = e^{-At} g(0)(1-i\tau)^{-1-\mu t}$, etc. Then, from the Fourier inversion formula,

$$h_t(x) = \lim_{T_1\to\infty} \frac{1}{2\pi} \int_{-T_1}^{T_1} e^{-i\tau x} (g_1 + t g_2 + t g_3 + t^2 g_4) \, d\tau. \tag{2}$$

We recognize $g_1 + t g_2$ as the Fourier transform of the terms involving gamma functions on the right side of (6.5), and hence this part of (6.5) is accounted for.

Let us next consider the function

$$\varphi(x, t) = \frac{\Gamma(\mu t + 1)}{2\pi} \int_{-\infty}^{\infty} e^{-i\tau x} \frac{\xi(\tau) \, d\tau}{(1-i\tau)^{1+\mu t}}. \tag{3}$$

Since $\xi(\tau)$ is the Fourier transform of $e^{-x} g'(x)$ and $\Gamma(\mu t+1)(1-i\tau)^{-1-\mu t}$ is the transform of $e^{-x} x^{\mu t}$, the product of these two transforms is the transform of the convolution

$$\int_0^x e^{-(x-y)} (x-y)^{\mu t} e^{-y} g'(y) \, dy. \tag{4}$$

The function in (4) has a continuous partial derivative in x for $x > 0$ and hence is of bounded variation in every positive x neighborhood. Hence from the Fourier inversion formula we see that the function in (4) is $\varphi(x, t)$, and $\varphi(x, 0) = e^{-x}(g(x) - g(0))$. From the definition of g and Conditions 3.2 we see that $|g'(x)| \leqq c_1 e^{c_2 x}$, $0 < c_2 < 1$. Hence $|\partial\varphi/\partial t| \leqq c_3 e^{-(1-c_2)x}$. Since $1/\Gamma(\mu t + 1)$ is analytic and bounded and has a bounded derivative for $t \geqq 0$, we have, from the law of the mean,

$$\frac{\varphi(x, t)}{\Gamma(\mu t + 1)} = e^{-x}(g(x) - g(0)) + t e^{-(1-c_2)x} R_2, \tag{5}$$

where R_2 is bounded. Hence the term in (2) involving g_3 gives us

$$t e^{-x}(g(x) - g(0)) + t^2 R_3,$$

where $\int_0^\infty |R_3|\, dx$ is a bounded function of t.

If we now examine g_4 in (2), we see that g_4, $\partial g_4/\partial\tau$, and $\partial^2 g_4/\partial\tau^2$ vanish at $\tau = \pm\infty$, and are absolutely integrable in τ, the integrals of the absolute values being bounded functions of t. Hence, integrating twice by parts, we see that

$$\int_{-\infty}^{\infty} e^{-i\tau x} g_4\, d\tau = -\frac{1}{x^2}\int_{-\infty}^{\infty} e^{-i\tau x}\frac{\partial^2 g_4}{\partial\tau^2}\, d\tau, \qquad x > 0. \tag{6}$$

The left side of (6) can be seen directly to be a bounded function of t and x. On the other hand, from the right side of (6) we see that the bound is $\leqq c_4/x^2$. Hence the term in (2) involving g_4 has the form $R_4 t^2$, where $\int_{-\infty}^{\infty} |R_4|\, dx$ is a bounded function of t.

We complete the proof of Theorem 6.2 by collecting the terms in the above estimates. □

Theorem 6.3 is deduced by writing

$$\int_0^\infty F(x)\, h_t(x)\, dx = F(0) + \int_0^\infty (F(x) - F(0))\, h_t(x)\, dx$$

and then using Theorem 6.2 to express $h_t(x)$. In evaluating the integrals we make use of the law of the mean to write $x^{\mu t} = 1 + \mu t x^{\mu\bar{t}} \log x$, $0 < \bar{t} < t$. The details of the proof are straightforward and need not be included here.

Appendix 2

Proof of (11.1). The purpose of this appendix is to derive (11.1), which is basic to establishing the equation (11.14b). We shall take $f_1(s, E, t) = 1$ when $t < 0$.

Lemma 1. *With probability* 1 *we have*

$$\left.\begin{aligned} &\mathscr{E}_2\{s^{N(E,t)} | t_1, u_1, t_2, u_2, \ldots\} \\ &= s^{J[\varepsilon(t;\, t_1, u_1, t_2, u_2, \ldots) - E]} \prod_{i=1}^{\infty} f_1\left[s, \frac{E}{u_i \varepsilon(t_i - 0;\, t_1, u_1, \ldots)}, t - t_i\right], \end{aligned}\right\} \tag{1}$$

where $J(x) = 1$, $x > 0$ *and* $J(x) = 0$, $x \leqq 0$.

Remark. Hence $f_2(s, E, t)$ is equal to the expectation of the right side of (1). This will be used later.

Note that $u_i \varepsilon(t_i - 0)$ is the energy of the photon $\langle i\rangle$, and (1) means that the electrons at t are the original electron $\langle 0\rangle$ plus those descended from the first-generation photons born before t. Since $u_i \to 0$ with probability 1, only finitely many factors on the right side of (1) differ from 1.

Proof. We can write

$$\left.\begin{aligned} N(E, t; t_1, u_1, \ldots) &= J\left[\varepsilon(t; t_1, u_1, t_2, u_2, \ldots) - E\right] \\ &+ \sum_{i=1}^{\infty} N'\left(\frac{E}{u_i \varepsilon(t_i - 0; t_1, u_1, \ldots)}, t - t_i; \ell_i, u_{i1}, u_{i2}, \ldots\right), \end{aligned}\right\} \quad (2)$$

where $N'(E, t) = N'(E, t, \omega)$ is a random function, constructed analogously to $N(E, t)$, denoting the number of electrons of energy $> E$ at t, if initially there is one *photon* of energy 1. If $u_1, u_2, \ldots$ and $t_1, t_2, \ldots$ are fixed numbers, then the i-th term and the j-th term on the right side of (2), $i \neq j$, are independent. Let $\overline{P}$ be the probability measure on those coordinates in ω (Sec. 7.3) other than $t_1, t_2, \ldots$ and $u_1, u_2, \ldots$. Then $\overline{P}$ is a product measure $\overline{P}_1 \times \overline{P}_2 \times \cdots$, where $\overline{P}_i$ is the probability measure on coordinates (other than $t_1, u_1, t_2, u_2, \ldots$) appearing in the i-th term of the sum on the right side of (2). Using (2) to evaluate $\int s^N d\overline{P}$, we obtain Lemma 1. □

Now consider the set V of Def. 5.1. We shall refer to a denumerable set ξ of points in V as a *point-distribution* if it has the following properties. (1) No two points of ξ have the same ordinate u. (2) If $c > 0$, only finitely many points of ξ have ordinates $> c$. (3) For each interval (a, b), $0 < a < b < T$, infinitely many points of ξ have abscissae between a and b. Let Ξ be the set of all point-distributions ξ. Let $\xi(A)$ denote the number of points of ξ in the set A. We allow $\xi(A) = \infty$.

We can define a probability measure P_0 on Ξ, based on the Poisson law that follows Def. 5.1. The procedure is similar to that used in Chapter III to define measures on point-distributions and will not be discussed here. The measurable sets in Ξ correspond to those generated by cylinder sets of the form $\{\xi(A_1) = r_1, \ldots, \xi(A_n) = r_n\}$, where $A_1, A_2, \ldots$ are rational intervals in V that are bounded away from the t-axis.

Now let Ω^* be the set of sequences $\omega^* = (t_1, u_1, t_2, u_2, \ldots)$, $0 < t_i < T$, $0 < u_i < 1$. Let σ denote the mapping taking each point-distribution ξ into a sequence $\omega^* = \sigma\xi = (t_1, u_1, t_2, u_2, \ldots)$, where the (t_i, u_i) are the coordinates of the points in ξ, ordered so that $u_1 > u_2 > \cdots$. The measurable sets in Ω^* will be those generated in the usual way by cylinder sets on a finite number of coordinates. It can be verified that if A is a measurable set in Ω^*, then $\sigma^{-1}A$ is a measurable set in Ξ. Hence we can define a measure P^* on Ω^* by putting $P^*(A) = P_0(\sigma^{-1}A)$.

Now let Δ be a positive number, $0 < \Delta < T$, and let $V = V' \cup V''$, where V' is the part of V for which $t < \Delta$ and V'' is the remainder of V. A point-distribution ξ determines a pair of sequences $\omega' = (t_1', u_1', \ldots)$ and $\omega'' = (t_1'', u_1'', \ldots)$, where ω' corresponds to the points of ξ in V' and ω'' to the points of ξ in V'' and the u-coordinates are in decreasing order. Thus we have a mapping S of Ξ into the product space $\Omega' \times \Omega''$, where Ω' is the space of sequences $(t_1', u_1', \ldots)$ with $0 < t_i' < \Delta$ and $0 < u_i' < 1$, and Ω'' is the space of sequences $(t_1'', u_1'', \ldots)$ with $\Delta \leq t_i'' < T$ and $0 < u_i'' < 1$. This mapping determines a probability measure Q on $\Omega' \times \Omega''$. It is important to notice that Q is a product measure, $Q = P' \times P''$, where P' is the measure induced on Ω', and P'' is the measure induced on Ω''. This corresponds to the fact that the points of ξ in the region $0 < t < \Delta$, $0 < u < 1$ are independent of the points of ξ in the region $\Delta \leq t < T$, $0 < u < 1$. We shall assume that none of the t_i (and hence none of the t_i' or t_i'') is equal to Δ; this is true with probability 1.

The t_i' and the t_i'' are all independent and are uniformly distributed on the intervals $(0, \Delta)$ and (Δ, T), respectively. The random variables $u_1', u_2', \ldots$ are independent of the t_i' and the t_i'' and are distributed like $u_1, u_2, \ldots$ in (5.1) except that

T is replaced by Δ. The random variables $u_1'', u_2'', \ldots$ are independent of $u_1', u_2', \ldots$ and of the t_i' and the t_i'', and are distributed as in (5.1) except that T is replaced by $T-\Delta$.

The mapping S has a unique inverse for all (ω', ω''), except a set of Q-measure 0.

Now let $f(\omega^*)$ be an integrable function on Ω^*. Then

$$\left.\begin{aligned} \int_{\Omega^*} f(\omega^*)\, dP(\omega^*) &= \int_{\Xi} f(\sigma\xi)\, dP_0(\xi) \\ &= \int_{\Omega'\times\Omega''} f[\sigma S^{-1}(\omega', \omega'')]\, dQ(\omega', \omega'') \\ &= \int_{\Omega'} dP'(\omega') \int_{\Omega''} f[\sigma S^{-1}(\omega', \omega'')]\, dP''(\omega''). \end{aligned}\right\} \tag{3}$$

Let us apply (3) to Lemma 1. From the definition of the random function ε we see that if $t > \Delta$, then

$$\left.\begin{aligned} &\varepsilon(t; t_1, u_1, \ldots) \\ &= \varepsilon(\Delta; t_1', u_1', t_2', u_2', \ldots)\, \varepsilon(t-\Delta; t_1''-\Delta, u_1'', t_2''-\Delta, u_2'', \ldots). \end{aligned}\right\} \tag{4}$$

Next, we can write the infinite product on the right side of (1) as the product of two products, one containing factors with $t_i < \Delta$, the other containing factors with $t_i > \Delta$. Using (4) and noting that $J(ax) = J(x)$ if $a > 0$, we can write the right side of (1), with t replaced by $t+\Delta$, as

$$\left.\begin{aligned} &s^{J[\varepsilon(t;\, t_1''-\Delta,\, u_1'', \ldots) - E/\varepsilon(\Delta;\, t_1', \ldots)]} \\ &\times \prod_{i=1}^{\infty} f_1\left(s, \frac{E}{u_i''\,\varepsilon(t_i''-\Delta-0;\, t_1''-\Delta, u_1'', \ldots)\,\varepsilon(\Delta;\, t_1', u_1', \ldots)}, t-(t_i''-\Delta)\right) \\ &\times \prod_{i=1}^{\infty} f_1\left(s, \frac{E}{u_i'\,\varepsilon(t_i'-0;\, t_1', u_1', \ldots)}, t+\Delta-t_i'\right) = s^J \Pi_1 \Pi_2 . \end{aligned}\right\} \tag{5}$$

Now let us consider what happens if we calculate the expectation of (5) using (3). We first integrate with respect to $dP''(\omega'')$; in this integration, Π_2 is constant. Reference to the remark following Lemma 1 shows that

$$\int s^J \Pi_1\, dP'' = f_2\left(s, \frac{E}{\varepsilon(\Delta)}, t\right). \tag{6}$$

Hence we have

$$f_2(s, E, t+\Delta) = \int \left\{ f_2\left(s, \frac{E}{\varepsilon(\Delta)}, t\right) \Pi_2 \right\} dP'$$

and this is just (11.1). □

Bibliography

References marked with an asterisk have not been cited in the text but are included in the Bibliography because of their general relevance.

AM: Annals of Mathematics
AMS: The Annals of Mathematical Statistics
BAMS: Bulletin of the American Mathematical Society
Berk. Symp.: Proceedings of the Berkeley Symposium on Mathematical Statistics and Probability
BMB: The Bulletin of Mathematical Biophysics
CJM: Canadian Journal of Mathematics
CPAM: Communications on Pure and Applied Mathematics
Doklady: Doklady Akademii Nauk SSSR[1]
IJM: Illinois Journal of Mathematics
JMAA: Journal of Mathematical Analysis and Applications
JMM: Journal of Mathematics and Mechanics
JRSS: Journal of the Royal Statistical Society
JSIAM: Journal of the Society for Industrial and Applied Mathematics
PAMS: Proceedings of the American Mathematical Society
PCPS: Proceedings of the Cambridge Philosophical Society
PIAS: Proceedings of the Indian Academy of Sciences
PLMS: Proceedings of the London Mathematical Society
PNAS: Proceedings of the National Academy of Sciences, U.S.A.
PPS: Proceedings of the Physical Society (London)
PR: The Physical Review
PRS: Proceedings of the Royal Society
PSAM: Proceedings of Symposia in Applied Mathematics, **11**: Nuclear Reactor Theory
PTRS: Philosophical Transactions of the Royal Society of London
RMP: Reviews of Modern Physics
TAMS: Transactions of the American Mathematical Society
TV: Teoriya Veroyatnostei i ee Primeneniya[2]

ABEL, N. H.: (1881) Oeuvres, vol 2. Christiania, 36—39.

AMALDI, EDOARDO: (1959) The production and slowing down of neutrons. Handbuch der Physik **38**, Berlin.

AMBARTSUMIAN, V. A.: (1943) Diffuse reflection of light by a foggy medium. Doklady **38**, 229—232.

ANSCOMBE F. J.: (1953) Sequential estimation. JRSS B **15**, 1—21.

ARLEY, NIELS: (1943) On the theory of stochastic processes and their application to the theory of cosmic radiation. Copenhagen: G. E. C. Gads Forlag.

—, and VIBEKE BORCHSENIUS: (1945) On the theory of infinite systems of differential equations and their application to the theory of stochastic processes and the perturbation theory of quantum mechanics. Acta Math. **76**, 261—322.

[1] Recent issues translated regularly as *Soviet Mathematics* by the American Mathematical Society.

[2] Translated regularly as *Theory of Probability and Its Applications* by the Society for Industrial and Applied Mathematics.

ARMITAGE, P.: (1952) The statistical theory of bacterial populations subject to mutation. JRSS B **14**, 2—40.

BAILEY, NORMAN T. J.: (1957) The mathematical theory of epidemics. New York: Hafner Publishing Company.

BANACH, STEFAN: (1932) Théorie des opérations linéaires. Warsaw.

BARTHOLOMAY, ANTHONY F.: (1958) Stochastic models for chemical reactions, I: Theory of the unimolecular reaction process. BMB **20**, 175—190.

BARTLETT, M. S.: (1946) Stochastic processes, lecture notes. University of North Carolina.

— (1949) Some evolutionary stochastic processes. JRSS B **11**, 211—229.

*— (1951) The dual recurrence relation for multiplicative processes. PCPS **47**, 821—825.

— (1954) Processus stochastiques ponctuels. Ann. Inst. H. Poincaré **14**, 35—60.

— (1955) An introduction to stochastic processes with special reference to methods and applications. Cambridge University Press.

*— (1956) Deterministic and stochastic models for recurrent epidemics. 3d Berk. Symp. **4**, 81—109.

— (1960) Stochastic population models in ecology and epidemiology. New York: John Wiley and Sons, Inc.

—, and DAVID G. KENDALL: (1951) On the use of the characteristic functional in the analysis of some stochastic processes occurring in physics and biology. PCPS **47**, 65—76.

BATEMAN, H.: (1943) Some simple differential difference equations and the related functions. BAMS **49**, 494—512.

*BELLMAN, RICHARD: (1952) The iteration of power series in two variables. Duke Math. J. **19**, 339—347.

— (1953) Stability theory of differential equations. New York: McGraw-Hill Book Company, Inc.

—, and KENNETH L. COOKE: (1963) Differential-difference equations. Princeton University Press.

—, and THEODORE HARRIS: (1948) On the theory of age-dependent stochastic branching processes. PNAS **34**, 601—604.

— — (1952) On age-dependent binary branching processes. AM **55**, 280—295.

—, R. E. KALABA, and M. C. PRESTRUD: (1962) Invariant imbedding and radiative transfer in slabs of finite thickness. Report R-388-ARPA, Santa Monica, Calif.: The RAND Corporation.

— — and G. MILTON WING: (1957) On the principle of invariant imbedding and one-dimensional neutron multiplication. PNAS **43**, 517—520.

— — — (1958) On the principle of invariant imbedding and neutron transport theory, I—one-dimensional case. JMM **7**, 149—162.

— — — (1960) Invariant imbedding and neutron transport theory, V—diffusion as a limiting case. JMM **9**, 933—944.

BETHE, H., and W. HEITLER: (1934) On the stopping of fast particles and on the creation of positive electrons. PRS London A **146**, 83—112.

BHABHA, H. J.: (1950) On the stochastic theory of continuous parametric systems and its applications to electron cascades. PRS London A **202**, 301—322.

*—, and S. K. CHAKRABARTY: (1942) Calculations on the cascade theory with collision loss. PIAS A **15**, 464—476.

— — (1943) The cascade theory with collision loss. PRS London A **181**, 267—303.

—, and W. HEITLER: (1937) The passage of fast electrons and the theory of cosmic showers. PRS London A **159**, 432—458.

BHABHA, H. J., and A. RAMAKRISHNAN: (1950) The mean square deviation of the number of electrons and quanta in the cascade theory. PIAS A **32**, 141—153.

*BHARUCHA-REID, A. T.: (1953) An age-dependent stochastic model of population growth. BMB **15**, 361—365.

*— (1954) Age-dependent branching stochastic processes in cascade theory. PR **96**, 751—753.

*— (1956) On the stochastic theory of epidemics. 3d Berk. Symp. **4**, 111—119.

— (1958) Comparison of populations whose growth can be described by a branching stochastic process—with special reference to a problem in epidemiology. Sankhyā **19**, 1—14.

— (1960) Elements of the theory of Markov processes and their applications. New York: McGraw-Hill Book Company, Inc.

— (1961) Markov branching processes and semigroups of operators, I. Wayne State University, Technical Report No. 3.

—, and H. RUBIN: (1958) Generating functions and the semigroup theory of branching Markov processes. PNAS **44**, 1057—1060.

BIRKHOFF, GARRETT: (1957) Extensions of Jentzsch's theorem. TAMS **85**, 219—227.

— (1961) Positivity and criticality. PSAM **11**, 116—126.

—, and RICHARD S. VARGA: (1958) Reactor criticality and nonnegative matrices. JSIAM **6**, 354—377.

—, and EUGENE P. WIGNER (eds.): (1961) Nuclear reactor theory. PSAM **11**.

BLANC-LAPIERRE, A., and ROBERT FORTET: (1953) Théorie des fonctions aléatoires. Paris: Masson et Cie.

BOCHNER, S.: (1947) Stochastic processes. AM **48**, 1014—1061.

— (1955) Harmonic analysis and the theory of probability. University of California Press.

BROADBENT, S. R., and J. M. HAMMERSLEY: (1957) Percolation processes, I: Crystals and mazes. PCPS **53**, 629—641.

BUSBRIDGE, I. W.: (1960) The mathematics of radiative transfer. Cambridge Tracts in Mathematics and Mathematical Physics, No. 50. Cambridge University Press.

CARLSON, J. F., and J. R. OPPENHEIMER: (1937) On multiplicative showers. PR **51**, 220—231.

*CASESNOVES, DARÍO MARAVALL: (1958) La adición de variables aleatorias en número aleatorio y el proceso estocástico de la descendencia de un mismo progenitor. Rev. Mat. Hisp. Amer. (4) **18**, 214—243.

*— (1959) Algunos nuevos procesos estocásticos y sus aplicaciones. Rev. Acad. Ci. Madrid **53**, 435—489, 659—726.

CHAKRABARTY, S. K., and M. R. GUPTA: (1956) Calculations on the cascade theory of showers. PR **101**, 813—819.

CHANDRASEKHAR, S.: (1950) Radiative transfer. Oxford University Press. New York: Dover Publications, Inc., 1960.

*CHIANG, CHIN LONG: (1957) An application of stochastic processes to experimental studies of flour beetles. Biometrics **13**, 79—97.

CHUNG, K. L.: (1951) Notes on Markov chains, lecture notes. Columbia University.

— (1960) Markov chains with stationary transition probabilities. Berlin-Göttingen-Heidelberg: Springer.

ČISTYAKOV, V. P.: (1957) Local limit theorems of the theory of branching random processes. TV **2**, 360—374 [Russian].

— (1959a) Generalization of a theorem of the theory of branching random processes. TV **4**, 109—113 [Russian].

ČISTYAKOV, V. P.: (1959b) Two limit theorems for branching processes with n types of particles. TV **4**, 477—478 [Russian].

— (1960) Transition phenomena in branching processes with n types of particles (abstract). Doklady **131**, 522—524 [Russian].

— (1961) Transition phenomena in branching processes with n types of particles. TV **6**, 31—46 [Russian].

—, and N. P. MARKOVA: (1962) On some theorems for inhomogeneous branching processes. Doklady **147**, 317—320 [Russian].

CONNER, HOWARD E.: (1961) A limit theorem for a position-dependent branching process. JMAA **3**, 560—591.

*CONSAEL, R., and A. LAMENS: (1960) Processus Markoviens d'embranchement en démographie. Bull. Inst. Int. Stat. (The Hague) **37**, 271—289.

COURANT, E. D., and P. R. WALLACE: (1947) Fluctuations of the number of neutrons in a pile. Report from the Chalk River Laboratories of the National Research Council (Canada).

COURANT, R.: (1937) Differential and integral calculus. New York: Nordeman Publishing Company, Inc.

CZERWONKO, JERZY: (1957) Asymptotic behaviour of higher factorial moments of electron-photon cascades at large depths. Acta Phys. Polon. **16**, 305—325.

D'ANCONA, UMBERTO: (1954) The struggle for existence. Series D of Bibliotheca Biotheoretica, vol. VI. Leiden: E. J. Brill.

DARWIN, CHARLES: (1859, 1st ed.) The origin of species, Modern Library Edition. New York: Random House.

*DARWIN, J. H.: (1953) Population differences between species growing according to simple birth and death processes. Biometrika **40**, 370—382.

— (1956) The behaviour of an estimator for a simple birth and death process. Biometrika **43**, 23—31.

DAVISON, B.: (1957) Neutron transport theory. Oxford University Press.

DE CANDOLLE, ALPHONSE: (1873) Histoire des sciences et des savants depuis deux siècles, 384—396. Geneva-Basel-Lyons.

*DELBRÜCK, M.: (1944) A statistical problem. J. Tennessee Acad. Sci. **19**, 177—178.

DERMAN, C.: (1954) A solution to a set of fundamental equations in Markov chains. PAMS **5**, 332—334.

— (1955) Some contributions to the theory of denumerable Markov chains. TAMS **79**, 541—555.

DOEBLIN, W.: (1940) Éléments d'une théorie générale des chaînes simples constantes de Markoff. Ann. Sci. École Norm. Sup. (3) **57**, 61—111.

DOOB, J. L.: (1945) Markoff chains—denumerable case. TAMS **58**, 455—473.

— (1953) Stochastic processes. New York: John Wiley and Sons, Inc.

DUBINS, LESTER E.: (1962) Rises and upcrossings of nonnegative martingales. IJM **6**, 226—241.

ECCLES, JOHN CAREW: (1952) The neurophysiological basis of mind. Oxford University Press.

*ENGELBERG, J.: (1961) A method of measuring the degree of synchronization of cell populations. Experimental Cell Research **23**, 218—227.

EVERETT, C. J., and S. ULAM: (1948a) Multiplicative systems in several variables, I. Los Alamos Scientific Laboratory, LA-683.

— — (1948b) Multiplicative systems in several variables, II. Los Alamos Scientific Laboratory, LA-690.

— — (1948c) Multiplicative systems in several variables, III. Los Alamos Scientific Laboratory, LA-707.

— — (1948d) Multiplicative systems, I. PNAS **34**, 403—405.

*FATOU, P.: (1910) Sur une classe remarquable de séries de Taylor. Ann. Sci. École Norm. Sup. **27**, Ser. III, 43—53.

— (1919) Sur les équations fonctionnelles. Bull. Soc. Math. France **47**, 161—271.

— (1920) Sur les équations fonctionnelles (Troisième mémoire). Bull. Soc. Math. France **48**, 208—314.

FELLER, WILLIAM: (1939) Die Grundlagen der Volterraschen Theorie des Kampfes ums Dasein in wahrscheinlichkeitstheoretischer Behandlung. Acta Biotheoretica **5**, 11—40.

— (1940) On the integro-differential equations of purely discontinuous Markoff processes. TAMS **48**, 488—515.

— (1941) On the integral equation of renewal theory. AMS **12**, 243—267.

— (1949) Fluctuation theory of recurrent events. TAMS **67**, 98—119.

— (1951) Diffusion processes in genetics. 2d Berk. Symp., 227—246.

— (1957) An introduction to probability theory and its applications, second edition. New York: John Wiley and Sons, Inc.

FERMI, E.: (1936) Sul moto dei neutroni nelle sostanze idrogenate. Ricerca Sci. **7** (2), 13—52.

*FEYNMAN, R. P.: (1947) The calculation of critical masses including the effects of the distribution of neutron energies. Los Alamos Scientific Laboratory, Ser. B, LA-524.

—, F. DE HOFFMAN, and R. SERBER: (1956) Dispersion of the neutron emission in U-235 fission. J. Nuclear Energy **3**, 64—69.

FILIPPOV, A. F.: (1961) On the distribution of the sizes of particles in grinding. TV **6**, 299—318 [Russian].

FISHER, R. A.: (1922) On the dominance ratio. PRS Edinburgh **42**, 321—341.

— (1930a) The genetical theory of natural selection. Oxford University Press. New York: Dover Publications, Inc., 1958.

— (1930b) The distribution of gene ratios for rare mutations. PRS Edinburgh **50**, 204—219.

FISZ, M., and V. S. VARADARAJAN: (1963) A condition for absolute continuity of infinitely divisible distribution functions. Z. für Wahrscheinlichkeitstheorie und Verwandte Gebiete **1**, 335—339.

FROBENIUS, G.: (1908) Über Matrizen aus positiven Elementen. Sitz. Preuss. Akad., 471—476.

— (1909) Über Matrizen aus positiven Elementen II. Sitz. Preuss. Akad., 514—518.

— (1912) Über Matrizen aus nicht negativen Elementen. Sitz. Preuss. Akad., 456—477.

FROST, A. A., and R. G. PEARSON: (1961) Kinetics and mechanism, second edition. New York: John Wiley and Sons, Inc.

FURRY, W. H.: (1937) On fluctuation phenomena in the passage of high energy electrons through lead. PR **52**, 569—581.

GALTON, FRANCIS: (1889) Natural inheritance, second American edition (an earlier edition appeared in 1869), App. F., 241—248. London: Macmillan and Co.

— (1891) Hereditary genius, second American edition. New York: D. Appleton and Co.

GANTMACHER, F. R.: (1959) The theory of matrices. New York: Chelsea Publishing Co.

GNEDENKO, B. V.: (1962) The theory of probability. New York: Chelsea Publishing Co.

GOOD, I. J.: (1949) The number of individuals in a cascade process. PCPS **45**, 360—363.

GOOD, I. J.: (1951) Random motion on a finite Abelian group. PCPS **47**, 756—762.
— (1955) The joint distribution for the sizes of the generations in a cascade process. PCPS **51**, 240—242.
— (1960) Generalizations to several variables of Lagrange's expansion with applications to stochastic processes. PCPS **56**, 367—380.
*— (1963) Cascade theory and the molecular weight averages of the sol fraction. PRS London A **272**, 54—59.

*GREISEN, KENNETH, and BRUNO ROSSI: (1941) Cosmic-ray theory. RMP **13**, 240—309.

HADAMARD, J.: (1944) Two works on iteration and related questions. BAMS **50**, 67—75.

HALDANE, J. B. S.: (1927) A mathematical theory of natural and artificial selection, part V: Selection and mutation. PCPS **23**, 838—844.
— (1939) The equilibrium between mutation and random extinction. Ann. of Eugenics **9**, 400—405.

HALMOS, PAUL R.: (1950) Measure theory. New York: D. Van Nostrand Company, Inc.

HAMMERSLEY, J. M.: (1957a) Percolation processes, II: The connective constant. PCPS **53**, 642—645.
— (1957b) Percolation processes: Lower bounds for the critical probability. AMS **28**, 790—795.
— (1959) Bornes supérieures de la probabilité critique dans un processus de filtration. Le calcul des probabilités et ses applications, 17—37. Paris: Centre National de la Recherche Scientifique.

HARDY, G. E., J. E. LITTLEWOOD, and G. PÓLYA: (1952) Inequalities, second edition. Cambridge University Press.

HARRIS, T. E.: (1947) Some theorems on the Bernoullian multiplicative process. Thesis, Princeton University.
— (1948) Branching processes. AMS **19**, 474—494.
— (1951) Some mathematical models for branching processes. 2d Berk. Symp. 305—328.
— (1952) First passage and recurrence distributions. TAMS **73**, 471—486.
— (1956) The existence of stationary measures for certain Markov processes. 3d Berk. Symp. **2**, 113—124.
— (1957a) Transient Markov chains with stationary measures. PAMS **8**, 937—942.
— (1957b) The random functions of cosmic-ray cascades. PNAS **43**, 509—512.
— (1958) A stationary measure for the multiplicative process. Abstracts of short communications and scientific programme, 121. International Congress of Mathematicians, Edinburgh.
— (1959a) A theorem on general branching processes. AMS Notices **6**, 55.
— (1959b) A mathematical model for multiplication by binary fission: The kinetics of cellular proliferation, 368—381. New York: Grune and Stratton, Inc.
— (1960a) A lower bound for the critical probability in a certain percolation process. PCPS **56**, 13—20.
— (1960b) On one-dimensional neutron multiplication. Research Memorandum RM-2317, Santa Monica, Calif.: The RAND Corporation.
— (1960c) Probability one convergence for age-dependent branching processes. AMS Notices **7**, 486.
—, and HERBERT ROBBINS: (1953) Ergodic theory of Markov chains admitting an infinite invariant measure. PNAS **39**, 860—864.

HAWKINS, D., and S. ULAM: (1944) Theory of multiplicative processes, I. Los Alamos Scientific Laboratory, LADC-265.

HEISENBERG, W. (ed.): (1943) Kosmische Strahlung. Berlin: Springer.

*HEITLER, W., and L. JÁNOSSY: (1949) On the absorption of meson-producing nucleons. PPS A **62**, 374—385.

*— — (1949) On the size-frequency distribution of penetrating showers. PPS A **62**, 669—683.

HINSHELWOOD, C. N.: (1946) The chemical kinetics of the bacterial cell. Oxford University Press.

HOBSON, E. W.: (1927) The theory of functions of a real variable, third edition. New York: Dover Publications, Inc.

HOFFMAN, JOSEPH G., NICHOLAS METROPOLIS, and VERNA GARDINER: (1956) Digital computer studies of cell multiplication by Monte Carlo methods. J. Nat. Cancer Inst. **17**, 175—188.

HOSTINSKY, B.: (1931) Méthodes générales du calcul des probabilités, fasc. 52. Mémor. Sci. Math., Paris.

JÁNOSSY, L.: (1950a) Cosmic rays, second edition. Oxford University Press.

— (1950b) Note on the fluctuation problem of cascades. PPS A **63**, 241—249.

*— (1952) Studies on the theory of cascades. Acta Phys. Acad. Sci. Hungar. **2**, 289—333.

—, and H. MESSEL: (1950) Fluctuations of the electron-photon cascade—moments of the distribution. PPS A **63**, 1101—1115.

JENTZSCH, ROBERT: (1912) Über Integralgleichungen mit positivem Kern. J. Reine Angew. Math. **141**, 235—244.

JIŘINA, MILOSLAV: (1957) The asymptotic behavior of branching stochastic processes. Czechoslovak Math. J. **7**, 130—153 [Russian].

— (1958) Stochastic branching processes with continuous state space. Czechoslovak Math. J. **8**, 292—313.

JÖRGENS, KONRAD: (1958) An asymptotic expansion in the theory of neutron transport. CPAM **11**, 219—242.

*JOHN, P. W. M.: (1961) A note on the quadratic birth process. J. London Math. Soc. **36**, 159—160.

KAC, M.: (1945) Random walk in the presence of absorbing barriers. AMS **16**, 62—67.

KAMKE, E.: (1944) Differentialgleichungen, Lösungs-Methoden und Lösungen, II. Leipzig: Akademische Verlagsgesellschaft.

KARLIN, SAMUEL: (1959) Positive operators. JMM **8**, 907—938.

*—, and JAMES MCGREGOR: (1955) Representation of a class of stochastic processes. PNAS **41**, 387—391.

— — (1957a) The classification of birth and death processes. TAMS **86**, 366—400.

— — (1957b) The differential equations of birth and death processes and the Stieltjes moment problem. TAMS **85**, 489—546.

— — (1958) Linear growth, birth, and death processes. JMM **4**, 643—662.

— — (1959) Random walks. IJM **3**, 66—81.

— — (1960) Classical diffusion processes and total positivity. JMAA **1**, 163—183.

— — (1962) On a genetics model of Moran. PCPS **58**, 299—311.

KELLY, C. D., and OTTO RAHN: (1932) The growth of individual bacterial cells. J. of Bacteriology **23**, 147—153.

KEMPERMAN, J. H. B.: (1950) The general one-dimensional random walk with absorbing barriers. Thesis, Excelsior, The Hague.

— (1961) The passage problem for a stationary Markov chain. University of Chicago Press.

KENDALL, DAVID G.: (1948a) On the generalized "birth-and-death" process. AMS **19**, 1—15.

KENDALL, DAVID G.: (1948b) On the role of variable generation time in the development of a stochastic birth process. Biometrika **35**, 316—330.

— (1949) Stochastic processes and population growth. JRSS B **11**, 230—264.

*— (1950) An artificial realization of a simple "birth-and-death" process. JRSS B **12**, 116—119.

— (1951) Some problems in the theory of queues. JRSS B **13**, 151—185.

— (1952) Les processus stochastiques de croissance en biologie. Ann. Inst. H. Poincaré **13**, 43—108.

*— (1953) Stochastic processes and the growth of bacterial colonies. Symposia of the Society for Experimental Biology, No. 7, 55—65.

— (1956) Deterministic and stochastic epidemics in closed populations. 3d Berk. Symp. **4**, 149—165.

— (1958) Integral representations for Markov transition probabilities. BAMS **64**, 358—362.

*— (1960) Birth-and-death processes and the theory of carcinogenesis. Biometrika **47**, 13—21.

KHINCHIN, A. I.: (1949) Mathematical foundations of statistical mechanics. New York: Dover Publications, Inc.

KIMURA, MOTOO: (1957) Some problems of stochastic processes in genetics. AMS **28**, 882—901.

KOENIGS, G.: (1884) Recherches sur les intégrales de certaines équations fonctionnelles. Ann. Sci. École Norm. Sup. **1**, Supplement S. 2—S. 41.

KOLMOGOROV, A.: (1933) Grundbegriffe der Wahrscheinlichkeitsrechnung, Springer-Verlag, Berlin; translated as Foundations of the theory of probability (1956). New York: Chelsea Publishing Company.

— (1935) La transformation de Laplace dans les espaces linéaires. C. R. Acad. Sci. Paris **200**, 1717—1718.

— (1938) Zur Lösung einer biologischen Aufgabe, Izvestiya nauchno-issledovatelskogo instituta matematiki i mechaniki pri Tomskom Gosudarstvennom Universitete **2**, 1—6.

— (1941) Über das logarithmisch normale Verteilungsgesetz der Dimensionen der Teilchen bei Zerstückelung. Doklady **31**, 99—101.

—, and N. A. DMITRIEV: (1947) Branching stochastic processes. Doklady **56**, 5—8.

—, and B. A. SEVAST'YANOV: (1947) The calculation of final probabilities for branching random processes. Doklady **56**, 783—786 [Russian].

KONWENT, HENRYK, and JAN ŁOPUSZAŃSKI: (1956) Some remarks on the asymptotic behaviour of the electron-photon cascade for large depth of the absorber. Acta Phys. Polon. **15**, 191—203.

KRAMER, H. P.: (1959) Symmetrizable Markov matrices. AMS **30**, 149—153.

KREIN, M. G., and M. A. RUTMAN: (1948) Linear operators leaving invariant a cone in a Banach space. Uspehi Matemat. Nauk **3**, 1—95 [Russian]; American Mathematical Society Translation No. 26, 1950.

*KRISHNA IYER, P. V.: (1949) The first and second moments of some probability distributions arising from points on a lattice and their application. Biometrika **36**, 135—141.

— (1955) Random association of points on a lattice. Nature **176**, 40.

*LAMENS, A.: (1957) Sur le processus non homogène de naissance et de mort à deux variables aléatoires. Acad. Roy. Belg. Bull. Classe Sciences, 5 Sér., **43**, 711—719.

*LANDAU, L.: (1944) On the energy loss of fast particles by ionization. J. Physics **8**, 201—205.

*LEAU, L.: (1897) Sur les équations fonctionnelles à une ou à plusieurs variables. Ann. Fac. Sci. Univ. Toulouse **11**, E. 1—E. 110.

LECAM, L.: (1947) Un instrument d'étude des fonctions aléatoires: La fonctionnelle charactéristique. C. R. Acad. Sci. Paris **224**, 710—711.

LEDERMAN, W., and G. E. H. REUTER: (1954) Spectral theory for the differential equations of simple birth and death processes. PTRS A **246**, 321—369.

LEHNER, JOSEPH, and G. MILTON WING: (1955) On the spectrum of an unsymmetric operator arising in the transport theory of neutrons. CPAM **8**, 217—234.

— — (1956) Solution of the linearized Boltzmann transport equation for the slab geometry. Duke Math. J. **23**, 125—142.

*LEIBOWITZ, MARTIN A.: (1960) A rigorous derivation of Fermi age theory. Report prepared for the Office of Naval Research.

LEONTOVIČ, M. A.: (1935) (Title not known). Zhurnal Eksper. Teoret. Fiz. **5**, 211—231.

LESLIE, P. H.: (1948) Some further notes on the use of matrices in population mathematics. Biometrika **35**, 213—245.

LEVINSON, NORMAN: (1959) Limiting theorems for Galton-Watson branching process. IJM **3**, 554—565.

— (1960) Limiting theorems for age-dependent branching processes. IJM **4**, 100—118.

LÉVY, PAUL: (1948) Processus stochastiques et mouvement Brownien. Paris: Gauthier-Villars.

LOÈVE, MICHEL: (1960) Probability theory, second edition. New York: D. Van Nostrand Company, Inc.

ŁOPUSZAŃSKI, J.: (1955a) Some remarks on the asymptotic behaviour of the cosmic ray cascade for large depth of the absorber, II: Asymptotic behaviour of the probability distribution function. Il Nuovo Cimento **2** (Supplement), Ser. X, 1150—1160.

— (1955b) Some remarks on the asymptotic behaviour of the cosmic ray cascade for large depth of the absorber, III: Evaluation of the distribution function. Il Nuovo Cimento **2** (Supplement), Ser. X, 1161—1167.

— (1956) Some remarks on the theory of the electron-photon cascade. Acta Phys. Polon. **15**, 177—180.

LOTKA, A. J.: (1922) The stability of the normal age distribution. PNAS **8**, 339—345.

— (1931a) The extinction of families—I. J. Wash. Acad. Sci. **21**, 377—380.

— (1931b) The extinction of families—II. J. Wash. Acad. Sci. **21**, 453—459.

— (1934) Théorie analytique des associations biologiques. Actualités Scientifiques et Industrielles **187**. Paris: Hermann.

— (1939a) Théorie analytique des associations biologiques, deuxième partie. Actualités Scientifiques et Industrielles **780**. Paris: Hermann.

— (1939b) A contribution to the theory of self-renewing aggregates, with special reference to industrial replacement. AMS **10**, 1—25.

— (1948) Application of recurrent series in renewal theory. AMS **19**, 190—206.

*LUNDBERG, O.: (1940) On random processes and their application to sickness and accident statistics. Thesis, Uppsala.

LURIA, S. E., and M. DELBRÜCK: (1943) Mutations of bacteria from virus sensitivity to virus resistance. Genetics **28**, 491—511.

MASLOV, K. V., and A. YA. POVZNER: (1958) The infinitesimal operators of a class of Markov processes. TV **3**, 70—83 [Russian].

MCKENDRICK, A. G.: (1914) Studies on the theory of continuous probabilities, with special reference to its bearing on natural phenomena of a progressive nature. PLMS **13**, Ser. II, 401—416.

McKENDRICK, A. G.: (1926) Applications of mathematics to medical problems. Proc. Edinburgh Math. Soc. **44**, part I, 98—130.

MENZEL, DONALD H.: (1960) Fundamental formulas of physics. New York: Dover Publications, Inc.

MESSEL, H.: (1956) On the solutions of the fluctuation problem in cascade showers. Il Nuovo Cimento **4**, Ser. X, 1339—1348.

*—, and R. B. POTTS: (1952) Note on the fluctuation problem in cascade theory. PPS A **65**, 854—856.

MORAN, P. A. P.: (1951) Estimation methods for evolutive processes. JRSS B **13**, 141—146.

— (1953) The estimation of the parameters of a birth and death process. JRSS B **15**, 241—245.

*— (1958) Random processes in genetics. PCPS **54**, 60—71.

*— (1961) The survival of a mutant under general conditions. PCPS **57**, 304—314.

— (1962) The statistical processes of evolutionary theory. Oxford University Press.

MOYAL, J. E.: (1949) Stochastic processes and statistical physics. JRSS B **11**, 150—210.

— (1956) Theory of the ionization cascade. Nuclear Phys. **1**, 180—195.

*— (1957) Discontinuous Markoff processes. Acta Math. **98**, 221—264.

— (1961 a) Multiplicative population processes. Applied Math. and Stat. Lab. Stanford University, Technical Report No. 7.

— (1961 b) The general theory of stochastic population processes. Applied Math. and Stat. Lab., Stanford University, Technical Report No. 41. Published in Acta Math. **108**, 1—31 (1962).

— (1962) Multiplicative population chains. PRS London A **266**, 518—526.

MULLIKIN, T. W.: (1961 a) Criticality estimates for spheres and slabs. PNAS **47**, 349—351.

— (1961 b) Principles of invariance in transport theory. JMAA **3**, 441—454.

*— (1962) Estimates of critical dimensions of spherical and slab reactors. JMAA **5**, 184—199.

— (1963) Limiting distributions for critical multitype branching processes with discrete time. TAMS **106**, 469—494.

NEVEU, JACQUES: (1961) Une généralisation des processus à accroissements positifs indépendants. Abhandlungen Math. Seminar, University of Hamburg **25**, 36—61.

NEY, PETER: (1961) Some contributions to the theory of cascades. Thesis, Columbia University.

— (1962a) Generalized branching processes, I: Existence and uniqueness theorems (unpublished report). Cornell University.

— (1962b) Generalized branching processes, II: Asymptotic theory (unpublished report). Cornell University.

*NEYMAN, JERZY: (1955) Sur la théorie probabiliste des amas de galaxies et la vérification de l'hypothèse de l'expansion de l'univers. Ann. Inst. H. Poincaré **14**, 201—244.

— (1956) (ed.) Contributions to biology and problems of health. 3d Berk. Symp. **4**.

— (1961) (ed.) Contributions to biology and problems of medicine. 4th Berk. Symp. **4**.

*— (1961) A two-step mutation theory of carcinogenesis. Bull. Int. Stat. Inst. (Tokyo) **38**, part III, 123—135.

*—, THOMAS PARK, and ELIZABETH L. SCOTT: (1956) Struggle for existence. The tribolium model: Biological and statistical aspects. 3d Berk. Symp. **4**, 41—79.

NEYMAN, JERZY, and ELIZABETH L. SCOTT: (1957a) On a mathematical theory of populations conceived as conglomerations of clusters. Cold Spring Harbor Symposia on Quantitative Biology **22**, 109—120.

— (1957b) Birth and death random walk process in *s* dimensions (abstract). AMS **28**, 1071.

*— (1958) On certain stochastic models of population dynamics, abstract of paper presented at the meeting of the Society for General Systems Research.

NORDSIECK, A., W. E. LAMB, Jr., and G. E. UHLENBECK: (1940) On the theory of cosmic-ray showers, I: The Furry model and the fluctuation problem. Physica **7**, 344—360.

OTTER, RICHARD: (1948) The number of trees. AM **49**, 583—599.

— (1949) The multiplicative process. AMS **20**, 206—224.

PALM, C.: (1943) Intensitätsschwankungen im Fernsprechverkehr, Ericsson Technics, No. 44. Stockholm: Telefonaktiebolaget LM Ericsson.

PERRON, O.: (1907a) Grundlagen für eine Theorie des Jacobischen Kettenbruchalgorithmus. Math. Ann. **64**, 1—76.

— (1907b) Zur Theorie der Matrizen. Math. Ann. **64**, 248—263.

PETROVSKIĬ, I. G.: (1957) Lectures on the theory of integral equations. Rochester, New York: Graylock Press.

PÓLYA, G.: (1937) Kombinatorische Anzahlbestimmungen für Gruppen, Graphen und chemische Verbindungen. Acta Math. **68**, 145—254.

POWELL, E. O.: (1955) Some features of the generation times of individual bacteria. Biometrika **42**, 16—44.

PREISENDORFER, RUDOLPH W.: (1958) Invariant imbedding relation for the principles of invariance. PNAS **44**, 320—323.

*PRÉKOPA, A.: (1956) On stochastic set functions, I. Acta Math. Acad. Sci. Hungar. **7**, 215—263. (N.B. These are set functions with "independent increments".)

PROHOROV, YU. V.: (1961) The method of characteristic functionals. 4th Berk. Symp. **2**, 403—419.

RAMACHANDRAN, B.: (1962) On the order and the type of entire characteristic functions. AMS **33**, 1238—1255.

RAMAKRISHNAN, ALLADI: (1950) Stochastic processes relating to particles distributed in a continuous infinity of states. PCPS **46**, 595—602.

*— (1951) Some simple stochastic processes. JRSS B **13**, 131—140.

— (1962) Elementary particles and cosmic rays. London: Pergamon Press.

—, and P. M. MATHEWS: (1954) Studies on the stochastic problem of electron-photon cascades. Progr. Theoret. Phys. **11**, 95—117.

— — (1956) Straggling of the range of fast particles as a stochastic process. PIAS **41**, 202—209.

*—, and S. K. SRINIVASAN: (1954) Two simple stochastic models of cascade multiplication. Progr. Theoret. Phys. **11**, 595—603.

— — (1955) Fluctuations in the number of photons in an electron-photon cascade. Progr. Theoret. Phys. **13**, 93—99.

— — (1956) A new approach to the cascade theory. PIAS **44**, 263—273.

*RIGAS, DEMETRIOS A.: (1958) Kinetics of isotope incorporation into the desoxyribonucleic acid (DNA) of tissues: life span and generation time of cells. BMB **20**, 33—70.

ROSSI, BRUNO: (1959) High energy cosmic rays. Scientific American **201**, 134—146.

SAVIN, A. A., and V. P. ČISTYAKOV: (1962) Some theorems for branching processes with several types of particles. TV **7**, 95—104 [Russian].

SCHROEDER, E.: (1871) Über iterirte Funktionen. Math. Ann. **3**, 296—322.

SCHROEDINGER, ERWIN: (1945) Probability problems in nuclear chemistry. Proc. Roy. Irish Acad. A **51**, 1—8.

SCOTT, W. T., and G. E. UHLENBECK: (1942) On the theory of cosmic-ray showers. PR **62**, 497—508.

SEMAT, HENRY: (1946) Introduction to atomic physics, revised edition. New York: Rinehart and Company, Inc.

SEMENOFF, N.: (1935) Chemical kinetics and chain reactions. Oxford University Press.

SEVAST'YANOV, B. A.: (1948) On the theory of branching random processes. Doklady **58**, 1407—1410 [Russian].

— (1951) The theory of branching random processes. Uspehi Matemat. Nauk **6**, 47—99 [Russian].

— (1957a) Transition phenomena in branching stochastic processes (abstract). TV **2**, 136—138 [Russian].

— (1957b) Limit theorems for branching stochastic processes of special form. TV **2**, 339—348 [Russian].

— (1958) Branching stochastic processes for particles diffusing in a restricted domain with absorbing boundaries. TV **3**, 121—136 [Russian].

— (1959) Transition phenomena in branching stochastic processes. TV **4**, 121—135 [Russian].

— (1961) The extinction conditions for branching processes with diffusion. TV **6**, 276—286 [Russian].

*SHAPIRO, MAURICE M.: (1962) Supernovae as cosmic-ray sources. Science **135**, 175—193.

SHOCKLEY, W., and J. R. PIERCE: (1938) A theory of noise for electron multipliers. Proc. Inst. Radio Engineers **26**, 321—332.

SIKORSKI, ROMAN: (1960) Boolean algebras. Ergebnisse der Mathematik, Neue Folge **25**. Berlin-Göttingen-Heidelberg: Springer.

SILVERMAN, R. J., and TI YEN: (1959) Characteristic functionals. PAMS **10**, 471—477.

SINGER, K.: (1953) Application of the theory of stochastic processes to the study of irreproducible chemical reactions and nucleation processes. JRSS B **15**, 92—106.

SMITH, WALTER L.: (1954) Asymptotic renewal theorems. PRS Edinburgh A **64**, 9—48.

— (1955) Regenerative stochastic processes. PRS London A **232**, 6—31.

— (1958) Renewal theory and its ramifications. JRSS B **20**, 243—302.

SNOW, R. N.: (1959a) N-dimensional age dependent branching processes, preliminary report, I: Formulation. AMS Notices **6**, 616.

— (1959b) N-dimensional age dependent processes, preliminary report, II: Asymptotic behavior, irreducible case. AMS Notices **6**, 616—617.

STACHOWIAK, HENRYK: (1956) Some properties of distribution functions of electron-photon cascades at large absorber depth. Acta Phys. Polon. **15**, 181—190.

STEFFENSEN, J. F.: (1930) Om sandsynligheden for at afkommet uddør. Matematisk Tidsskrift B **1**, 19—23.

— (1932) Deux problèmes du calcul des probabilités. Ann. Inst. H. Poincaré **3**, 319—344.

STOKES, G. G.: (1862) On the intensity of the light reflected from or transmitted through a pile of plates. PRS London Jan. 23, 1862. Reprinted in Mathematical and Physical Papers, vol. IV, Cambridge University Press, 1904, 145—156.

*STRUTT, J. W.: (1917) On the reflection of light from a regularly stratified medium. Scientific Papers, vol. VI (1911—1919), Cambridge University Press, 1920, 492—503. Reprinted in PRS London A **93**, 565—577.

*TÄCKLIND, SVEN: (1944) Elementare Behandlung vom Erneuerungsproblem für den stationären Fall. Skand. Aktuarietidskr. **27**, 1—15.

— (1945) Fourieranalytische Behandlung vom Erneuerungsproblem. Skand. Aktuarietidskr. **28**, 68—105.

*TAKACS, LAJOS: (1956) On some probabilistic problems in the theory of nuclear reactors (in Hungarian, English summary). Publications of the Mathematical Institute of the Hungarian Academy of Sciences **1**, 55—66.

TER-MIKAELYAN, M. L.: (1955) Radiation and dispersion of fast particles in a medium. Izv. Akad. Nauk SSSR, Ser. Fiz. **19**, 657—660 [Russian].

TIMOFEEV, G. A.: (1961) Effect of polarization of the medium on the development of electron-photon showers. Zhurnal Eksper. Teoret. Fiz. **41**, 1487—1492. Translated in Soviet Physics, JETP **14**, 1062—1065.

TITCHMARSH, E. C.: (1937) Introduction to the theory of Fourier integrals. Oxford University Press.

— (1939) The theory of functions, second edition. Oxford University Press.

TOBIAS, CORNELIUS A.: (1961) Quantitative approaches to the cell division process. 4th Berk. Symp. **4**, 369—385.

*TUCKER, HOWARD G.: (1961) A stochastic model for a two-stage theory of carcinogenesis. 4th Berk. Symp. **4**, 387—403.

— (1962) Absolute continuity of infinitely divisible distributions. Pacific J. Math. **12**, 1125—1129.

URBANIK, K.: (1955a) On a stochastic model of a cascade. Bull. Acad. Polon. Sci. **3**, 349—351.

— (1955b) Some remarks on the asymptotic behaviour of the cosmic ray cascade for large depth of the absorber, I: Estimation of the factorial moments. Il Nuovo Cimento **2** (Supplement), Ser. X, 1147—1149.

— (1955c) Bemerkungen über die mittlere Anzahl von Partikeln in gewissen stochastichen Schauern. Studia Math. **15**, 34—42.

— (1956) On a problem concerning birth and death processes. Acta Math. Acad. Sci. Hungar. **7**, 99—106 [Russian].

— (1958) On a stochastic model of a cascade. Studia Math. **16**, 237—267.

VOLTERRA, V.: (1931) Leçons sur la théorie mathématique de la lutte pour la vie. Cahiers scientifiques **VII**. Paris: Gauthier-Villars.

WALKER, P. M. B.: (1954) The mitotic index and interphase processes. J. of Exp. Biology **31**, 8—15.

WATSON, H. W.: (1874, 1889) See WATSON and GALTON (1874); GALTON (1889).

—, and FRANCIS GALTON: (1874) On the probability of the extinction of families. J. Anthropol. Inst. Great Britain and Ireland **4**, 138—144.

WAUGH, W. A. O'N.: (1955) An age-dependent birth and death process. Biometrika **42**, 291—306.

*— (1958) Conditioned Markov processes. Biometrika **45**, 241—249.

— (1961) Age-dependence in a stochastic model of carcinogenesis. 4th Berk. Symp. **4**, 405—413.

WEINBERG, ALVIN M.: (1961) Reactor types. PSAM **11**, 1—19.

—, and EUGENE P. WIGNER: (1958) The physical theory of neutron chain reactors. University of Chicago Press.

*WHITTLE, P.: (1952) Certain nonlinear models of population and epidemic theory. Skand. Aktuarietidskr. **35**, 211—222.

WIDDER, DAVID VERNON: (1941) The Laplace transform. Princeton University Press.

WIELANDT, H.: (1950) Unzerlegbare, nicht negative Matrizen. Math. Z. **52**, 642—648.

Wilson, J. G. (ed.): (1952, 1954, 1956) Progress in cosmic ray physics, vol. 1, 1952, vol. 2, 1954, vol. 3, 1956. New York: Interscience Publishers, Inc., and Amsterdam: North-Holland Publishing Co.

—, and S. A. Wouthuysen (eds.): (1958) Progress in elementary particle and cosmic ray physics, vol. 4. New York: Interscience Publishers, Inc., and Amsterdam: North-Holland Publishing Co.

Wing, G. Milton: (1961) Transport theory and spectral problems. PSAM **11**, 140—150.

Wold, H.: (1949) Sur les processus stationnaires ponctuels. Le calcul des probabilités et ses applications, 75—86. Paris: Centre National de la Recherche Scientifique.

*Woods, W. Max, and A. T. Bharucha-Reid: (1958) Age-dependent branching stochastic processes in cascade theory, II. Il Nuovo Cimento **10**, 569—578.

Woodward, P. M.: (1948) A statistical theory of cascade multiplication. PCPS **44**, 404—412.

*Wright, Sewall: (1945) The differential equation of the distribution of gene frequencies. PNAS **31**, 382—389.

Yaglom, A. M.: (1947) Certain limit theorems of the theory of branching random processes. Doklady **56**, 795—798 [Russian].

Yule, G. Udny: (1924) A mathematical theory of evolution based on the conclusions of Dr. J. C. Willis, F.R.S. PTRS B **213**, 21—87.

Zolotarev, V. M.: (1954) On a problem in the theory of branching processes. Uspehi Matemat. Nauk **9**, 147—156 [Russian].

— (1957) More exact statements of several theorems in the theory of branching processes. TV **2**, 256—266 [Russian].

Index

See also the Bibliography. This index includes only names and subjects cited in the text.

Selected RAND Books

ARROW, KENNETH J., and MARVIN HOFFENBERG: *A time series analysis of interindustry demands.* Amsterdam: North-Holland Publishing Company 1959.

BELLMAN, RICHARD: *Introduction to matrix analysis.* New York: McGraw-Hill Book Company, Inc. 1960.

BELLMAN, RICHARD, and KENNETH L. COOKE: *Differential-difference equations.* New York: Academic Press 1963.

BELLMAN, RICHARD, and STUART E. DREYFUS: *Applied dynamic programming.* Princeton, N. J.: Princeton University Press 1962.

BELLMAN, RICHARD E., ROBERT E. KALABA and MARCIA C. PRESTRUD: *Invariant imbedding and radiative transfer in slabs of finite thickness.* New York: American Elsevier Publishing Company, Inc. 1963.

DANTZIG, G. B.: *Linear programming and extensions.* Princeton, N. J.: Princeton University Press 1963.

DORFMAN, ROBERT, PAUL A. SAMUELSON, and ROBERT M. SOLOW: *Linear programming and economic analysis.* New York: McGraw-Hill Book Company, Inc. 1958.

DRESHER, MELVIN: *Games of strategy: Theory and applications.* Englewood Cliffs, N. J.: Prentice-Hall, Inc. 1961.

EDELEN, DOMINIC G. B.: *The structure of field space: An axiomatic formulation of field physics.* Berkeley and Los Angeles: University of California Press 1962.

FORD, L. R., Jr., and D. R. FULKERSON: *Flows in networks.* Princeton, N. J.: Princeton University Press 1962.

GRUENBERGER, F. J., and D. D. MCCRACKEN: *Introduction to electronic computers.* New York: John Wiley & Sons, Inc. 1963.

HASTINGS, CECIL, Jr.: *Approximations for digital computers.* Princeton, N.J.: Princeton University Press 1955.

MCKINSEY, J. C. C.: *Introduction to the theory of games.* New York: McGraw-Hill Book Company, Inc. 1952.

WILLIAMS, J. D.: *The compleat strategyst: Being a primer on the theory of games of strategy.* New York: McGraw-Hill Book Company, Inc. 1954.